AQA Geography

AS

SERIES EDITORS: ROSS ▸ DIGBY

CHAPMAN ▸ COWLING

OXFORD

UNIVERSITY PRESS

OXFORD
UNIVERSITY PRESS

Great Clarendon Street, Oxford OX2 6DP

Oxford University Press is a department of the University of Oxford.
It furthers the University's objective of excellence in research,
scholarship, and education by publishing worldwide in

Oxford   New York

Auckland   Cape Town   Dar es Salaam   Hong Kong   Karachi
Kuala Lumpur   Madrid   Melbourne   Mexico City   Nairobi
New Delhi   Shanghai   Taipei   Toronto

With offices in

Argentina   Austria   Brazil   Chile   Czech Republic   France   Greece
Guatemala   Hungary   Italy   Japan   Poland   Portugal   Singapore
South Korea   Switzerland   Thailand   Turkey   Ukraine   Vietnam

Oxford is a registered trade mark of Oxford University Press
in the UK and in certain other countries

© Oxford University Press 2011

Authors: Simon Ross, Bob Digby, Russell Chapman, Dan Cowling

Database right Oxford University Press (maker)

First published 2011

British Library Cataloguing in Publication Data

Data available

ISBN 978-0-19-913544-8

10 9 8 7 6 5 4

Printed in Malaysia by Vivar Printing Sdn Bhd.

Paper used in the production of this book is a natural, recyclable product
made from wood grown in sustainable forests. The manufacturing process
conforms to the environmental regulations of the country of origin

Simon Ross would like to thank Dan Moncrieff of the Field Studies Council,
Martyn Knappett, and Elspeth Goate for their helpful contributions, and Susi
Ross who, as an AQA geography student, offered useful advice from a student's
perspective.

Russell Chapman would like to thank Maria Dunn for passing on news items
at every opportunity and Linda Chapman for her support and constructive
criticism.

The authors would also like to thank Chloe Asker and Katie O'Gorman for
their help in identifying the books to read, music to listen to, and films to see.

Acknowledgements
The publisher and authors would like the thank the following for permission
to use photographs and other copyright material:

Reuters/Corbis **P6**; Curved-Light/Alamy **P8**; Elisa Cane **P11**(t); Rob Atherton
P11(r); Imagebroker/Photolibrary **P14**; Lloyd Cuff/Corbis **P15**; Meeta Sinha/
Alamy **P18**; Simon Ross **P19**; Jeremy Bright/Photolibrary **P20**; Thilo Brunner/
Corbis **P22**; Simon Ross **P28**(t); Pulp Foto/Alamy **P29**(b); Daniel Moncrieff
P30(l&r); Tony Waltham Geophotos **P34**(t); Simon Ross **P34**(b); Lasting Images/
Photolibrary **P35**; Loren McIntyre/Photolibrary **P36**(t); Simon Ross **P36**(b);
National Geographic Image Collection/Alamy **P38**; Planet Observor/Universal
Images Group/Getty Images **P39**(t&b); Tim Graham/Getty Images **P45**(t);
Barry Batchelor/Press Association **P45**(b); Seapix/Alamy **P46**; Environment
Agency **P47**(t&b); United National Photographers/Rex Features **P49**; British
Geological Survey **P50**; Pat Roque/Press Association **P52**; AFP/Getty Images
P53; David Lyon/Alamy **P55**(t); Camille Moirenc/Photolibrary **P55**(b); Ed
Brown **P56**; Mel Clarke **P58**; Environment Agency **P60**; John Schwieder/
Alamy **P64**; Felix Ausin Ordonez/Reuters Photo Agency **P66**; W Lynch/Arctic
Photo **P71**; Simon Ross **P72**; Tony Waltham Geophotos **P73**; Simon Ross
P74(t); Shutterstock **P74**(b); Roger Antrobus/Corbis **P75**(t); Robert Harding
Travel/Photolibrary **P80**; Tony Waltham Geophotos **P81**, **P82**, **P83**, **P85**, **P87**;
Imagebroker/Alamy **P88**(t); Bryan & Cherry Alexander/Arctic Photo **P88**(b);
Tony Waltham Geophotos **P90**; Karl Weatherly/Getty Images **P91**; Bryan and
Cherry Alexander/Arctic Photo **P92**; Enrique Aguirre/Photolibrary **P93**; Pixmac
P96; Mike Page **P98**; Simon Ross **P100**(t); Dave Caulkin/Press Association
P100(b); John Giles/Press Association **P102**; Shutterstock **P164**; Simon Ross
P165; Barry Bateman/Alamy **P187**; Mike Page **P108**; Pixmac **P111**(t); Tony
Waltham Geophotos **P111**(b), **P112**(t); Simon Ross **P112**(b); AFP/Virgilio
Rodrigues/Getty Images **P114**; Paul Glendall/Alamy **P115**; Tim Graham/Getty
Images **P116**; AFP/Getty Images **P117**; Tony Waltham Geophotos **P118**; Simon
Ross **P121**; Tony Waltham Geophotos **P122**; Nigel Cox **P123**; Abbotts Hall
Farm **P125**; Greg Vaugh/Alamy **P128**; Simon Ross **P130**; Frans Lemmens/
Getty Images **P131**; Joel Sartore/Getty Images **P132**; George Steinmetz/Corbis
P133(t); Rovingmagpie@flickr.com/Getty Images **P133**(b); Simon Ross **P134**;
National Geographic Image Collection/Alamy **P135**; Simon Ross **P136**(t); Tony
Waltham Geophotos **P136**(b); Getty Images **P137**(t); Michael Wald/Alamy
P137(c), Simon Ross **P137**(b); Tore Kjeilen **P139**; Tony Waltham Geophotos
P140(t); George Steinmetz/Science Photo Library **P140**(b); AFP/Getty Images
P141(t); Frans Lemmens/Getty Images **P142**; Tony Waltham Geophotos **P144**;
Eco Images/Universal Images Group/Getty Images **P146**; Thomas Mukoya/
Reuters Photo Agency **P147**(t); Getty Images **P147**(b); Simon Ross **P148**; Victor
Englebert **P149**; Mark Edwards/Still Pictures **P150**; Melanie Stetson Freeman/
The Christian Science Monitor/Getty Images **P151**; Tony Waltham Geophotos
P154; Jim West/Alamy **P153**; Getty Images **P155**(t); Pete Starman/Getty Images
P155(b); Red Brick Stock/Alamy **P158**; Science Photo Library **P162**(t); AFP/
Getty Images **P162**(b); Getty Images **P166**; Marcia Levy **P167**; Louise Batella
Duran **P168**; Marcia Levy **P169**; Martin Godwin **P172**; Caroline Maitland
Smith/Alamy **P173**; Camera Press **P177**; Shutterstock **P180**; Eightfish/Getty
Images **P182**; Xinhau Photo/Alamy **P183**(t); View Stock/Alamy **P183**(b); Jim
Hawthorne/Alamy **P184**; Press Association **P184**(t); Bob Digby **P186**(b); Bob
Digby **P187**(b); Kevin Allen/Alamy **P190**; Bob Digby **P191**; Canning Town
and Custom Regeneration Project **P193**; Jon Arnold Images Ltd/Alamy **P194**;
Angela Hampton Photo Library/Alamy **P198**; Justin Kase/Alamy **P201**(l);
Shutterstock **P202**; Roy Lane/Alamy **P204**; Bob Digby **P205**; Jill Archer/
Alamy **P208**; AFP/Getty Images **P210**; Eightfish/Getty Images **P211**; Bob Digby
P214; Inga Spence/Getty Images **P215**; Jose Cendon/Bloomberg **P217**; Still
Pictures **P218**; Cheryl Ravelo/Reuters Photo Agency **P220**; Ariana Cubillos/
Press Association **P222**(t); Daniel Morel/Press Association **P222**(b); Hans Deryk/
Reuters Photo Agency **P224**; Jose Goitia/Press Association **P225**(t); Fabienne
Fossez **P225**(b); Andre Penner/Press Association **P226**; Jaynata Dey/Reuters
Photo Agency **P227**; Reuters Photo Agency **P228**(t); Jose Cendon/Bloomberg/
Getty Images **P228**(b); Yves Herman/Reuters Photo Agency **P230**; Agrophotos/
Alamy **P232**; UNEP/DEWA/Grid-Europe **P233**; Shangara Singh/Alamy **P234**;
Peter Wiles **P235**; Prakash Singh/AFP/Getty Images **P236**; Imagestate
Media Partners Ltd-Impact Photos/Alamy **P238**; Bob Digby **P239**; 1971yes/
Shutterstock **P242**; NASA **P244**; Jason Hawkes/Getty Images **P245**(t); Nicholas
Gill/Alamy **P245**(b); Ashley Cooper Pictures/Alamy **P246**(tl); Bloomberg/Getty
Images **P246**(tr); Shutterstock **P246**(bl); Glow Images/Alamy **P246**(br); Russell
Chapman **P247**(l); Shutterstock **P247**(r); Image Bank/Getty Images **P249**; Press
Association **P250** (t); Getty Images **P250**(b); Caroline Penn/Photolibrary **P254**;
Eamon MacMahon/Press Association **P255**; Sergei Chuzavkov/Press Association
P259(b); Misha Japaridze/Press Association **P259**(t); Bullit Marquez/Press
Association **P261**; Gerald Herbert/Press Association **P262**(t); Russell Chapman
P262(b); Ahmed Jadallah **P263**; Gareth Fuller/Press Association **P266**; Patrick
Forget/Saga Photo.com **P267**; Eddie Stobart **P268**(t); Adnams Brewery **P268**;
Practical Action/Karen Robinson **P269**; Jeff Moore/Empics/Press Association
P272(t); Alan Schein/Alamy **P272**(b); JLI Images/Alamy **P273**; Fancy/Corbis
P276; Mario Anzuoni/Reuters Photo Agency **P278**(t); Greg E Mathieson/Press
Association **P278**(b); Wire/Press Association **P279**; Phototake Inc/Alamy **P280**;
Rex Features **P281**(t); Sean Sprague Photos **P281**(b); Worldmapper **P282**;
Picture Contact BV/Alamy **P283**(t); Worldmapper **P283**(b); Science Photo
Library **P285**(t); Boris Roesoker/Press Association **P285**(b); Phototake Inc/
Alamy **P287**(t); Steve Turner **P289**(b); Fredrich Stark/Alamy **P290**(t); Fedrik
Renaider/Alamy **P290**(b); Worldmapper **P290**; Steve Gschmeissner/Science
Photo Library **P291**; Peter DeJong/Press Association **P291**(b); Dr P Marazzi/
Science Photo Library **P293**; Keith Dannemiller/Alamy **P295**; Julien Behal/
Press Association **P297**; Phil Degginger/Alamy **P298**; Worldmapper **P300**,
Directphoto.org/Alamy **P301**; Nadia Bettega **P302**; Barcroft Media/Getty
Images **P303**; Worldmapper **P305**(t); A Glauberman/Science Photo Library
P306(t); Image Source/Alamy **P306**(b); Getty Images **P307**(t); Fredrik Renander/
Alamy **P307**(b); Pat Tuscon/Alamy **P312**; Truro Leisure Centre **P313**.

Ordnance Survey maps reproduced by permission of Ordnance Survey on
behalf of HMSO © Crown copyright. All rights reserved – OS Licence number
00000249.

Front Cover: Serp/Shutterstock

Illustrations are by Barking Dog Art.
Design by John Dickinson Design.

Every effort has been made to contact copyright holders of material
reproduced in this book. Any omissions will be rectified in subsequent
printings if notice is given to the publisher.

- Use this textbook. Read what's in the news as well. Some issues provoke different viewpoints. Read or surf the internet to get as many angles on a topic as you can. This book contains suggestions for your own research. Use journals, newspapers, and the internet to be aware of the most recent developments in your studies. Keep a record of sources from which you get information – you might want to check the book, article, or web page later.

- In class, take part in questions, answers, and discussions. Get involved in whatever goes on – map or stats analysis, number crunching, research in books or on the internet, debates, role plays. Question when you don't understand either what your teacher or class colleagues are saying.

- Sometimes your classwork will be in rough notes and won't need to be copied up neatly. However, do keep legible work. Don't leave it more than a week to tidy up your files!

- Keep up with work – essays, exercises, past questions, or prescribed notes – as it is set; don't leave it so that work builds up. Always meet deadlines.

Geography and careers

Geography can lead to careers in finance, management, the media, and advertising. Plenty of solicitors, accountants, and IT specialists have a Geography qualification. And some careers use Geography directly, like teaching, environmental work, land management, and planning.

Contents

See also: Skills boxes on pages 11, 16, 24, 29, 40, 44, 50, 60, 160, 169, 170, 179, 188, 191

This book uses:
LIC – low-income country
MIC – middle-income country
HIC – high-income country

Tewkesbury, Gloucestershire, 2007

Look at the photo. Where is the river channel?

What has caused this flood?

What time of year did this flood occur?

What effect has this flood had on people's lives?

What happens if this flood gets worse?

Introduction

Rivers provide tremendous opportunities for people – water for drinking and irrigating crops, power to generate electricity, and magnificent landscapes to enjoy. However, they can be unpredictable and unimaginably destructive.

Every year, flooding causes devastation and brings misery to millions of people across the world. In 2010 some 20 million people were affected by widespread floods in Pakistan, and in 2008 Cyclone Nargis devastated parts of the Irrawaddy delta in Myanmar (Burma) killing over 10 000 people and destroying farmland and rural settlements.

In this chapter you will learn about river processes and landforms, as well as the causes of flooding and associated management issues.

Books, music, and films

Books to read

Floods, when disaster strikes by Aleksandrs Rozens

Music to listen to

'When the levée breaks' by Led Zeppelin
'Natural Disaster' by Muse
'The Flood' by Take That

Films to see

Flood
2012
Blue Gold: World Water Wars (documentary, 2009)
Rising Waves (documentary, 2000)

About the specification

'Rivers, floods and management' is the compulsory Physical Geography topic in Unit 1 – you have to study this topic.

This is what you have to study:

- The drainage basin hydrological cycle and the water balance.
- The factors affecting river discharge and the storm hydrograph.
- The long profile. Changing processes: types of erosion, transportation, and deposition. Types of load. The Hjulstrom curve.
- Valley profiles. The long profile and changing cross-profiles downstream. The graded profile. Potential and kinetic energy.
- Changing channel characteristics: cross-profile, wetted perimeter, hydraulic radius, roughness, and efficiency. How these characteristics link to and influence velocity and discharge.
- Landforms of fluvial erosion and deposition: potholes, rapids, waterfalls, gorges, meanders, braiding, levees, flood plains, and deltas.
- The process and impact of rejuvenation: knick points, waterfalls, river terraces, and incised meanders.
- Magnitude-frequency analysis of flood risk.
- The physical and human causes of flooding – two case studies of recent flooding events from contrasting areas of the world. The impact of flooding – two case studies of recent flooding events from contrasting areas of the world.
- Flood management strategies:
  - hard engineering – dams, river straightening, building up of levees, and diversion spillways;
    soft engineering – forecasts and warnings, land use management on the flood plain, wetland and river bank conservation, and river restoration.

In this section you will learn about:
- the individual components of the drainage basin hydrological cycle and their interactions
- the concept of the water balance
- how to compare the water balance between two drainage basins

Skills

In this section you will:
- calculate water balance values
- draw a dispersion diagram
- use measures of dispersion when comparing data sets
- calculate standard deviation

What is a drainage basin?

When water falls as precipitation, it begins a long journey on its way to a river – and eventually to the sea. This journey might take just a few minutes, if it forms a flood on the surface. Or it might take hundreds – if not thousands – of years, as it passes through rocks deep underground. This watery journey takes place within an area of land called a drainage basin.

The **drainage basin** is the area of land that is drained by a river and its tributaries. The movement of water within the drainage basin is illustrated by the **drainage basin hydrological cycle** (Figure 1.2).

A drainage basin acts as a largely 'closed' system, because it has a very definite outer edge or boundary (called the **watershed**). This means that almost all aspects to do with water transfer, water quality and water supply are self-contained within a distinct area of land. There is little, if any, transfer of water between drainage basins. This explains why the drainage basin is the obvious unit for all aspects of water planning, such as pollution control and flood prevention.

Understanding how a drainage basin works is very important for all aspects of water management, e.g. knowing how water can be stored in a basin, and how it can be transferred between stores. It is also important to understand how transfers vary, and what factors encourage different forms of transfer. For example, flooding is more likely to occur if overland flow is the main form of transfer. This type of flow is more likely to occur if the rocks are **impermeable**, or the soil is saturated. You can begin to see how complex water movement is within a drainage basin!

Figure 1.1 *Cow Green Reservoir on the River Tees* ▼

Precipitation

The type (rain, snow, etc.), the total amount – and the intensity – of precipitation are key factors in determining the nature of water movement. Prolonged or heavy rainfall is more likely to lead to flooding. Snow acts as a store of water, which can lead to flooding when it melts.

Overland flow

This is a rapid form of water transfer over the surface of the ground. It is most likely to occur during periods of very heavy rainfall, or when the soil has become completely saturated.

Infiltration

This involves water moving from the ground surface into the soil. The rate of infiltration (the infiltration capacity) depends on the moisture content of the soil and its porosity (the number of air spaces contained within it).

Throughflow

This is the downhill transfer of water through the soil layer to the river. This shallow transfer can be quite rapid in very porous sandy soils.

Evapotranspiration

When water loss from the ground surface to the atmosphere (**evaporation**), combines with water given off from plants (**transpiration**) to form the main output from the system.

Interception

Vegetation, particularly trees, intercepts some precipitation on its way to the ground. Water is then lost back into the atmosphere by evapotranspiration. The intercepting plants also use some water for growth. Vegetation reduces and slows down water transfer.

Depression storage

When water is stored temporarily on the ground surface in the form of puddles.

Soil moisture

The existing moisture content of the soil is very important in determining whether precipitation will be absorbed, or be forced to flow as overland flow. Clay soils can be very wet and boggy (leading to overland flow), whereas sandy soils tend to be much drier, so they absorb more precipitation.

Baseflow, or groundwater flow

This is the very slow transfer of water through rocks. Only in limestone areas, where there are extensive underground channels, can the flow be faster.

River channel

The river itself is an important store of water. It also forms the 'exit' for water transferred through the drainage basin.

Percolation

This is the deeper transfer of water into **permeable** rocks – those with joints (**pervious**), or those that are **porous**.

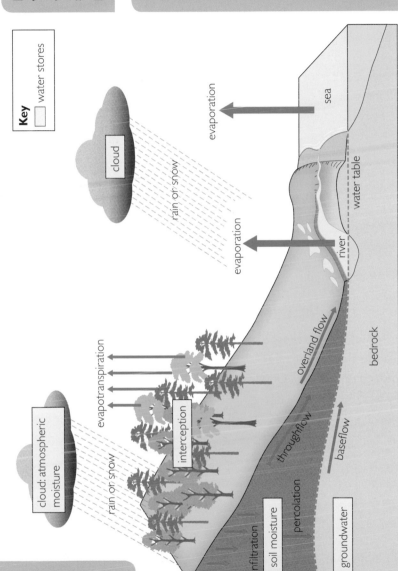

Key
☐ water stores

cloud

rain or snow

evaporation

sea

water table

river

overland flow

bedrock

throughflow

baseflow

percolation

groundwater

soil moisture

infiltration

interception

evapotranspiration

rain or snow

cloud: atmospheric moisture

evaporation

Figure 1.2 *The drainage basin hydrological cycle* ▲

The water balance

In order to gain a better understanding of water resources in a drainage basin, we use a simple equation called **the water balance**. This is expressed as follows:

P = O + E +/− S

P = precipitation
O = total runoff (streamflow)
E = evapotranspiration
S = storage (in soil and rock)

The water balance helps hydrologists to understand the unique hydrological characteristics of an individual drainage basin. This helps them to plan for future water supply and flood control.

An important aspect of the equation is the amount of runoff – expressed as a percentage of precipitation (Figure 1.3). It is a measure of the proportion of total precipitation that makes its way into streams and rivers.

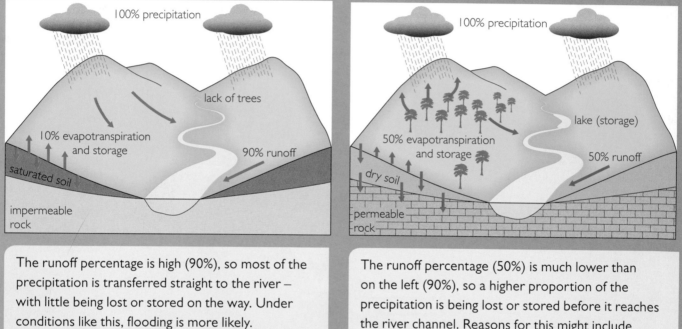

The runoff percentage is high (90%), so most of the precipitation is transferred straight to the river – with little being lost or stored on the way. Under conditions like this, flooding is more likely.

The runoff percentage (50%) is much lower than on the left (90%), so a higher proportion of the precipitation is being lost or stored before it reaches the river channel. Reasons for this might include a heavily forested river basin, or one which has **permeable rocks**. Under these conditions, flooding is much less likely.

Figure 1.3 *Variations in runoff and water balance between drainage basins* ▲

ACTIVITIES

1   Study Figure 1.1.
    a   Would you expect the amount of runoff as a proportion of precipitation to be high or low in this drainage basin? Explain your answer.
    b   If the area of moorland shown in the photo was afforested (had trees planted) with pine trees, how and why would this affect:
        i   the drainage basin hydrological cycle?
        ii   the water balance?
2   Study Figure 1.2. Construct a flow diagram, using a series of boxes and arrows, to describe how water is transferred from the atmosphere into river channels. Use a colour-coding system to separate *stores* from *transfers*.

Refer to the relative transfer speeds, either in the text or by using proportional symbols – such as different-sized arrows. Use simple diagrams, sketches or thumbnail photos to help you describe what is happening at each stage.
3   Consider how the drainage basin hydrological cycle would be different for an urbanised drainage basin. Work in pairs to draw a diagram similar to Figure 1.2, but for an urbanised basin with its houses, impermeable tarmac and concrete surfaces, and underground drains and sewers. Consider how the stores and flows might differ from those shown in Figure 1.2.

SKILLS BOX

How and why does the water balance of the River Wye compare with that of the River Cam?

The aim of this investigation is to compare the water balance of two contrasting drainage basins in the UK – the River Cam and the River Wye. It involves skills associated with describing and measuring dispersion.

The drainage basins of the River Wye and the River Cam are among a number of basins that have been studied by the Institute of Hydrology over several decades. These studies have helped hydrologists to find out more about the impact of factors such as geology, vegetation, and land use change on the water balance.

1 The studies of the River Wye and the River Cam have focused on their upper courses. Examine the background information below (Figure 1.4) to set the scene for your investigation. Also use an atlas, and any additional information from the Internet, to add to your background knowledge. This information will help you to explain any contrasts that you identify.

River Wye
Study Area: Upper Wye in Plynlimon Hills (320–780 m above sea level)

River Cam
Study Area: Upper Cam near Saffron Walden (30–120 m above sea level)

The River Wye has its source in the Plynlimon Hills of mid-Wales. The upper course of the Wye is characterised by steep slopes, with acidic soils and grassland. Some of the land has been improved by the digging of drainage ditches and the addition of fertilizer. The rocks are mostly impermeable mudstones, shales and grits. There are no extensive forests.

The upper course of the River Cam flows over flat agricultural land, from its source in the chalk hills south of Saffron Walden. The river flows over impermeable clays that lie on top of permeable chalk. In places, there are chalk outcrops on the surface, especially to the south. The vegetation is mainly arable crops.

Figure 1.4 *Comparing drainage basins: the River Wye and the River Cam in their upper courses* ▲

2 Make copies of the two tables in Figure 1.5, and then use the water balance equation on page 10 to help you complete the 'storage' columns.

    a Why are some storage values negative numbers?

    b Does there appear to be any seasonal trend with the positive and negative values? Can you explain these trends?

    c Why do you think there is a high positive storage value for the River Wye in September?

    d Suggest why the storage value then drops significantly in October.

    e Why might the total storage figure for the River Cam be of concern if it were to be replicated year on year?

River Wye

Month	Precipitation	Runoff	Evapotranspiration	Storage	Runoff as a percentage of precipitation
January	280.8	275.7	10.6	-5.5	98.2
February	191.7	145.6	12.1	34.0	76.0
March	491.0	440.2	35.9		
April	103.8	43.7	62.2		
May	168.9	126.4	65.3		
June	98.7	92.8	71.0		
July	142.2	83.0	76.8		
August	93.8	50.8	75.6		
September	285.1	199.5	46.6		
October	497.9	449.8	25.5		
November	279.4	264.8	12.1		
December	188.4	141.2	3.7		

River Cam

Month	Precipitation	Runoff	Evapotranspiration	Storage	Runoff as a percentage of precipitation
January	36.9	11.4	8.6	16.9	30.9
February	17.0	8.6	15.9	-7.5	50.6
March	90.9	24.5	33.4		
April	50.5	16.0	62.0		
May	70.6	16.6	80.1		
June	21.7	11.7	89.7		
July	54.5	8.2	86.2		
August	53.9	8.7	75.4		
September	81.0	7.4	46.9		
October	64.7	10.9	20.2		
November	32.0	9.6	11.0		
December	50.7	16.9	6.7		

Figure 1.5 *Comparing the water balance between the River Wye and the River Cam* ▲

3 Now complete the final column in your two tables to give runoff as a percentage of precipitation. To do this, you need to divide each runoff value by the precipitation value and multiply by 100. For example, for January in the River Wye basin, the calculation is:

$$\frac{275.7}{280.8} \times 100 = 98.2\%$$

    a Which drainage basin records the highest runoff values as a percentage of precipitation?

    b Use the information in Figure 1.4 to help you suggest reasons for the main differences between the two basins.

    c Look at the percentage runoff values for the River Wye. Suggest reasons for the very high values in January and June.

    d Look at the percentage runoff values for the River Cam. Why do you think the highest value was in June? Suggest reasons for the very low value in September.

4 Construct a **dispersion diagram** to compare the percentage runoff values for the drainage basins of the River Wye and the River Cam. This will enable you to look more closely at the differences between the two data sets.

To help you do this, look at Figure 1.6 – which shows an example dispersion diagram for a similar study.

a Draw two vertical lines on a sheet of graph paper and plot the values for your two data sets.

b Write a title and label each graph correctly. Don't forget to explain the vertical scale.

c Work out the **range** for each data set. This is the difference between the highest and lowest values.

d For each graph, draw a line to indicate the **median** value. This is the middle value, and will lie midway between the sixth and seventh ranked values.

e Now plot the position of the **upper quartile** and the **lower quartile**, and calculate the **inter-quartile range** for both data sets.

f Use Figure 1.6 to help you interpret your dispersion diagram. Is there a significant difference between the two data sets?

5 Calculate the **standard deviation** for each of the data sets. This is a recognised measure of the spread of a set of data. It is an effective method of comparing two sets of data. Study Figure 1.7 and then calculate the standard deviation for the percentage runoff values for the drainage basins of the River Wye and the River Cam. Note that the higher the standard deviation, the greater the spread of data around the mean.

6 Write a summary paragraph comparing the percentage runoff values. Refer to the measures of dispersion, such as the range and the inter-quartile range. Also refer to standard deviation. Does there appear to be a significant difference between the River Wye and the River Cam?

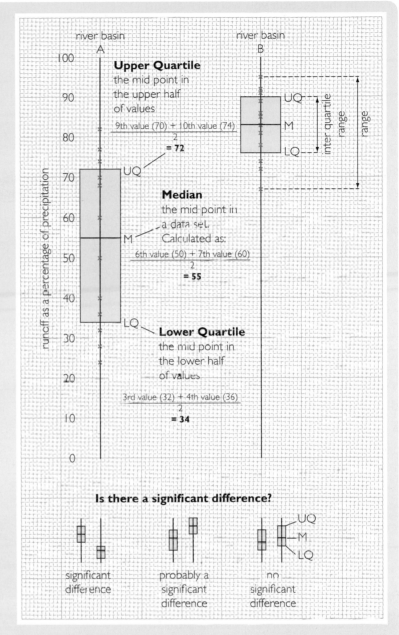

Figure 1.6 *A completed dispersion diagram, as a worked example* ▲

Step 1 Tabulate the values (x) and their squares (x^2)

Step 2 Find the mean of all the values ($\bar{x}$) and square it ($\bar{x}^2$)

Step 3 Calculate the standard deviation (σ) using the formula

$$\sigma = \sqrt{\left(\frac{\Sigma x^2}{n} - \bar{x}^2\right)}$$

σ = standard deviation
√ = square root of
Σ = sum of
n = number of values
$\bar{x}$ = mean of all values

Figure 1.7 *How to calculate standard deviation* ▲

River discharge and the storm hydrograph

In this section you will learn about:
- factors affecting river discharge
- the characteristics of the storm hydrograph
- factors affecting the shape of the storm hydrograph

What is river discharge?

Imagine standing on a bridge looking down onto a fast-flowing river (Figure 1.8). The volume of the water passing beneath you is the river's **discharge**. Values for river discharge are expressed in 'cumecs' (cubic metres per second). Discharge is most commonly calculated using the following equation:

Discharge (m³ per second) = cross-sectional area (m²) × velocity (metres per second)

What factors affect river discharge?

A number of different factors can affect the discharge of a river (Figure 1.9).

Figure 1.8 Examining a river's discharge ▲

> **Did you know?**
>
> The average discharge of the River Amazon is 219 000 m³/sec. By comparison, the discharge of the River Nile is 2830 m³/sec, and the River Thames is just 65.8 m³/sec. 1 m³/sec is roughly equivalent in weight to a medium-sized family car!

Figure 1.9 Factors affecting the discharge of a river ▼

Factor	Affect on river discharge
Distance downstream	In humid environments, such as the UK, river discharge increases with distance downstream. The main reason for this is the addition of water to the river channel as smaller tributaries join the main river.
Climatic characteristics	River discharge usually reflects precipitation, both in terms of its amount and its seasonal pattern. Snowfall results in a time delay, because the water is stored until the snow melts. When it melts, there is a surge in discharge. Other aspects of the climate, such as temperature, are important in affecting evaporation rates and vegetation growth.
Land use	Afforestation (tree planting) tends to reduce the discharge and make it more constant. Urbanisation often increases the discharge, because water is transferred rapidly over impermeable surfaces, such as tarmac and concrete – and through pipes and sewers.
Water abstraction (removal)	The abstraction of water for domestic use and the irrigation of crops reduces discharge. This can be seasonal, with a greater demand for water in the summer months.
Channel modifications	Constructing a reservoir regulates the discharge and makes it more constant. The river channel itself might also be modified to reduce the flood risk, e.g. channel straightening and enlargement tend to increase discharge, while river restoration and the creation of flood storage areas can reduce discharge because they encourage small-scale flooding in a river's headwaters.

CASE STUDY

The River Nile

With a total length of 6825 km , the River Nile is one of the world's longest rivers (Figure 1.10). Throughout history, the Nile has experienced tremendous fluctuations in its discharge. Heavy rainfall in the Ethiopian Highlands in June and July often resulted in large-scale flooding of the lower Nile in August and September. This flooding brought fertile silt downstream to Egypt's farmers, but it also led to widespread destruction. Also, during the drier months, the discharge would often be very low, which caused problems with river navigation and the water supply.

The Aswan High Dam (Figure 1.11) was constructed in 1968 to respond to these concerns. It cost over US$1 billion to build. The dam and its associated reservoir, Lake Nasser, is an example of hard engineering (see page 54). It now regulates the discharge of the river below the dam – preventing the peaks and troughs of the past. The navigation of the river has also improved, and flooding is no longer such an issue.

However, it has also caused a number of problems. The high rate of evaporation in such a hot, dry climate has led to an increasing problem with **salinisation**. The still water in the irrigation channels has become a breeding ground for insects linked with malaria and sleeping sickness. Also, Egypt's farmers no longer receive their annual top-up of free fertile soil!

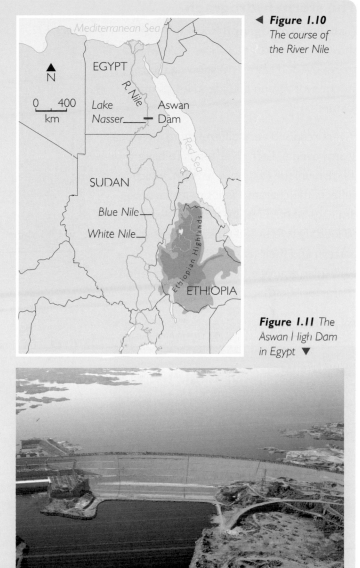

◀ *Figure 1.10*
The course of the River Nile

Figure 1.11 The Aswan High Dam in Egypt ▼

ACTIVITIES

1 Study Figures 1.8 and 1.9. Work in pairs or small groups to discuss questions (**a**) and (**b**) before attempting question (**c**) on your own.
   a Look at Figure 1.8. Under what circumstances might you expect the discharge of the river passing beneath the bridge to increase?
   b Discuss the factors identified in Figure 1.9, to clarify how they operate. Try to add some other factors that might affect a river's discharge.
   c Now produce a revision mind map/spider diagram to describe the factors affecting river discharge. Try to write in your own words and include simple illustrative sketches or thumbnail photographs to make your diagram more memorable.

2 Study Figure 1.11.
   a Why is the Aswan High Dam an example of hard engineering?
   b How does the Aswan High Dam reduce the peaks and troughs of discharge *downstream*?
   c Do you think that the construction of the dam and the reservoir is likely to have affected discharge patterns *upstream*?
   d Construct a summary table to identify the advantages and disadvantages of the construction of the dam to the people living *downstream*.

The storm hydrograph

A **storm hydrograph** is a graph that shows the discharge of a river following a storm event. Figure 1.12 shows a typical storm hydrograph and identifies its main features.

Rivers respond very differently to individual storm events. Each river tends to have its own unique response pattern, which reflects the characteristics of its particular drainage basin. Despite this, it is possible to identify two broad types of hydrograph pattern and their association with particular drainage basin and precipitation characteristics (Figure 1.13).

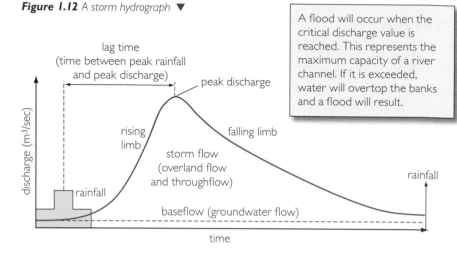

Figure 1.12 *A storm hydrograph* ▼

> A flood will occur when the critical discharge value is reached. This represents the maximum capacity of a river channel. If it is exceeded, water will overtop the banks and a flood will result.

Figure 1.13 *Characteristics that affect hydrographs* ▼

Drainage basin and precipitation characteristics	'Flashy' hydrograph with a short lag time and high peak	Low, flat hydrograph with a low peak
Basin size	Small basins often lead to a rapid water transfer.	Large basins result in a relatively slow water transfer.
Drainage density	A high density speeds up water transfer.	A low density leads to a slower transfer.
Rock type	Impermeable rocks encourage rapid overland flow.	Permeable rocks encourage a slow transfer by groundwater flow.
Land use	Urbanisation encourages rapid water transfer.	Forests slow down water transfer, because of interception.
Relief	Steep slopes lead to rapid water transfer.	Gentle slopes slow down water transfer.
Soil moisture	Saturated soil results in rapid overland flow.	Dry soil soaks up water and slows down its transfer.
Rainfall intensity	Heavy rain may exceed the infiltration capacity of vegetation, and lead to rapid overland flow.	Light rain will transfer slowly.

SKILLS BOX

1 Figure 1.14 provides discharge data for the River Easan Biorach, on the Isle of Arran in Scotland. Figure 1.15 is an OS map extract showing the drainage basin of the river. The data in Figure 1.14 was collected at grid reference 938503.

   a Make a copy of Figure 1.16 on a separate sheet of graph paper.

   b Now carefully plot the discharge data from Figure 1.14 onto your graph – to construct a hydrograph for the river.

   c Use Figure 1.12 to help you add labels to your hydrograph to describe its main features.

   d Calculate the lag time to the nearest hour.

   e With the aid of the OS map extract in Figure 1.15, attempt to explain the following features of the hydrograph:

- the steep rising limb
- the short time lag to the first peak
- the presence of two distinct peaks
- the second peak being higher than the first peak
- the gentle falling limb

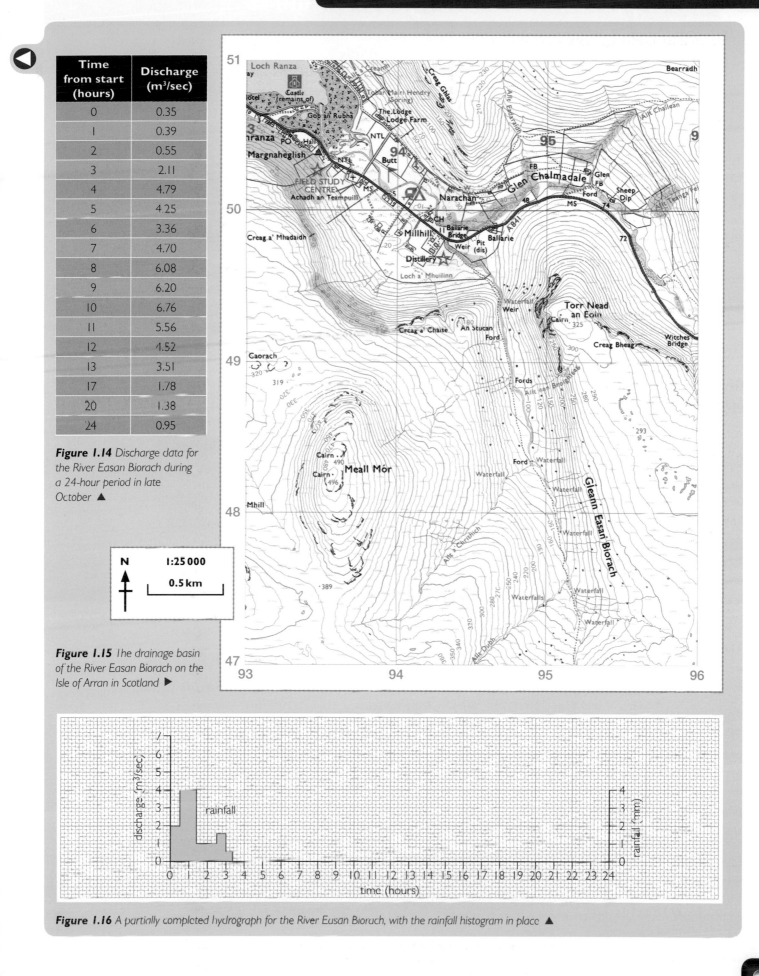

Time from start (hours)	Discharge (m³/sec)
0	0.35
1	0.39
2	0.55
3	2.11
4	4.79
5	4.25
6	3.36
7	4.70
8	6.08
9	6.20
10	6.76
11	5.56
12	4.52
13	3.51
17	1.78
20	1.38
24	0.95

Figure 1.14 *Discharge data for the River Easan Biorach during a 24-hour period in late October* ▲

Figure 1.15 *The drainage basin of the River Easan Biorach on the Isle of Arran in Scotland* ▶

Figure 1.16 *A partially completed hydrograph for the River Easan Bioruch, with the rainfall histogram in place* ▲

In this section you will learn about:
- the processes of erosion, transportation and deposition in a river
- the Hjulstrom curve
- changes in river processes with distance downstream

Skills
In this section you will:
▶ interpret a logarithmic graph

Energy in a river

Look at Figure 1.17. It shows a typical river during **low-flow** conditions. The water is clear and that the riverbed is visible. There is very little happening in this river apart from the downhill transfer of water. Yet, there is a mystery to be solved! How did the large rocks get there?

For much of the time, low-flow conditions (such as those in Figure 1.17) dominate in a river. However, occasionally – after periods of heavy rain – much larger quantities of water flow down a river's channel. It is under these **flood** conditions that the river becomes powerful enough to transport large amounts of sediment, including the large rocks shown in Figure 1.17. Water has the potential to be tremendously destructive. Armed with rocks and boulders, rivers are capable of carrying out severe erosion of the landscape – forming features such as V-shaped valleys, waterfalls and gorges.

Figure 1.17 *A river in low-flow conditions* ▲

The amount of energy available in a river to do 'work' depends on:
- the height the water has to descend (and its steepness) – this is essentially gravity
- the amount (or mass) of water available.

A still body of water at any point above sea level has a certain amount of stored energy. This is called **potential energy**. When the water starts to move downhill, the potential energy is converted into **kinetic energy**. However, some energy is lost overcoming friction, so the amount of kinetic energy generated is less than the amount of potential energy. If the gradient is steep and there is a lot of water in the channel, the amount of energy in a river will be high. This explains why fast-flowing large rivers transport more sediment and carry out more erosion than small, slow-moving rivers.

Much of the energy in a river is used up overcoming friction with the bed and banks. This is particularly true if the river flows in a wide, shallow channel, with lots of rocks and boulders protruding into the flow. It is only when the river has surplus energy that it is able to transport material and actively erode its channel.

River transportation

The sediment carried by a river is known as its **load**. It is possible to identify three main types of load:

- **Dissolved load** is the invisible transport of chemicals dissolved in the water. A common example is calcium carbonate, which is dissolved when rainwater flows over (or through) limestone or chalk.
- **Suspended load** is usually very fine-grained mud and silt, which is carried within the main body of the water. It is this form of sediment transport that makes some rivers look dark and murky.
- **Bedload** is larger sediment transported along the riverbed, which is too heavy to be picked up and carried as suspended load. Material will either be rolled along the riverbed (**traction**), or move in a series of small bounces (**saltation**). Look at Figure 1.18 to see how this operates.

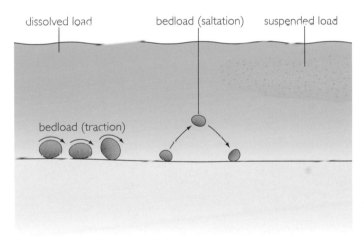

Figure 1.18 *The transportation of sediment* ▲

The type and amount of sediment transported by a river depends on several factors:

- The flow of the river is clearly very important, with most transportation occurring during high-flow conditions.
- The nature of the riverbed and banks is also important. If the river is flowing through loose material, such as sands and gravels, it is more likely to pick up and carry sediment than if it is flowing over solid rock.
- Human intervention, such as lining riverbanks with concrete, will reduce the amount of available sediment for transport.

River erosion

Erosion is the picking up and removal of material. It can involve picking up individual rocks from the riverbed, or gradually wearing away a rocky outcrop. Erosion occurs when a river has surplus energy available.

There are three main types of erosion:

- **Corrasion** is where particles of rock carried by the river scrape away at the riverbed and banks. This can dislodge rock particles to add to the load of a river. If a river is flowing over bare rocks, a sandpapering effect may occur – resulting in a smooth rock surface. This particular process is called **abrasion**.
- **Hydraulic action** is where the sheer power of moving water is able to dislodge loose particles of rock from the riverbed or banks. Hydraulic action is most effective during times of high flow, when the water forces itself strongly against the banks, particularly on the outside bends of meanders. Hydraulic action is also active when water plunges over a waterfall onto the rocks below (Figure 1.19).
- **Solution** is the dissolving of chemicals when a river flows over rocks such as limestone and chalk.

Figure 1.19 *A river in high-flow conditions* ▲

Attrition is an extra process of erosion. It involves the bashing together of rock particles as they are carried downstream by the river. Over time, and with the increasing distance downstream, the constant collisions cause individual rocks to become smaller and more rounded (Figure 1.20).

River deposition

Deposition is when material carried by a river is dumped or deposited. It occurs when a river no longer has enough energy to transport its load. This is usually when the flow slows down. Friction with the riverbed and banks leads to considerable deposition within the river channel itself. This explains why most rivers flow over sediment, rather than over bare rock. Large amounts of sediment are also deposited when a river enters the sea, or a lake, where the rate of flow is suddenly reduced. This explains why vast mud flats are often found at the mouths of rivers (Figure 1.21).

The Hjulstrom curve

The main factor that controls transportation, erosion and deposition is the speed or **velocity** of a river. The relationship between river processes and velocity is shown by a graph called the **Hjulstrom curve** (Figure 1.22). This graph resulted from research carried out during the 1930s.

Look at the graph's axes. The horizontal x-axis shows the size of rock particles, and the vertical y-axis shows velocity. Both axes are logarithmic, which means that there is a ten-fold increase between each of the equally spaced points on the axes. A logarithmic scale is commonly used when a wide range of values has to be plotted on a single graph.

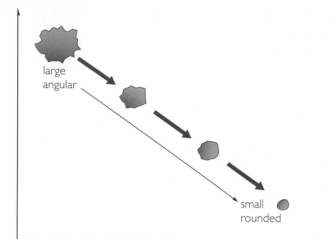

large angular

small rounded

distance downstream (increased attrition)

Figure 1.20 *The effects of attrition on particle size and shape* ▲

Figure 1.21 *The estuary of the River Blackwater at Maldon in Essex, showing the effects of deposition* ▲

Examine the **critical erosion velocity line**. This indicates the velocity needed to pick up (erode) particles of different sizes. In general, the larger the particle, the higher the velocity needed to pick it up. This is logical, because a large boulder will clearly need much more velocity to pick it up than a small particle of sand! However, there is an exception to this general trend – involving the very smallest particles. Clay particles are microscopic, flat platelets that tend to stick to each other. They need a relatively high velocity to pick them up.

Now examine the **critical deposition velocity line**. This shows the velocity below which particles of a particular size can no longer be carried and have to be deposited. As the velocity drops, successively smaller particles become deposited.

There is a gap between the two critical lines because water will suspend particles once they are picked up, even if the velocity drops slightly – a bit like floating in a swimming pool! It is only when the velocity drops significantly that material will start to be deposited.

The two critical velocity lines can be used to split the graph into three sections – erosion, transportation and deposition.

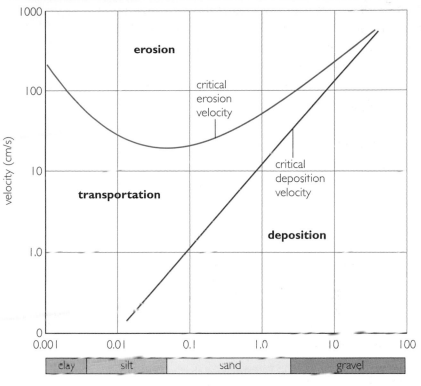

Figure 1.22 *The Hjulstrom curve* ▲

In this section you will learn about:
- valley long profiles and the concept of grade
- changes in valley cross profiles along the course of a river

Skills

In this section you will:
- interpret a 1:25 000 OS map
- draw valley cross profiles

If you paddled down a river from its source to its mouth, you would notice a number of changes. High in the mountains, the river would be flowing down a steep gradient – tumbling over rapids and waterfalls (Figure 1.23). The mountain valley in Figure 1.23 is very steep-sided and quite narrow.

If you continued paddling downstream, the valley would eventually open out and become much flatter. The land on either side might be used for farming or for settlements. The river itself would start to meander, and there would be less white-water. Paddling would become a lot easier!

Valley long profiles

It is the **gradient** (or slope) of the land that accounts for the above changes. High in the mountains there is a steep gradient. This results in the river tumbling over rapids and waterfalls. On reaching the lowlands, the gradient becomes gentler. These changes can be shown on a graph as the valley, or river, **long profile** (Figure 1.24), where the height – or altitude – of the river is plotted against distance downstream. As you can see, the upper section of the river has a steeper profile than the lower section.

In theory, as river erosion smoothes out the irregularities in the valley floor, a concave profile will result. This is called a **graded profile**. It represents a theoretical state of equilibrium, where the river is in balance with its environment. In reality, a perfectly smooth profile is very unlikely to occur. This is because the river often has to flow over outcrops of relatively resistant rock, which form steps in the long profile. Changes in sea level may also affect the long profile, as you will learn in Section 1.7 (page 41).

Figure 1.23 A river in its upper course ▲

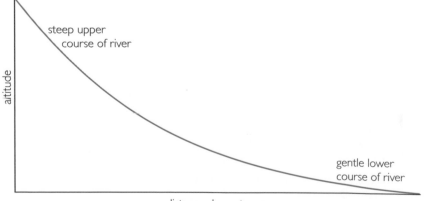

steep upper course of river

altitude

gentle lower course of river

distance downstream

Figure 1.24 A graded long profile ▶

Valley cross profiles

The cross profile of a valley also changes with distance downstream (Figure 1.25). Close to its source, a river usually flows through a narrow, steep-sided valley (Figure 1.23). With increasing distance downstream, the valley becomes wider and less steep. The river also tends to increase in size – both in width and depth – because it is carrying more water and has more energy. Even further downstream, and close to the river's mouth, the landscape is generally very flat. It may be hard to identify the edge of the valley at all!

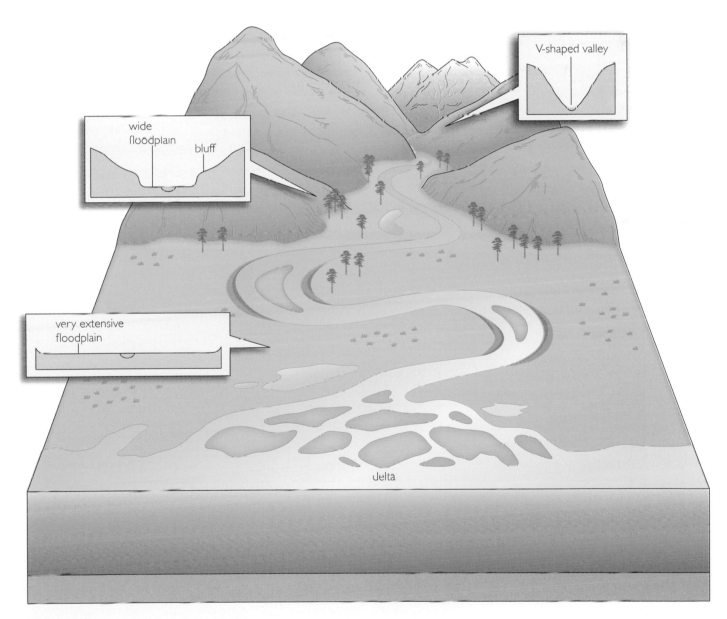

wide floodplain

bluff

V-shaped valley

very extensive floodplain

delta

Figure 1.25 *Changes in a river's cross profile* ▲

SKILLS BOX

This activity uses the River Horner – a small river in southwest England (Figure 1.27). The river flows northwards for about 10 km from its source on Exmoor to the Bristol Channel, on the north coast of Somerset. Despite its relatively short length, it exhibits many of the features and characteristics typical of rivers in the UK.

Look at Figure 1.26. It plots the long profile of the River Horner. The profile is broadly concave in shape, but is not perfectly smooth. High up on Exmoor, the valley is very steep and quite irregular in its profile – as it flows over the tough quartzite rock. Its profile becomes much smoother as it reaches the coast – and flows over the weaker marl (a form of clay). Rock type clearly affects both the landscape and the profile of the river.

You are going to draw three cross profiles for the River Horner – to see how the valley changes with the distance downstream.

1 Look closely at the long profile (Figure 1.26) and the map (Figure 1.27) to locate the three sites for the cross profiles. Write some observations about each site. Consider: the height and steepness of the land, the approximate height of the river above sea level, and the land use. Predict how you think the three cross profiles will look.

2 Draw accurate cross profiles for the three sites. (Each profile starts and finishes at either a spot height or a dark brown contour.)
  a Lay a piece of paper along the line of the profile on the map
  b Carefully mark off the contour lines onto your piece of paper. Write the values. If the contours are very close together, mark only the dark brown contours.
  c Now draw the axes for your graph. Use the length of your cross profile to draw the horizontal axis. For the vertical axis, a scale of 1 cm = 100 metres works well.
  d Draw the three profiles one under the other on the same sheet of graph paper. This will enable you to make comparisons.
  e Label the axes and write a title.
3 Add labels to each cross profile to identify its main features. Comment on the shape (width, steepness, etc.) of the valley. Label the river.
4 Briefly describe the changes in the cross profile of the River Horner as it flows from Exmoor to the coast.

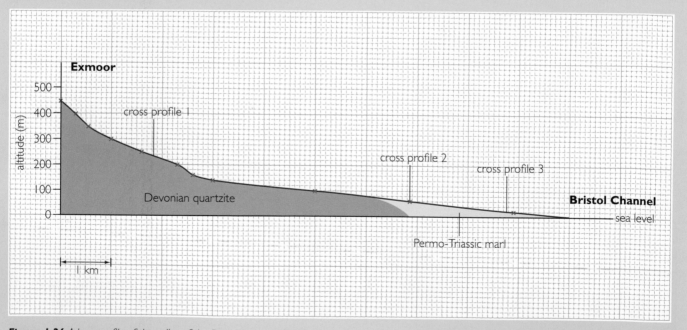

Figure 1.26 *A long profile of the valley of the River Horner in Somerset* ▲

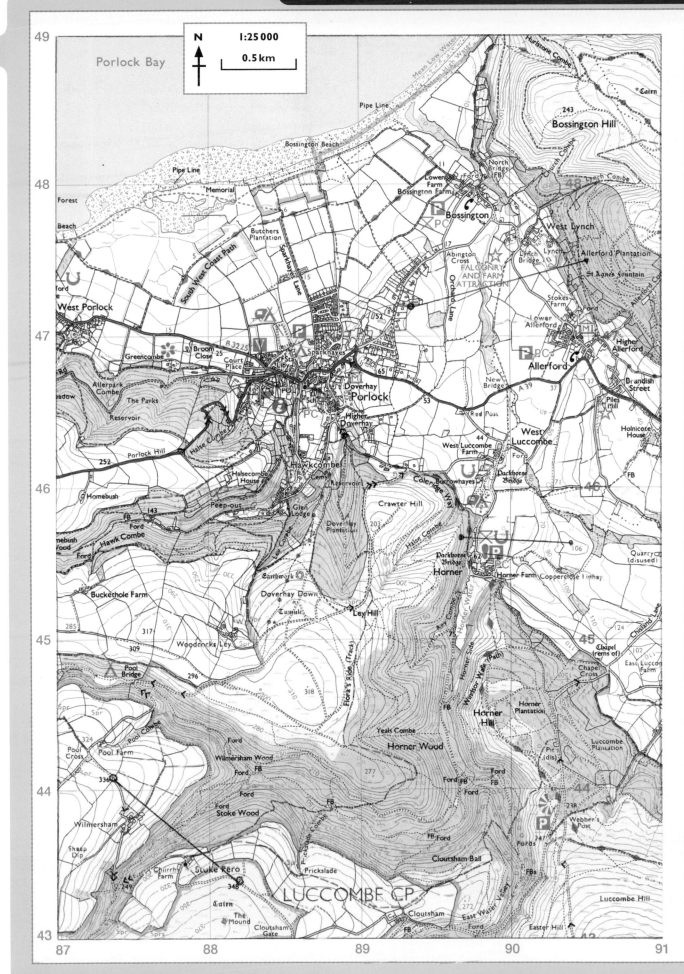

Figure 1.27 *A 1:25 000 OS map extract of the River Horner* ▲

In this section you will learn about:

- the changes to velocity and discharge downstream
- the links between velocity and river channel characteristics
- the factors affecting river channel cross profiles

Skills

In this section you will:

- investigate the changes to channel characteristics downstream, using actual data
- construct and interpret scattergraphs
- carry out and interpret the meaning of a simple statistical test of association

Downstream changes to velocity and discharge

As they tumble over rapids and waterfalls, rivers in mountain areas give the impression that they are flowing quickly. Those keen on adventure sports love white-water rafting or canoeing in these challenging environments. By contrast, the same rivers further downstream often appear to be sluggish and slow flowing. However, despite appearances, measurements suggest that both the **velocity** (the speed of flow) and the **discharge** (the volume of water) actually increase with the distance downstream. This can be explained as follows:

- Because the river's gradient decreases with the distance downstream, the velocity might be expected to decrease too. However, the reduction in gradient is more than offset by an increase in the mass of water, as smaller tributaries join the main river. It is this increase in the amount of water that is largely responsible for the overall slight increase in velocity and discharge with the distance downstream.

- The '**roughness**' of the river channel also decreases downstream. In a river's upper course, boulders and rocks create a 'rough' channel (Figure 1.28). This leads to disturbed turbulent flow. Further downstream, finer deposited sediments form a smoother, less 'rough' lining to the riverbed and banks. This promotes a less-disturbed laminar flow within the river, which is generally faster than the turbulent flow experienced upstream.

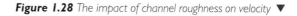

Figure 1.28 *The impact of channel roughness on velocity* ▼

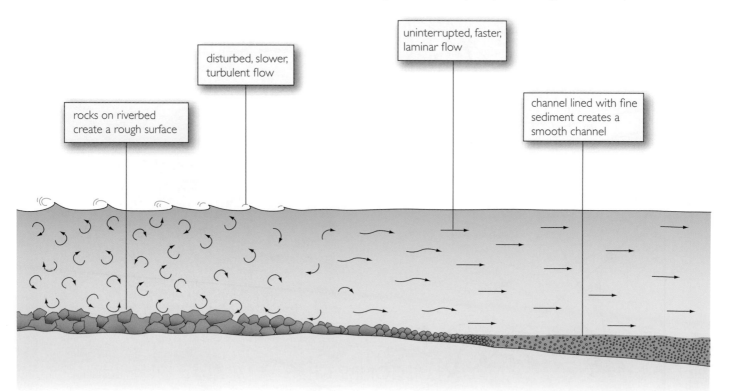

rocks on riverbed create a rough surface

disturbed, slower, turbulent flow

uninterrupted, faster, laminar flow

channel lined with fine sediment creates a smooth channel

- The channel further downstream also adopts more of a semi-circular shape, compared with the wide, shallow shape more typical upstream (Figure 1.29). This semi-circular shape is much more efficient for transferring water, because less energy is lost overcoming friction with the riverbed and banks.

The **hydraulic radius** measures the **efficiency** of a channel. It is calculated as the cross-sectional area divided by the wetted perimeter. (The **wetted perimeter** is the length of the cross-section in contact with the water.) The higher the hydraulic radius, the more efficient the channel is in carrying water.

Look at Figure 1.30. It summarises the changes to river flow and channel characteristics with the distance downstream. For example, the velocity and discharge both increase with the distance downstream. By contrast, some characteristics – such as gradient and size of load – decrease with the distance downstream.

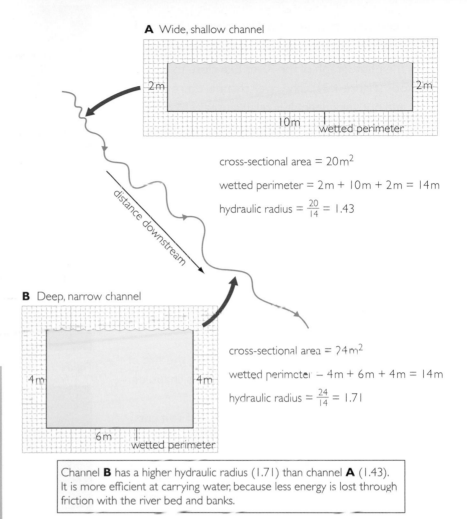

A Wide, shallow channel

cross-sectional area = $20\,m^2$

wetted perimeter = 2m + 10m + 2m = 14m

hydraulic radius = $\frac{20}{14}$ = 1.43

B Deep, narrow channel

cross-sectional area = $24\,m^2$

wetted perimeter = 4m + 6m + 4m = 14m

hydraulic radius = $\frac{24}{14}$ = 1.71

Channel **B** has a higher hydraulic radius (1.71) than channel **A** (1.43). It is more efficient at carrying water, because less energy is lost through friction with the river bed and banks.

Figure 1.29 Channel efficiency and hydraulic radius ▲

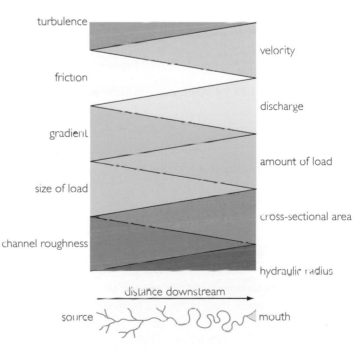

Figure 1.30 Changes to river channel characteristics with the distance downstream ▲

Downstream changes to the river channel cross profile

Look at the photograph of the River Exe close to its source on Exmoor (Figure 1.31). The river channel there is quite narrow and not very deep. It is relatively easy to cross the river at this point. Further downstream, however, the channel of the River Exe widens considerably (Figure 1.32). The wide, deep channel there carries a large amount of water.

The volume of water carried by a river affects the size of its channel. As more and more tributaries join the main river, it has to carve an ever-larger channel for itself to cope with the increasing volume. Only occasionally – during periods of exceptionally high discharge – will the river channel be unable to contain the water within its banks. A flood will occur when this happens.

While discharge is the primary factor affecting the size of a river channel, other factors can also affect its shape or profile. For example:

- a river might adopt a narrow, deep channel as it cuts through a resistant band of rock, or flows along a line of weakness – such as a fault or a joint
- human engineering can alter the shape of a channel – perhaps where it's lined with concrete, or has been dredged to deepen it
- farm animals like cattle can trample riverbanks and alter the channel profile.

Figure 1.31 *The River Exe near its source on Exmoor* ▲

Figure 1.32 *The River Exe close to Exeter* ▶

ACTIVITIES

1 Study Figure 1.30.
  a Does friction increase or decrease with the distance downstream?
  b How can downstream changes to channel roughness be used to explain your answer to question (**a**)?
  c Why does the velocity increase with the distance downstream, despite the fact that the gradient decreases?
  d Describe and attempt to explain the changes to the amount and physical size of the load carried by rivers downstream.
  e Both cross-sectional area and hydraulic radius increase with the distance downstream. Are the two connected?

2 Study Figure 1.31.
  a Describe the characteristics of the river channel shown in the photograph.
  b What evidence is there that turbulent flow is occurring?
  c Why is turbulent flow occurring?
  d If you calculated this channel's hydraulic radius, would you expect a high or a low value? Explain your answer.
  e Further downstream the riverbed is mostly fine-grained silt. How do you think this would affect the type and speed of the flow?

How do river channel characteristics change with distance downstream: River Conwy, North Wales?

The Afon (River) Conwy flows about 43 km from its source high up in the mountains of Snowdonia to its mouth at Conwy, on the north coast of Wales (Figure 1.33). For much of its lower course, the valley of the River Conwy reflects its recent glacial history – taking the form of an impressive glacial trough. The river here is wide and its channel characteristics do not change much. Therefore, the lower course of the River Conwy is not an ideal stretch of river for studying changes to channel characteristics. By contrast, however, the upper course of the River Conwy shows considerable changes to its channel characteristics with the increasing distance downstream.

In 2010, a group of A-level students on a residential field course (run by the Field Studies Council at Rhyd-y-creuau) studied the changing river channel characteristics along the upper course of the River Conwy. They took measurements at ten sites along the river – gradually moving further downstream. The six measurement sites near the source of the river are marked on Figure 1.34. Figures 1.35 and 1.36 are photographs of two measurement sites (about 8 km apart). The data collected by the students are given in Figure 1.37.

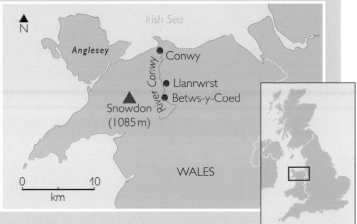

Figure 1.33 The course of the River Conwy ▲

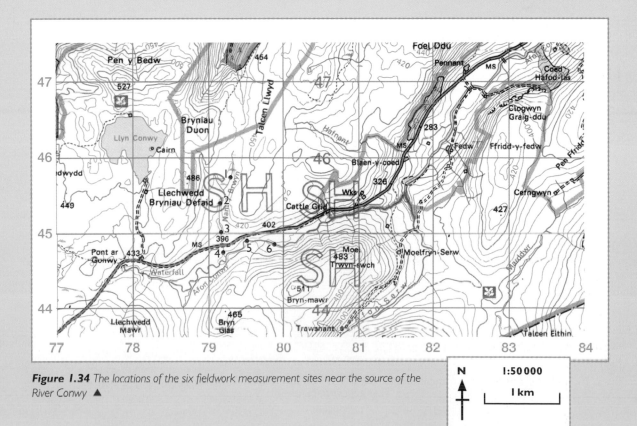

Figure 1.34 The locations of the six fieldwork measurement sites near the source of the River Conwy ▲

Figure 1.35 *Measurement site 3 (very near the source of the Conwy)* ▲

Figure 1.36 *Measurement site 8 (8.46 km downstream from site 1)* ▲

Site	1	2	3	4	5	6	7	8	9	10
Distance downstream from site 1 (km)	0.00	0.43	0.79	1.14	1.50	1.86	8.36	8.46	9.66	22.86
Water width (m)	1.80	1.42	2.30	5.80	6.56	5.40	14.50	10.22	12.80	24.60
Wetted perimeter (m)	2.00	1.60	2.90	7.70	6.63	6.60	15.67	10.54	14.80	26.70
Mean depth (m)	0.10	0.12	0.17	0.21	0.19	0.20	0.33	0.32	0.30	0.28
Average velocity (m/s)	0.23	0.31	0.30	0.41	0.48	0.60	0.48	0.80	0.74	0.76
Cross-sectional area (m²)										
Discharge (m³/sec)										
Hydraulic radius	0.09	0.10	0.13	0.16	0.19	0.17	0.31	0.31	0.26	0.26

Figure 1.37 *Channel characteristics for the River Conwy (data collected on 21 February 2010)* ▲

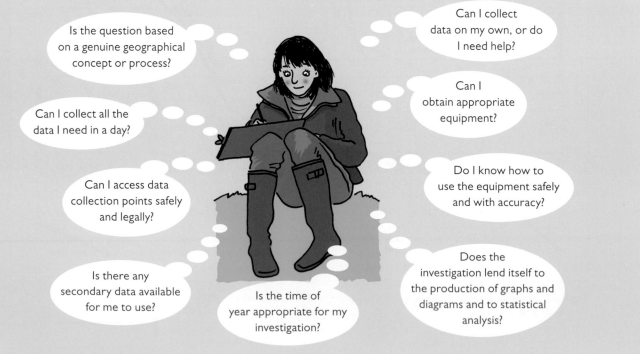

Figure 1.38 *Considerations when choosing a question for a geographical investigation* ▲

ACTIVITIES

1   Work in pairs for this activity. First study Figure 1.38. Then look closely at Figures 1.33 – 1.36.

   a   Why is the title question of this skills box suitable as a geographical investigation?

   b   With reference to the two photographs, suggest why the upper course of the River Conwy is a good location for the collection of data.

2   Study Figures 1.35 and 1.36 again. A Risk Assessment needs to be carried out when conducting fieldwork. The purpose of a Risk Assessment is to identify potential risks and then minimise them. What potential risks can you identify in the two photographs that should be minimised?

3   What can photographs like Figures 1.35 and 1.36 be used for when conducting a geographical investigation?

4   Study Figure 1.37. Copy the table and complete the two blank rows:

   a   Cross-sectional area of the channel (width x mean depth)

   b   Discharge (cross-sectional area x average velocity)

5   Work in pairs for this activity. Suggest how and why you might expect the following channel characteristics to change with the distance downstream:

   a   Cross-sectional area of the channel

   b   Average velocity

   c   Discharge

   d   Hydraulic radius

6   Draw scattergraphs to plot the data for the four characteristics listed in Activity **5**. Plot each of the sets of values (y-axis) against distance downstream (x-axis). If you are able to work in pairs, you could split this task between you. Use the example scattergraphs in Figure 1.39 to help you draw a best-fit line on each graph. Comment on any relationship that you observe.

7   A statistical technique – called the Spearman's Rank Correlation Coefficient Test (Figure 1.40) – can be used to investigate the possible relationship between two sets of data. It might look complicated, but it's quite straightforward really – if you check your results as you go and follow the steps laid out in Figure 1.40.

   a   Carry out a Spearman's Rank Correlation Coefficient Test for one or more of the data sets from Activities **5** and **6**.

   b   Interpret the Spearman's Rank value using the critical values table in Figure 1.40.

   c   Comment on the relationship between your chosen data set and the distance downstream.

   d   Now repeat the exercise – but this time look for a relationship between average velocity and a channel characteristic like mean depth.

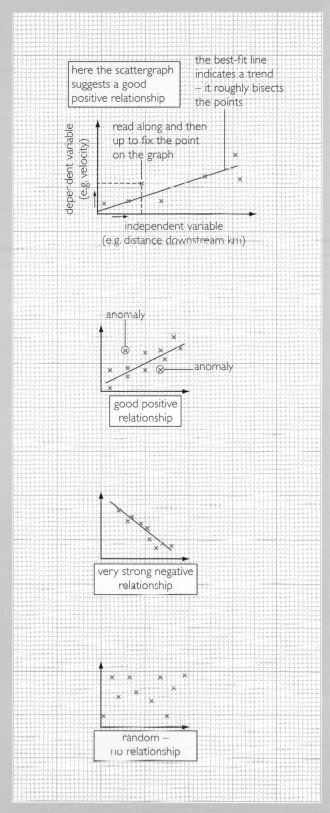

Figure 1.39 *Drawing and interpreting scattergraphs* ▲

Example question: Is there a relationship between average velocity and hydraulic radius?

Site	Average velocity	Rank of average velocity	Hydraulic radius	Rank of hydraulic radius	Difference between the ranks (d)	Difference squared (d^2)
1	0.23	10	0.09	10	0	0
2	0.31	8	0.10	9	1	1
3	0.30	9	0.13	8	1	1
4	0.41	7	0.16	7	0	0
5	0.48	5.5	0.19	5	0.5	0.25
6	0.60	4	0.17	6	2	4
7	0.48	5.5	0.31	1.5	4	16
8	0.80	1	0.31	1.5	0.5	0.25
9	0.74	3	0.26	3.5	0.5	0.25
10	0.76	2	0.26	3.5	1.5	2.25
					Sum d^2 = 25	

(**NB** Where two or more sites have the same value, calculate an average rank value. Sum the total ranks and divide by the number of values. In the table on the left, the values for average velocity for sites 5 and 7 are the same (0.48). These values are ranked 5 and 6. Therefore, sum ranks $\frac{5 + 6}{2} = 5.5$)

Spearman's Rank Correlation Coefficient:

$$R = 1 - \left(\frac{6\Sigma d^2}{n^3 - n}\right)$$

$$R = 1 - \left(\frac{150}{990}\right)$$

$$R = 1 - 0.15$$

$$R = 0.85$$

How to interpret the Spearman's Rank Correlation Coefficient

The result of the above test (0.85) can be crudely interpreted as follows. A value in excess of +/− 0.7 indicates a good relationship. The closer the R value is to +/− 1.0, the stronger the relationship will be. The above value of 0.85 suggests a strong positive relationship between velocity and hydraulic radius. Is this what you would expect? Why?

However, to make a more accurate interpretation, you need to take into consideration the number of data sets (in our case 10) – plus the likelihood that any relationship has occurred by chance. In geographical investigations, it's reasonable to accept a relationship if there is only a 5% probability (0.05) that it has occurred by chance. To evaluate this, a critical values table (see on the right) is used to look up the R value.

The number of data sets (n) used in our example is 10. In the table on the right, there are two critical values – one for a 5% probability that a relationship has occurred by

chance, and the other for a 1% probability. Our value of 0.85 is above both of these critical values. This means that there is less than a 1% probability that the relationship has occurred by chance. In other words, there is over a 99% probability that there is a significant relationship between velocity and hydraulic radius.

Essentially, you need to remember that your R value should be higher than the critical value in the table at 5% if you are going to accept that there is a relationship between two variables.

BUT!

A statistical acceptance is only as good as the geography! You have to judge if geographically it makes sense for a causal relationship to exist (i.e. one variable influences the other). Spearman's Rank is just a statistical test. High levels of acceptance can occur in two data sets that have absolutely no causal relationship between them. As a geographer, you have to decide if a relationship between two sets of variables makes sense before you start to carry out a statistical analysis.

n	5%	1%
4	1.000	*
5	0.900	1.000
6	0.829	0.943
7	0.714	0.893
8	0.643	0.833
9	0.600	0.783
10	0.564	0.745
11	0.536	0.709
12	0.503	0.678
13	0.484	0.648
14	0.464	0.626
15	0.446	0.604

Figure 1.40 *Spearman's Rank Correlation Coefficient Test* ▲

In this section you will learn about:
- the characteristics and formation of fluvial landforms (potholes, rapids, waterfalls, meanders, braiding, levees, floodplains, and deltas)
- how to recognise landforms on photographs, maps and diagrams

Skills

In this section you will:
- draw a sketch from a photograph
- draw a sketch map using an OS map (1:25 000)
- interpret and carry out measurements on an OS map (1:25 000)

Fluvial landforms: an overview

Rivers are responsible for forming some of the world's most spectacular landforms – such as Niagara Falls on the US-Canadian border and the Grand Canyon in the USA. The sweeping meanders of the River Mississippi and the River Ganges can be seen clearly from space. Both the River Nile and the River Mississippi form extensive deltas where they meet the sea. In fact, much of the country of Bangladesh is actually a huge delta – formed at the mouths of the Ganges and Brahmaputra Rivers.

Rivers are constantly shaping the landscape as they carry water and sediment along their courses (Figure 1.41). Where a river has excess energy, it will erode its channel – carving gorges and creating waterfalls. Elsewhere, when energy levels fall, deposition will occur – forming extensive floodplains. Often processes of erosion and deposition will combine to form features such as meanders, which are ever-changing as they migrate downstream.

This section will examine some of the common landforms associated with fluvial erosion and deposition.

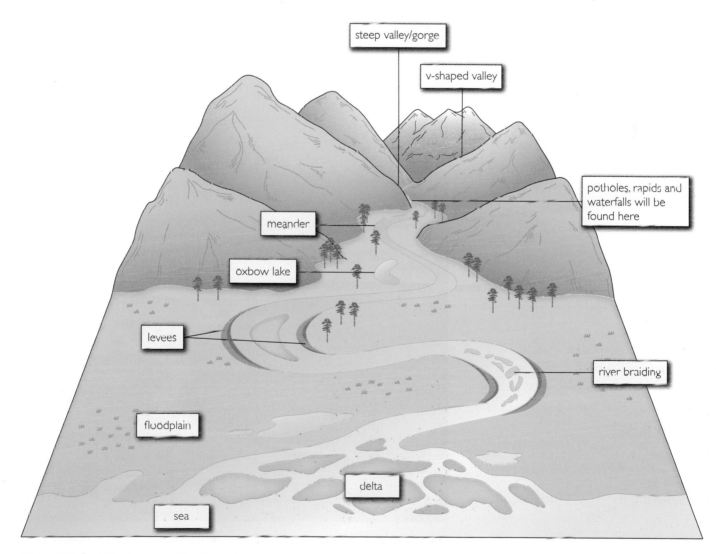

Figure 1.41 *Fluvial landscapes and landforms* ▲

Landforms of fluvial erosion and deposition

Potholes

Potholes are small circular depressions in the riverbed, carved out of solid rock (Figure 1.42). They are commonly found in upland areas, close to the source of a river, where the channel flows directly over the bedrock.

Potholes are formed when rock fragments are 'drilled' into holes and cavities by the turbulent whisking-like action of the river. This type of erosion is called corrasion (see page 19). Potholes are commonly found in the same locations as rapids and waterfalls. Over time, the holes are enlarged to form potholes such as the one shown in Figure 1.42.

Rapids and waterfalls

A river often flows over a variety of different rock types as it makes its way from its source to its mouth. Tougher, more-resistant rocks are less easily eroded than weaker rocks, and they will often form irregular steps in the long profile of a river. On a small scale, these irregularities may create a very turbulent stretch of river, called **rapids** (Figure 1.43). Rapids often form spectacular stretches of white-water, as the river plunges over jagged rocks and mini-waterfalls to form dangerous whirlpools and fast-flowing tubes of water. Canoeists often prize these challenging river landscapes.

By contrast, a **waterfall** is a single – rather more pronounced – 'step' in the long profile of a river. It is most commonly formed when a river flows over a relatively tough band of rock.

Figure 1.42 *A pothole and the rock fragments that helped to form it* ▲

Figure 1.43 *Rapids on the River Orchy in Scotland* ▲

Look at Figure 1.44, which shows the Bridal Veil Falls in North Island, New Zealand. Here the River Pakota plunges over a vertical wall of basalt – a tough igneous rock exposed by glacial action during the last Ice Age. Beneath the basalt is a relatively weak layer of sandstone. This has been eroded by hydraulic action and corrasion to form a deep **plunge pool**. Gradually, erosion of the underlying sandstone is causing the waterfall to be undercut. This explains why it is possible to walk behind the waterfall! In the future, if erosion continues in this way, a narrow, steep-sided **gorge** will be formed here.

CASE STUDY

High Force waterfall on the River Tees

High Force (Figure 1.45) is one of a number of waterfalls formed along the upper course of the River Tees in northern England (Figure 1.46). At High Force, the river plunges down about 20 metres over the edge of a tough outcrop of dark igneous rock, called dolerite. Beneath the dolerite are weaker bands of sedimentary rocks, mainly limestones and shales. These sedimentary rocks have been eroded to form the plunge pool. Erosion of these weaker rocks has undercut the waterfall at its base, causing the overlying dolerite to collapse into the plunge pool below. Over thousands of years, this cycle of undercutting and collapse has caused the waterfall to retreat upstream to form a 1.5-km gorge.

Figure 1.44 *The Bridal Veil Falls, New Zealand* ▲

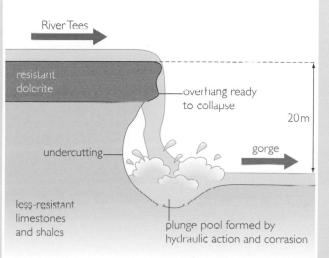

Figure 1.45 *High Force waterfall in cross-section* ▲

Figure 1.46 *The location of High Force waterfall* ▲

Meanders

Meanders are sweeping bends made by a river, and are most commonly found in lowland areas in the river's middle course (Figure 1.47). They are dynamic features, which slowly change their shape and position in response to the processes of erosion and deposition taking place in the river channel.

Look at Figure 1.48. It shows some of the main features and processes operating within a meandering river. Locate the **thalweg** – the line of greatest velocity or fastest flow within the river. As you can see, it swings from side to side, causing erosion on the outside bend and deposition on the inside bend. This explains why meanders gradually migrate across the valley floor. The flow pattern within a meandering channel is, however, much more complex than it appears in the diagram. It is more like a giant 3-D corkscrew, spiralling its way down the river. This type of flow is called **helicoidal flow**.

Figure 1.47 A meandering river in Brazil ▲

One of the main features of meandering channels is an alternating sequence of shallow, fast-flowing sections – called **riffles** – and deeper, slower-moving sections – called **pools** (Figures 1.48 and 1.49). Riffles and pools commonly form in channels during low-flow conditions, and they are closely linked to the very complex pattern of helicoidal flow. Indeed, experiments in laboratories have shown how alternating riffles and pools in straight channels can set up patterns of helicoidal flow that start to create meanders. It is all very complex!

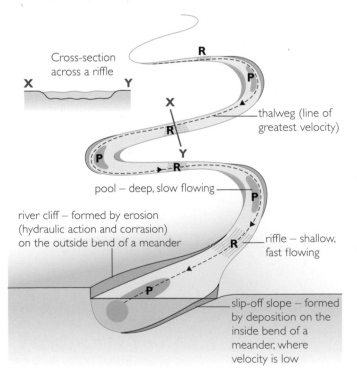

Cross-section across a riffle

thalweg (line of greatest velocity)

pool – deep, slow flowing

river cliff – formed by erosion (hydraulic action and corrasion) on the outside bend of a meander

riffle – shallow, fast flowing

slip-off slope – formed by deposition on the inside bend of a meander, where velocity is low

◀ **Figure 1.48** The main features and processes in a meandering river

Riffle

Pool

Figure 1.49 A riffle and a pool in a small stream on Exmoor ▲

Over time, as meanders migrate across the valley floor, opposite meander bends may start to erode towards each other (Figure 1.50). Gradually, the neck of the meander narrows until it's completely broken through (usually during a flood), to form a new straighter channel. The old meander loop is eventually sealed off by deposition to form an **oxbow lake**.

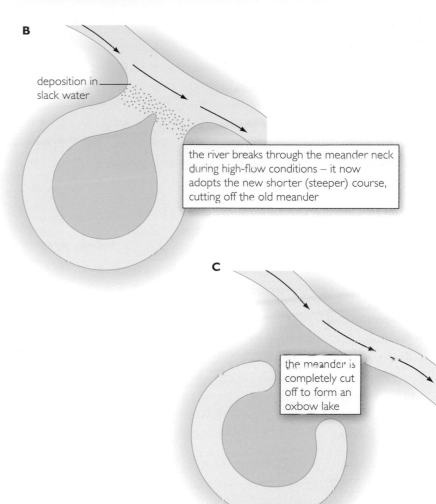

B

deposition in slack water

the river breaks through the meander neck during high-flow conditions – it now adopts the new shorter (steeper) course, cutting off the old meander

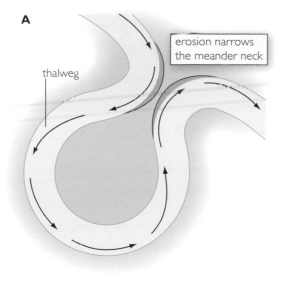

A

erosion narrows the meander neck

thalweg

C

the meander is completely cut off to form an oxbow lake

Figure 1.50 *The formation of an oxbow lake* ▲

ACTIVITIES

1 Study Figure 1.42.
   a Describe the shape of the pothole shown in the photograph.
   b Describe the processes responsible for the formation of potholes. Use a simple diagram to illustrate your answer.
   c Suggest reasons why potholes have formed in this stretch of the river.

2 Study Figure 1.44.
   a Draw a simple sketch of the waterfall in the photo.
   b Write detailed labels to describe the main features and processes associated with the development of the waterfall. Use the text to help you do this. Remember that the vertical cliff over which the water plunges is basalt and that the rock underneath is sandstone.

3 Study Figures 1.47 and 1.48.
   a What is the name of the feature at A?
   b Suggest how the feature at A has been formed.
   c Would you expect to find a pool or a riffle at B?
   d If you were on the ground at B, what supporting evidence would you look for to confirm your answer to (c) above?
   e Locate the line of cross section X-Y. Draw a simple profile across the river at this point to show how you would expect the river channel to look. Add labels to your sketch to identify the thalwag, the slip-off slope and the river cliff.
   f What is the feature at C?
   g With the aid of fully labelled diagrams, suggest how the feature at C has been formed.
   h If you were to see a photo of this river in 100 years' time, how and why might it have changed?

Braiding

A river is described as being braided when it becomes sub-divided into many separate channels (Figure 1.51). The main reason why this occurs is because the river becomes overloaded, which results in sediment being dumped in the channel to form islands – like the ones you can see in Figure 1.51. Around these islands, smaller channels are cut into the river's sediments – steepening the gradient and increasing their efficiency.

Braiding is common in rivers that experience large variations in discharge, and where the load is relatively large, e.g. rivers that flow from the snouts of glaciers. Similarly, rivers in tropical and semi-arid environments can have wildly fluctuating discharges – leading to braiding.

Floodplains and levees

A floodplain is an extensive, flat area of land on either side of a river, which periodically becomes flooded. River floodplains are most extensive in lowland areas, where they can be several kilometres wide. The silt deposits that form floodplains are fertile, which explains why they are often used for farming.

Look at Figure 1.52, which shows the main features of a floodplain. The river here flows over sediments that have been deposited on the valley floor by previous floods – every time the river floods, it deposits a fresh layer of silt on top of the existing floodplain. Over thousands of years, as meanders slowly migrate across the floodplain, they cut into the valley edge (the bluff). This gradually widens the valley and extends the floodplain.

During high-flow conditions – for instance after a period of heavy rain, or when a large amount of snow has melted –

Figure 1.51 A braided river channel ▲

the river channel might be unable to contain the increased volume of water flowing into it. As a result, the water starts to overflow the banks and spill onto the floodplain. When this happens, the larger, heavier sediment is deposited on top of the banks. This is because the water velocity at this point is much less than in the main river channel. The coarse sediment traps smaller sediment and gradually raises the height of the banks. These raised banks are called levees (Figure 1.52).

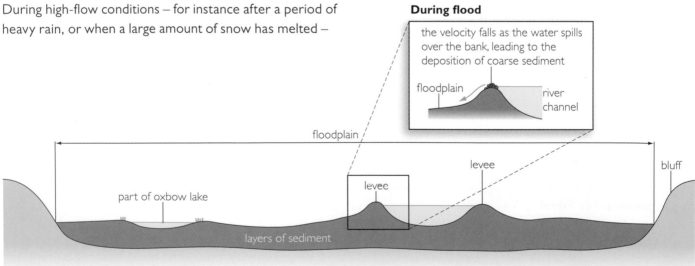

Figure 1.52 The features of a floodplain ▲

Deltas

A river loses energy very rapidly when it enters the sea (or a lake). As a result, vast amounts of silt and clay are deposited in a fan shape where the two meet. Over time, the build-up of deposited material breaks through the water surface and forms new land – called a delta (Figure 1.53). Often the river in a delta is forced to split into separate channels, called distributaries. This is simply the river's response to the massive deposition of sediment and the reduction in the gradient.

There are two main types of delta:

- An arcuate delta is a gently curved delta (Figure 1.54).
- A bird's foot delta is rather more complex in its shape, and represents deposition along the edges of several distributaries (Figure 1.55).

Figure 1.54 The Nile Delta ▲

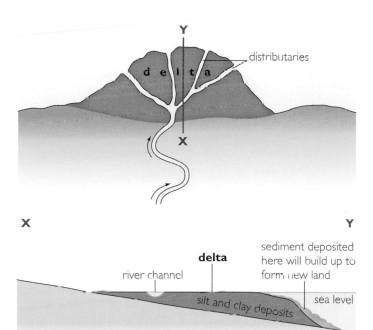

Figure 1.53 The features of a delta ▲

Figure 1.55 The Mississippi Delta ▲

ACTIVITIES

1 Study Figure 1.51.
   a Does the photograph show low-flow or high-flow conditions in the river? Explain your answer.
   b This river is a meltwater river – emerging from the snout of a glacier. What time of year do you think the photo was taken?
   c Will the pattern of braiding within the river channel constantly change? Explain your answer.
2 Why do you think braided rivers are very hazardous environments for people walking or camping?

Internet research

In your exam you will need to be able to identify, describe and explain the formation of river landforms. Even though you've already seen a number of photographs and diagrams in this section, you will benefit enormously from developing your own 'scrapbook' of photos and diagrams.

For each of the landforms listed in your specification (potholes, rapids, waterfalls, meanders, braiding, floodplains, levees and deltas) use the Internet to search for and select one really impressive photo that you think shows the main characteristics of the landform. Add labels to identify the main features. Give each photograph a title which names the landform and its location.

SKILLS BOX

Figure 1.56 shows a short stretch of the River Swale near Northallerton in North Yorkshire. The river is flowing from north to south across the extract.

1   The aim of this activity is to draw a sketch map to identify the main features of the river and its valley.

   a   Using a pencil, draw a faint grid on a sheet of plain paper – to represent the gridlines shown on the map extract. Use the same scale as the map (4 cm = 1 km). Write the grid numbers alongside each gridline.

   b   Now carefully draw in the course of the river, using the gridlines to help you. Draw a few arrows alongside the river to indicate the direction of flow.

   c   Try to add on the 25-metre and 30-metre contour lines. This is quite difficult in places, but do the best you can. Also locate a selection of spot heights, to indicate the general relief of the area.

   d   Identify the embankments that run alongside the river. Where they are right next to the river, they are probably levees.

   e   Label as many features as you can – choosing clear examples from the map. You should be able to label meanders, levees and the floodplain. Can you suggest the location of a bluff?

   f   Complete your sketch map by adding a title, a scale and a north point.

2   Locate the river as it flows through grid squares 3386 and 3486.

   a   Make a large sketch of the river channel in those two grid squares – ensuring that it's wide enough to add in the thalweg.

   b   Try to draw in the thalweg. Look back at Figure 1.48 for help, if you need it.

   c   Use two separate colours to suggest where you might expect to find pools and riffles along this section of river. Explain your colours in a key.

   d   Use a broken line to show where you might expect two oxbow lakes to form in the future.

   e   Carefully measure the length of the river on the map, as it passes through these two grid squares. Now measure it again as it would be when the two oxbow lakes form. How much shorter will its course be?

   f   Once the river flows along a shorter course, what effect will this have on the gradient of the river, the available energy, and the processes operating in the river? Consider how the river channel characteristics will change.

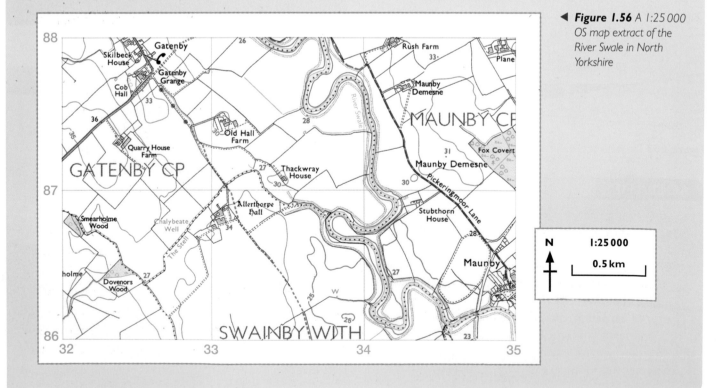

◄ *Figure 1.56 A 1:25 000 OS map extract of the River Swale in North Yorkshire*

In this section you will learn about:
- the processes of rejuvenation
- the impact of rejuvenation
- the characteristics and formation of knick points, waterfalls, river terraces, and incised meanders

What is rejuvenation?

Over a long period, a river assumes a generally smooth long profile (Figure 1.57). This is the most efficient profile for a river to have in order to transport water and sediment. It represents a state of equilibrium (balance) with the environment. In theory, a river will always be trying to achieve this smooth concave profile.

Occasionally an event occurs that de-stabilises the situation, and causes the river to actively erode its channel in order to re-establish its smooth long profile. This renewed period of erosion is called **rejuvenation**.

How does a river become rejuvenated?

The sudden and rapid increase in erosion is due to a fall in **base level** (Figure 1.57). Base level is the height or altitude to which the river flows before it either joins another river or reaches the sea. If base level falls, a 'kink' or step called a **knick point** is formed in the river's long profile (Figure 1.57).

The increase in the gradient caused by the fall in base level means the river now has more energy available. This is used to actively erode the irregularity, so that the river once again achieves a state of equilibrium by creating a smooth long profile. Like a normal waterfall, erosion will cause a knick point to retreat upstream.

There are two main causes of rejuvenation:
- **Sea-level change.** If sea-level falls faster than the rate of vertical river erosion, a knick point will form close to the coast. There are examples of this along the Antrim coast of Northern Ireland (Figure 1.58).
- **River capture.** Over hundreds of years, rivers gradually cut backwards at their source. This is called headward erosion. Occasionally, as a river cuts back, it can break into an adjacent valley and 'capture' the tributaries of a nearby river. There is often a considerable height difference at the point of river capture and a waterfall is formed. This is a knick point and, as the river cuts into its former valley, features of rejuvenation will be formed.

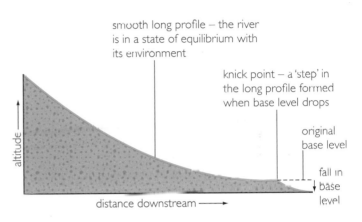

smooth long profile – the river is in a state of equilibrium with its environment

knick point – a 'step' in the long profile formed when base level drops

original base level

fall in base level

altitude

distance downstream

Figure 1.57 *Rejuvenation and a river's long profile* ▲

More about sea-level change

Since the end of the last ice age, about 8000 years ago, sea-levels have fluctuated considerably. Melting ice initially caused sea-levels to rise (eustatic change), but this was followed by a much slower rise in the land (isostatic recovery) as the weight of the ice cover was removed. In the south and east of the UK, the trend has been for a slight rise in relative sea-level, whereas in the north and the west there has been a slight drop in sea-level, creating features like the rapids shown in Figure 1.58.

Figure 1.58 *Rapids at a knick point* ▼

Rejuvenation of the River Rheidol, Ceredigion

The River Rheidol is in mid-Wales. It flows from the Plynlimon uplands to the coast at Aberystwyth, and it exhibits some superb features of rejuvenation.

Look at Figure 1.59, which shows how the River Rheidol captured the headwaters of the River Teifi at Devil's Bridge, at an altitude of about 250 m above sea-level. With a shorter distance to the sea and a much steeper gradient, the headwaters of the Teifi were diverted along the course of the River Rheidol. Immediately, the headwaters started to erode vertically at a rapid rate to form a deep and distinctive gorge within the old valley of the Teifi. Falling some 550 m in just 45 km , the River Rheidol is one of the UK's fastest-flowing rivers.

There are several **waterfalls** north of Devil's Bridge, formed by the rapid vertical erosion of the river. The original knick point that would have existed at Devil's Bridge has retreated upstream to form a series of **rapids** close to Ponterwyd (Figure 1.60). The River Mynach, a small tributary, forms the spectacular Mynach Falls where it joins the Rheidol at Devil's Bridge (Figure 1.61).

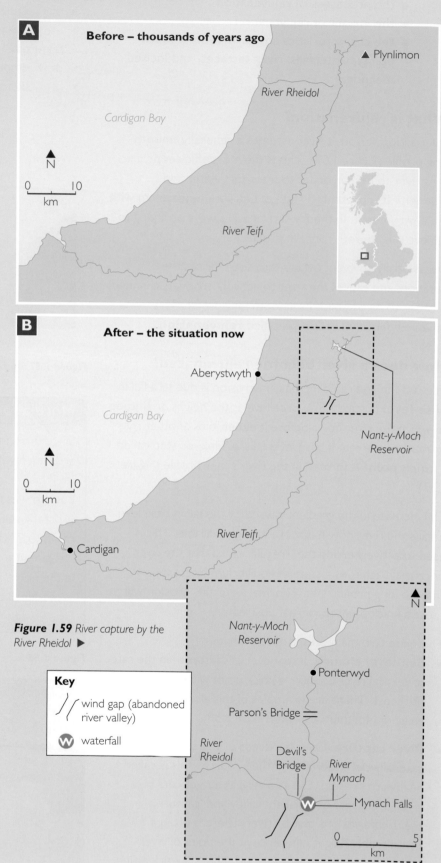

A Before – thousands of years ago

B After – the situation now

Figure 1.59 *River capture by the River Rheidol* ▶

Key

⫽ wind gap (abandoned river valley)

Ⓦ waterfall

Figure 1.60 *Rapids formed at the knick point on the River Rheidol close to Ponterwyd* ▶

Figure 1.61 *Mynach Falls at Devil's Bridge* ▶

At Parson's Bridge (Figure 1.59), there is a spectacular **incised meander** (Figure 1.62). This is a steep gorge cut into a meandering channel. An incised meander is formed when rapid vertical erosion caused by rejuvenation far exceeds the normal rate of lateral erosion that occurs within a meandering channel. As you can see, they really are very dramatic and unusual river features.

Another common feature formed by rejuvenation is a **river terrace** (Figure 1.62). This is the old floodplain, left perched up on the valley side as rapid vertical erosion has cut down into the valley floor. River terraces are common settlement sites, because they offer protection from flooding.

Figure 1.62 *Incised meander on the River Rheidol at Parson's Bridge* ▶

ACTIVITIES

1 The word rejuvenation means 'to restore to youth'.
   a Why do you think this is the geographical term used to describe the river changes covered in this section?
   b Draw your own version of Figure 1.57 and add labels to describe the changes that take place in a river's long profile when base level falls. Consider making your diagram more complex by imagining that base level falls more than once!

2 Study Figure 1.58. The stream joins the sea by dropping steeply over a series of rapids and small waterfalls. How and why will this feature change in the future?

3 Study Figure 1.59. Using simple sketch maps, describe the cause of the rejuvenation of the River Rheidol. Explain clearly what is meant by river capture.

4 Study Figure 1.60. Imagine that you were to take a walk upstream and downstream from these rapids. How would you expect the river and its valley to differ in both directions? Give reasons for your answer.

5 Study Figure 1.62.
   a Use a pencil to draw a large sketch to show the main features of the river and its valley.
   b Add boxed annotations to **identify**, **describe**, and **explain** the following features:
   - incised meander
   - turbulent river flow
   - river terrace
   - land uses

SKILLS BOX

Figure 1.63 A 1:50 000 OS map extract of the River Rheidol ▶

Figure 1.63 is a 1:50 000 OS map extract of the River Rheidol between Devil's Bridge and Ponterwyd (see Figure 1.59).

1 What is the six-figure grid reference of the confluence of the River (Afon) Rheidol and the Afon Mynach?

2 Give a four-figure grid reference to locate the 'wind gap' shown in Figure 1.59.

3 Locate grid square 7477. Look closely at the River Rheidol and its valley. What evidence is there that this stretch of the river has been affected by rejuvenation? Use a simple sketch map and a sketch cross-section to support your answer.

4 Locate grid square 7579. This is an excellent example of an incised meander. It is the same one shown in Figure 1.62. Draw an enlarged version of this grid square to show the main features of the meander. Mark on some contours to indicate how the meander has become a winding gorge. Add labels to describe the main features.

5 What evidence is there on the map to suggest that the present day position of the knick point is just south of the bridge at Ponterwyd?

6 With reference to Figures 1.60–1.63, suggest why there is no public footpath running alongside the river, as might be expected given its stunning features.

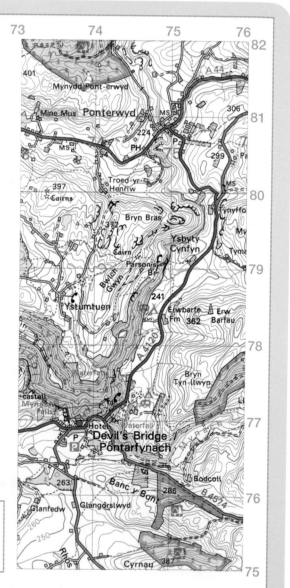

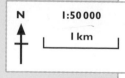

In this section you will learn about:
- the physical and human causes of flooding
- the location of areas in the UK with a high flood risk
- the causes, impacts and responses of flooding in contrasting localities

What is river flooding?

On 19 November 2009, a remote mountain weather station at Seathwaite, in the Lake District, recorded an astonishing 314.4 mm of rain in just 24 hours! This was the wettest day ever recorded in the UK, and it unleashed a devastating flood that inundated farmland, washed away bridges, and tore apart towns and villages in western Cumbria. This flood event is explained in more detail on pages 48–50.

Flooding is where land that is not normally underwater becomes inundated. A river flood occurs when the channel can no longer accommodate the amount of water flowing within it. Water overtops the banks and floods the adjacent land – hence the use of the word 'floodplain' to describe the flat land that borders a river.

River floods often occur after a period of prolonged rainfall. The volume of water steadily increases, causing river levels to rise. Eventually, the river may overtop its banks to cause a flood. These slow build-up river floods are common in the UK in late winter and early spring, when periods of heavy rainfall can coincide with melting snow (Figure 1.64).

Occasionally, more dramatic floods occur – following torrential storms. These are called **flash floods**, and they are often associated in the UK with extreme rainfall events in the summer months. The sheer intensity of the rainfall exceeds the capacity of the river to cope with that amount of water, and flooding results. In 2004, a flash flood tore through the village of Boscastle in Cornwall, causing considerable damage to property (Figure 1.65).

> **Did you know?**
>
> During the height of the 2004 flash flood in Boscastle, an equivalent of 21 petrol tanker loads of water surged through the village every second!

Figure 1.64 *A British river in spring flood* ▲

Figure 1.65 *The aftermath of the flash flood in Boscastle in 2004* ▶

Causes of river floods

Both physical and human factors contribute towards river floods.

Physical

- **Rock type** – Impermeable rocks encourage overland flow, which transfers water rapidly to river channels.

- **Steep slopes** – In mountain environments, steep slopes encourage a rapid transfer of water, both by overland flow and throughflow.

- **Long profile** – If a river has a steep upper course and a flat lower course, water is transferred rapidly to the lowlands, where the gentle gradient causes it to slow down. This can lead to a build up of water, which might ultimately cause flooding.

- **Drainage density** – A high drainage density means that water reaches river channels quickly. This speeds up the transfer of water and can make flooding more likely.

- **Snowmelt** – Snow and ice form a store of water. When they eventually melt, this store is rapidly released and, if the ground is still frozen, the meltwater flows quickly into rivers and increases the risk of flooding.

- **Intense and prolonged rainfall** – Intense rainfall events, like summer storms in the UK, tropical storms (hurricanes) in the Caribbean, and monsoons in India, have the potential to cause significant floods. Prolonged winter rainfall in Europe or the USA can also lead to river floods in early spring.

Human

- **Building construction** – Building on the floodplain creates an impermeable surface (tarmac roads, concrete driveways, slate roofs, etc.) that, together with the system of drains and sewers, transfers water quickly to nearby river channels – causing them to fill up rapidly and overtop their banks (Figure 1.66).

- **Deforestation** – Trees act as huge umbrellas that shelter the ground underneath them from the direct impact of rainfall. Much of the water that falls on trees is evaporated or stored temporarily on leaves and branches. Trees also use up water as they grow. Therefore, when trees are removed from an area, much more water is suddenly available to be transferred rapidly to river channels – thereby increasing the risk of flooding.

- **Agriculture** – With arable farming, there are periods when the soil is left bare and exposed to the elements. This can lead to an increase in runoff, which is made worse if the land is ploughed up and down a slope (instead of across it), because the water can then flow quickly downslope in the grooves left by the plough.

- **Ineffective flood management and warning systems** – The lack of appropriate flood defence schemes, or poor warning systems, can increase the risk of flooding and its impacts on people. If flood defences are allowed to fall into disrepair, or are not upgraded, flooding is more likely. The 2005 floods in New Orleans, USA, were partly blamed on the poor maintenance of river levees (raised river banks). Badly maintained culverts (underground channels) or bridges can also lead to blockages and flooding.

Figure 1.66 *The result of building on the floodplain in West Pulborough, Sussex* ▲

Flood risk in the UK

Flood risk in the UK is managed by three separate agencies, depending on where you live:

- England and Wales – the Environment Agency
- Scotland – the Scottish Environment Protection Agency
- Northern Ireland – the Rivers Agency

Using historic data and computer modelling, each agency has produced an interactive map – available on the Internet – which plots the possible extent of river and sea flooding, and locates flood defences. This service, which enables the public to find detailed information by postcode, is intended to encourage people living and working in flood risk areas to find out more and take appropriate action. In England and Wales, the Environment Agency estimates that over 5 million people live and work in properties at risk of flooding from rivers or the sea (Figure 1.67).

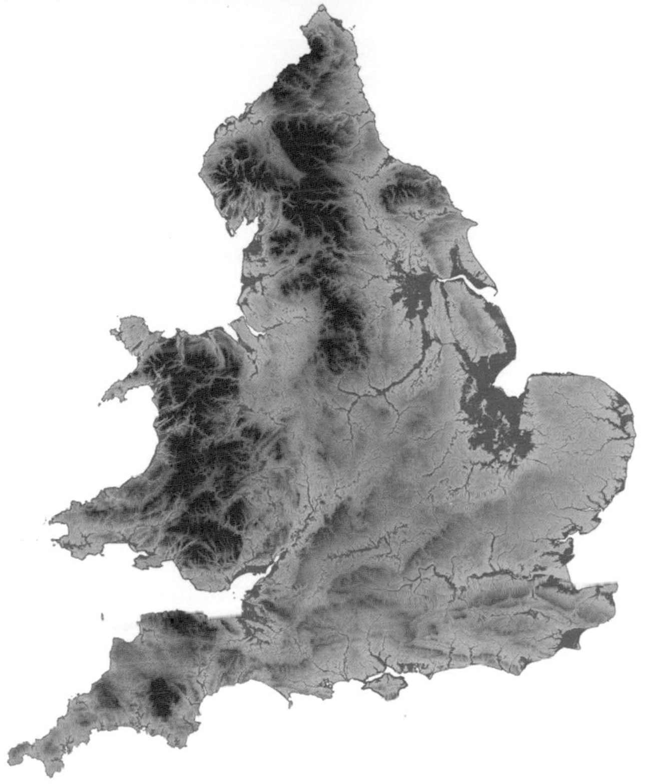

The severity or **magnitude** of a flood is generally related to its **frequency** (how often it occurs) – the higher the frequency, the lower the magnitude or severity. High-magnitude flood events are rare – only occurring, on average, once every few hundred years. Low-magnitude events are much more common – possibly several times a decade.

Look at Figure 1.68, which shows the detailed flood risk map for Cockermouth in the Lake District. The standard flood risk zone marked on the map is for a relatively low-magnitude event – occurring, on average, once every 100 years. The more extreme higher-magnitude flood event is described as occurring, on average, once every 1000 years.

Figure 1.67 Areas in England and Wales at risk from river and sea flooding, according to the Environment Agency. The areas shaded dark blue are the 'at risk' areas. The brown areas are uplands. ▲

Figure 1.68 The Environment Agency flood risk map for Cockermouth ▼

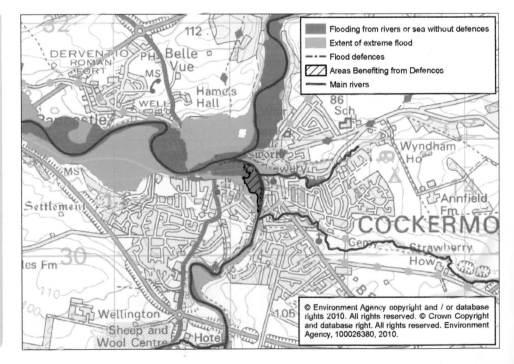

Flooding from rivers or sea without defences
Extent of extreme flood
— · — · Flood defences
Areas Benefiting from Defences
—— Main rivers

ACTIVITIES

1 Study Figures 1.64 and 1.65. Briefly summarise the differences between slow-build-up river floods and flash floods. Consider both causes and effects in your summary.

2 Complete a summary 'spider diagram', or 'mind map', to identify the physical and human factors that contribute to flooding. Use plenty of colour and include simple diagrams and photographs to illustrate your work. This will give you a useful overview to help with revision.

3 Should modern housing estates be built on floodplains? Consider both sides of the debate before arriving at your conclusion.

Internet research

1 Use the appropriate weblink below to access a flood risk map for your home area. To do this, you will need to enter a postcode or placename.

England and Wales:

http://www.environment-agency.gov.uk/homeandleisure/floods/31650.aspx

Scotland:

http://www.multimap.com/clients/places.cgi?client=sepa

Northern Ireland:

http://www.riversagencyni.gov.uk/index/stategic-flood-maps.htm

a Copy and paste the accessed map into a Word document and write a few sentences describing the flood risk and trying to explain its extent. If your home area does not have any flood risk (lucky you!), type in other postcodes or placenames until you find a suitable flood risk.

b Which groups of people might find this map useful?

c Can you think of any negative issues arising from the production of flood risk maps?

2 Make a critical comparison of the flood risk maps produced by the three agencies in the UK. Which agency do you think has produced the most useful and user-friendly map? Justify your choice by identifying advantages and disadvantages of the three maps.

CASE STUDY

The Cumbrian floods of November 2009

Following an extreme rainfall event on 19 November 2009, there was widespread flooding throughout the county of Cumbria. In Cockermouth, a torrent of water up to 2.5 metres high cascaded down the main road – flooding and severely damaging shops, offices and homes. In Workington, on the coast, a policeman died when a bridge collapsed. The main road bridge across the River Derwent in Workington had to be closed – effectively cutting the town in two.

Physical factors

A number of physical factors affected this flood event:

- The record-breaking rainfall came from a weather front that stalled over northwest England (Figure 1.69). This front led to a constant stream of warm, moist air all the way from the South Atlantic. The saturated air was forced to rise by the Cumbrian mountains, which resulted in the exceptionally high rainfall totals.
- The steep mountain slopes transferred the water rapidly to the much flatter lowlands.
- Cockermouth is sited at the mouth of the River Cocker, where it joins the larger River Derwent. The vast amounts of water at this natural river confluence led to serious flooding in the town.

- The rock in Cumbria is mostly impermeable and, with soils already saturated from earlier rainfall events, much of the rain that fell on 19 November flowed rapidly into the river channels as overland flow.

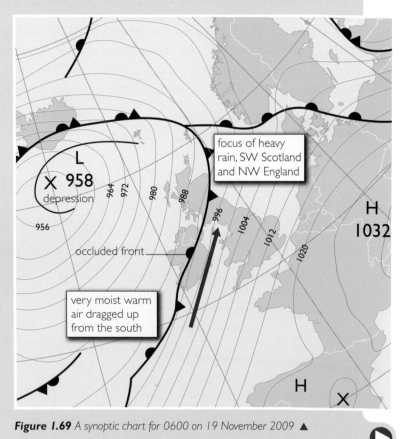

Figure 1.69 *A synoptic chart for 0600 on 19 November 2009* ▲

Impacts

This event was extreme – a 1 in 1000 year flood – and, despite being correctly forecast by the Met Office, it caused much more devastation than had been expected:

- Over 1300 people had their homes flooded.
- Power supplies were cut off and schools closed.
- A number of bridges collapsed and others were damaged – cutting off whole communities. All of Cumbria's 1800 bridges had to be inspected to make sure they were safe.
- Insurance companies estimated that the eventual cost of the damage was more than £100 million.
- Many farm livestock were killed, and pastures were covered with rocks carried downstream by the swollen rivers. Fences and walls were destroyed, farm buildings damaged and machinery ruined.
- Local businesses, such as shops, pubs and hotels faced months with little prospect of earning any money.

Responses

- In the immediate aftermath of the flood, the main focus was on search and rescue. About 200 people had to be airlifted from the roofs of their houses by RAF helicopters, or rescued by boats.
- In the following days, as the floodwaters receded, buildings that had been flooded were assessed for their safety by experts before residents and shopkeepers were allowed back. Electricity supplies were re-established and the process of repair and refitting began. For weeks, the streets were lined with skips overflowing with damaged shop stock and people's ruined personal belongings.
- Many residents had to be housed elsewhere for several months.
- Network Rail built a new station in Workington – on wasteland on the north side of the Derwent – to connect the two sides of the town cut off by the closure of the road bridges. The army also built a temporary footbridge across the river.
- In Cockermouth, some small businesses were provided with temporary trading accommodation in the town centre, while their premises were refitted. Several roads in the town centre were so severely broken up by the raging floodwaters, that they had to be completely re-surfaced.

Figure 1.70 *Economic damage in Cockermouth after the 19 November flood* ▲

ACTIVITIES

1 Study the text and the photograph (Figure 1.70). Draw up a table identifying the social, economic and environmental impacts of the 2009 Cumbrian floods.
2 What do you think the long-term impacts (over a period of months) of the Cumbrian floods would be on the following people:
   - Farmers
   - Local people living in Workington north of the River Derwent
   - Shop-owners in Cockermouth whose properties have been flooded and stock damaged
   - Villagers in Low Lorton, cut off from each other by the destruction of their bridge across the River Cocker.
   - Hotel owners who rely on tourism for their income
3 'Physical factors were largely responsible for the Cumbrian floods.' Do you agree with this statement? Why or why not?

The Cockermouth flood – an aerial photo study

Study Figure 1.71. It shows widespread flooding in the centre of Cockermouth following the record-breaking rainfall event of 19 November 2009. The photograph is taken looking upstream along the River Derwent.

Figure 1.71 *Cockermouth underwater on 20 November 2009* ▲

1 Study Figure 1.71.

  a Describe the extent of the flooding in the photograph. Try to identify the different land uses affected by the flood.

  b Use evidence from the photograph to suggest some social, economic and environmental impacts of the flooding.

  c Using evidence from the photograph, suggest some short- and longer-term responses to the flooding in Cockermouth.

2 Look closely at the flooding shown in the photo (Figure 1.71) and compare it with the Environment Agency's flood risk map (Figure 1.68).

  a Comment on the severity of the flooding that occurred in Cockermouth in 2009. Was the event, as some newspapers reported, really a 1-in-1000-year flood?

  b How accurate was the extent of the flooding in Cockermouth in 2009, as predicted by the Environment Agency's flood risk map?

Flooding in the Philippines in September 2009

On 26 September 2009, the island of Luzon in the Philippines was struck by a powerful typhoon. Typhoon Ketsana (local name Ondoy) moved slowly over the island, causing torrential rainfall – particularly on the capital city, Manila (Figure 1.72). According to the state weather bureau PAGASA, about 455 mm of rain fell in just 12 hours – the highest intensity rainfall ever recorded in the Philippines. Over 340 mm fell in a six-hour period! Around 250 people were killed by the floods and associated landslides.

One of the worst hit areas was Marikina City – part of the vast urban sprawl of Greater Manila. Here the massive rainfall caused the Marikina River to burst its banks – sending torrents of water down busy streets. Homes and businesses were destroyed as cars, swept up in the tumult, crashed into buildings. The floodwaters were so high that, when they receded, debris was left dangling from overhead power lines. Figure 1.73 describes some of the impacts of the flooding.

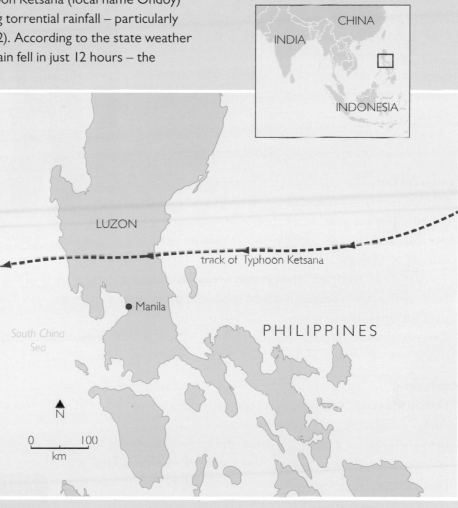

Figure 1.72 *The track of Cyclone Ketsana – just north of Manila, the capital of the Philippines* ▲

Economic	Social	Environmental
• Many homes and businesses were destroyed in Manila. • The international airport was closed for a day. • The estimated cost of the flooding was US$100 million. • 3% of future rice production in the Philippines was lost because of damage to farmland.	• About 250 people were killed. • Around 60 000 people were displaced from their homes. • Some people had their power supplies and water cut off. • A lot of mental distress was shown by Filipinos on Internet social networking sites.	• The flooding left thick mud on roads, fields, and in homes and businesses. • 100 000 litres of oil from Noah's Paper Factory in Marikina City leaked out because of the surge of floodwater. • Landslides were widespread, affecting rural areas and farmland.

Figure 1.73 *The impacts of the flooding* ▲

Both physical and human factors contributed to the flooding and its impacts:

Physical factors

- The heavy rain that fell on Luzon was due to the passage of an intense tropical storm that developed into Typhoon Ketsana as it passed over the South China Sea between Luzon and Vietnam.
- September is the height of the typhoon (hurricane) season in this part of the world, and several storms had already passed through the Philippines. The already saturated ground promoted rapid runoff into rivers, as well as encouraging landslides to occur.

Human factors

- Despite warnings being issued, there was some criticism that the government had underestimated the scale of the event, and that people had not been warned sufficiently.
- Torrential rain falling on urban Manila led to flash flooding in many parts of the city, because the storm drains could not cope with the huge quantities of runoff flowing over impermeable surfaces.
- Deforestation on hillsides also contributed to many landslides and increased runoff into rivers.

Responses

- During the event, Filipinos responded by seeking shelter from the heavy rain and rising floodwaters.
- In the first few hours after the event, the immediate priority for families, communities and the government (through its disaster relief agency the National Disaster Coordinating Council) was search and rescue. About 1000 soldiers were deployed, and the Red Cross and Coast Guard provided expert assistance using rubber boats (Figure 1.74). Social networking sites identified people in need of help, and their locations were plotted on Google Maps. Food, water, emergency shelter and medical help were all needed.
- The immediate clean-up operation took many weeks, but a lot of major work was still necessary to rebuild properties destroyed by the floods. Businesses also needed support to become re-established. Roads and bridges damaged by the floodwaters and landslides needed to be repaired or re-built. Farmers affected by the event also needed long-term support.
- The Philippines government welcomed international support from the United Nations, together with individual countries – such as the USA, Japan and China – and NGOs, such as AmeriCares.

Figure 1.74 *Rescue efforts in Marikina City, Greater Manila, in September 2009* ▲

ACTIVITIES

1 Work in pairs or small groups for this activity. Compare and contrast the two flooding case studies, considering both their similarities and their differences. Include the following aspects:
   - The type of flood event and the scale.
   - The importance of physical and human factors in contributing to the flooding.
   - The impacts of the two events
   - The similarities and differences between the responses?
2 Conduct a study of a recent flood event using the Internet to help you. Follow the same structure as above – considering the physical and human **causes**; the social, economic and environmental **impacts**; and the short- and long-term **responses**.

In this section you will learn about:
- flood management options
- hard engineering options (dams, river straightening, levees and embankments, diversion spillways)
- soft engineering (forecasts and warnings, land-use management, wetland and river bank conservation and river restoration)

Flood management options

River flooding is an entirely natural event – it's what rivers do. After all, that's why a floodplain is called that! However, flooding is starting to have an increasing impact on people and their activities (Figure 1.75). This is largely because, due to a lack of space for the increasing population, more and more decisions are being made to build houses, businesses and transport links on land at risk of flooding.

Often these new developments increase the flood risk by removing trees, replacing vegetation with tarmac surfaces, and installing drains and sewers – all of which direct water rapidly into rivers. People are increasingly putting themselves and their property in harm's way, which is why flooding and its risks need to be managed.

Figure 1.75 *Flooding in Carlisle in 2005* ▲

Approaches to flood management

- **Basin management** employs schemes that aim to reduce the amount and speed of water flowing towards a river. Large-scale tree planting (afforestation) increases interception and uses up some of the water. The construction of a reservoir to store and regulate water flow is another effective form of basin management.

- **Channel management** makes changes to the river channel itself, e.g. increasing the height of the banks, or straightening the channel.

- **Hard engineering** intervenes directly with the river's natural processes. It usually involves building things, e.g. a concrete embankment alongside a river channel.

- **Soft engineering** aims to work with the river's natural processes. It tends to have less impact on the environment than hard engineering. One example is the creation of wetlands on a floodplain to store water.

- **Warning and forecasting** is a behavioural response to flooding. By careful monitoring of precipitation and river levels, it's possible to issue warnings to people to prepare themselves for a possible flood, e.g. move their furniture upstairs and use sandbags to seal the outside doors.

- **Do nothing** – no surprise what this means! This approach is often adopted in locations where there is a low level of risk. Some people get used to regular, small floods. They just mop up and carry on with their lives.

Making decisions

Look at Figure 1.76. It shows a selection of flood prevention measures that can be adopted to reduce the risk of flooding. As you can see, there is a big choice! Choosing the right option for a particular place depends on many factors, including construction costs, maintenance costs (how sustainable a scheme might be), social, economic and environmental impacts, the views of local people, government policy, and so on. The planning and decision-making often takes many years.

Very often the main factor is cost, which relates to the potential severity or magnitude of a particular flood risk. Most flood defence schemes aim to protect places from a 1-in-100-year flood. This is a flood with a magnitude likely to occur, on average, only once every 100 years. Technically, it is perfectly possible to construct defences to protect against more severe floods, such as the 1-in-1000-year flood that struck Cockermouth and Workington in November 2009. However, the cost of this would be enormous, and the structures involved would completely alter the nature of the river environment. Decision-making is all about compromise between costs, levels of impact, and protection of people and property.

Hard-engineering options

Dams and reservoirs

Dams and reservoirs are widely used around the world to regulate river flow and reduce the risk of flooding. Most dam projects are **multi-purpose**, which means that they serve several functions, e.g. flood prevention, irrigation, water supply, hydroelectric power production, and recreation (sailing, fishing, etc.).

Dams can be very effective in regulating water flow. During periods of high rainfall, water can be stored in the reservoir. It can then be released slowly during low-flow conditions. This allows a steady flow of water to be maintained throughout the year. However, while dams and reservoirs are usually effective, they are very controversial. They cost a huge amount of money to build, and the reservoirs behind them often flood large areas of useful land (sometimes resulting in people being forcibly moved).

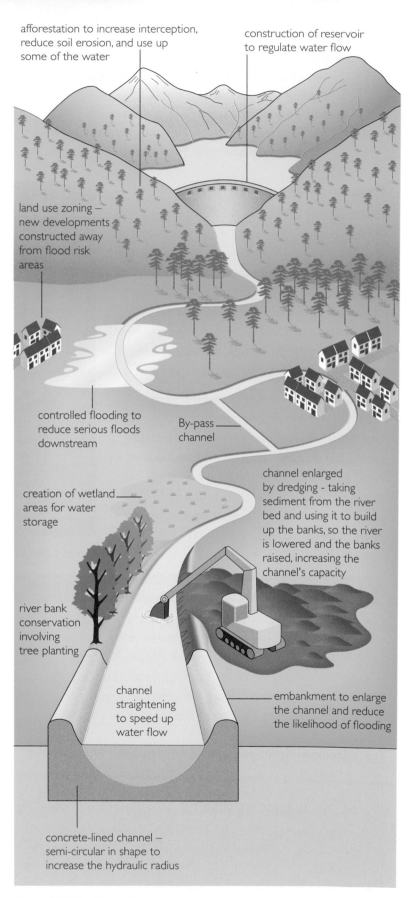

afforestation to increase interception, reduce soil erosion, and use up some of the water

construction of reservoir to regulate water flow

land use zoning – new developments constructed away from flood risk areas

controlled flooding to reduce serious floods downstream

By-pass channel

creation of wetland areas for water storage

channel enlarged by dredging - taking sediment from the river bed and using it to build up the banks, so the river is lowered and the banks raised, increasing the channel's capacity

river bank conservation involving tree planting

channel straightening to speed up water flow

embankment to enlarge the channel and reduce the likelihood of flooding

concrete-lined channel – semi-circular in shape to increase the hydraulic radius

Figure 1.76 *Flood prevention measures* ▲

River straightening

When a river cuts off a meander to form an oxbow lake, it straightens and shortens its course. This results in an increase in channel gradient, which in turn leads to an increase in velocity. Engineers apply this same principle when a river is artificially **straightened** – water flows faster and is transferred downstream more rapidly. However, while this might relieve the flood risk in one location, it increases it further downstream. So the problem is not really dealt with, just shifted somewhere else!

In some places, the straightened sections of river channel are lined with concrete to speed up the flow and reduce bank collapse, which can cause the channel to silt up. This technique is called **channelisation** (Figure 1.77). However, while channelisation improves the rate of flow and also benefits transportation, it does create a very unnatural-looking river environment, and damages local wetlands and wildlife habitats. In fact, in some places, rivers are being returned to their natural state by ripping out the concrete channels.

Levees and embankments

Raising the height of its banks increases the capacity of a river channel. This allows it to hold more water before flooding occurs. A raised riverbank is usually called an **embankment** in the UK, but in the USA the term **levee** is more common.

Raising the height of riverbanks can be done quite easily and cheaply. A dredger mounted on a barge, or on the riverbank, scoops sediment from the river channel (lowering the riverbed) and dumps it onto the banks to raise their height (Figure 1.76). This double approach is sustainable and has a minimal impact on the environment.

Alternative measures involve built structures like concrete walls. They are much more expensive to construct and are more intrusive on the environment. But they are generally regarded as a more-effective form of defence, and are widely used to protect towns and cities from potentially damaging floods.

Look at Figure 1.78. It shows part of the embankment next to the River Rhone in Lyon, France. The River Rhone has a long history of serious flooding, particularly in late spring (following ice and snowmelt in the Alps). The embankments have been imaginatively landscaped to provide green spaces and cycle ways.

Figure 1.77 *Channelisation in Damascus, Syria* ▲

Figure 1.78 *A landscaped embankment alongside the River Rhone in Lyon, France* ▼

Diversion spillways (by-pass channels)

Another option to reduce flooding in towns and cities is the construction of **diversion spillways** – to divert excess water away from built-up areas. During high-flow periods, sluice gates are opened to allow the excess water to flow along the new channel. In some places, the by-pass channel is a permanently flowing part of the river – although the flow does need to be controlled to stop the river abandoning one of the channels.

The Big Sioux River flows through the city of Sioux Falls in South Dakota, USA. In the spring, melting snow and ice create a major flood risk, particularly in the low-lying downtown district, so a diversion channel was constructed to cut off the giant 25-km meander that loops around the city (Figure 1.79).

In the UK, a by-pass channel – now named the Jubilee River – was constructed on the River Thames near Maidenhead in Berkshire (Figure 1.80). The 11km long channel was opened in 2002, and cost over £80 million to build. It's had a positive impact on the environment, by creating new wetlands. It's also popular for recreational activities, like walking and fishing.

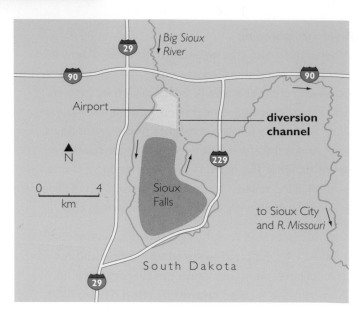

Figure 1.79 *The diversion channel around Sioux Falls in the USA* ▲

Figure 1.80 *The Jubilee River (by-pass channel) near Maidenhead* ▲

ACTIVITIES

1 Study Figure 1.76.
   a Briefly outline the differences between hard-engineering and soft-engineering solutions.
   b Sort the flood defence measures shown in Figure 1.76 into two groups: hard and soft engineering.
   c Are there any measures that do not fit comfortably into one or other category? Explain your answer.
   d Under what circumstances is the 'Do nothing' approach the most appropriate option?

2 Study Figure 1.79.
   a Why was Sioux Falls at risk from flooding? Try to refer to river characteristics (e.g. gradient and velocity) and landforms in your answer.
   b Use the scale to measure the length of the by-pass channel. How does this compare with the length of the natural channel?
   c Do you think water would flow faster or slower along the by-pass channel? Explain your answer.
   d In pairs, suggest why the authorities kept the original channel open, rather than replacing it with the by-pass channel? There are lots of possible reasons here!

Internet research

Write a short report about the Jubilee River by-pass channel on the River Thames at Maidenhead. There is plenty of information available on the Internet, e.g. http://www.thamesweb.co.uk/floodrelief/relief_bckgrnd.html

- Why was the decision taken to build the new channel?
- Were there any other alternatives?
- Briefly describe the project, and assess its success.
- Include annotated photos and maps.

Soft-engineering options

Soft engineering involves working with natural processes. Most schemes aim to reduce and slow down the movement of water to a river channel. This can be achieved in lots of ways.

Planting trees

Afforestation programmes, often in upland areas, increase interception and slow down water transfer. The trees also take up water for their growth – thus reducing the amount flowing into river channels.

Establishing wetlands

Wetland environments on river floodplains are highly efficient at storing water. They are also important habitats and breeding grounds. Increasingly, areas that have been drained and developed are being returned to their natural state, as wetland environments.

Riverbank conservation

This involves stabilising riverbanks to prevent their erosion and collapse – and the subsequent silting-up of the river channel. It often involves planting vegetation, like bushes and trees – the roots of which bind together the loose sediments forming the banks.

Land-use management and floodplain zoning

Careful planning can significantly reduce flood risk. Areas that are close to rivers and prone to flooding can be kept clear of expensive and vulnerable developments, like housing. Instead, they can be used for pasture, or for recreational uses (parkland, racecourses, etc.). Not only does this type of land use reduce the amount of water flowing into rivers (and its speed) but, should a flood event occur, the damage will be minimal. This approach is called land-use **floodplain zoning** (Figure 1.81).

Since the 1970s, the Federal Emergency Management Authority (FEMA) has prepared Flood Insurance Rate Maps (FIRMs) to show areas that are at high risk of flooding after intense storms. These maps are freely available on the Internet.

Each map identifies a **base line flood**, taken as a 1-in-100-year flood. All areas below this elevation are considered to be at 'high risk'. Homeowners must have house insurance. Any proposed developments in this zone require a Floodplain Development Permit. Each permit requires official surveys and a considerable amount of time and money.

Local authorities are encouraged to keep the floodplain below the base line as open natural space – to allow the river to flood. Appropriate land uses include parking, playgrounds, wetlands, recreational areas and gardens.

In the UK, similar maps are produced. For example, the Environment Agency produces maps for England and Wales which identify areas at risk from both the 1-in-100-year flood and also the more extreme 1-in-1000-year flood. These maps are considered by planning authorities when applications for developments are made, and by insurance companies when calculating insurance premiums. Look back at Figure 1.68 (page 47) to see the flood risk map for Cockermouth.

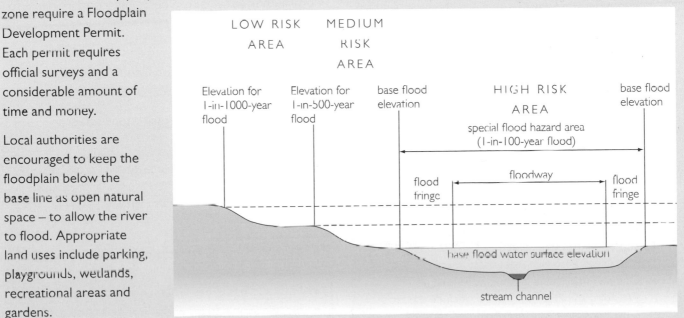

Figure 1.81 Land-use zoning in the USA ▲

River restoration

River restoration involves returning a river that has been artificially altered to its natural state. If left alone, rivers will flood occasionally – but they will mostly be small-scale floods. The natural processes and features of a river, such as meanders and wetlands, work to slow down river flow – reducing the likelihood of a major flood. River restoration projects often involve restoring features like meanders to rivers that have been artificially straightened.

A good example of river restoration in the UK is the River Skerne in Darlington. Here a 2-km stretch of river running through part of Darlington has been restored to create an attractive wetland environment (Figure 1.82). For 200 years, the river had been straightened to reduce the flood risk and improve drainage. It has now been transformed by the River Restoration Centre as a demonstration project from a polluted industrial wasteland into a thriving wildlife environment.

Figure 1.82 *The restoration of the River Skerne* ▲

ACTIVITIES

1 Study Figure 1.76 again.

   a Create a simple summary poster to describe a selection of soft-engineering measures. To do this, draw a large, simplified cross-section across a floodplain from one side of the valley to the other. Use Figure 1.25 on page 23 to guide you.

   b Now locate a selection of soft-engineering measures, using simple illustrations. Describe each one using an annotation.

   c Use the Internet to find some thumbnail photos to help you illustrate your poster.

2 Study Figure 1.81. This activity works well as a discussion activity with small groups or pairs.

   a What do you understand by the term floodplain zoning?

   b How useful is floodplain zoning in reducing flood risks?

   c What land uses are appropriate within the High Risk Area? Explain why they are appropriate.

   d Suggest appropriate land uses for Medium and Low Risk Areas. Also give reasons for these.

   e Outline the possible arguments for and against extending the High Risk Area to include the zone liable to a 1-in-1000-year flood.

Internet research

River restoration is becoming an extremely popular management option. In the UK, the River Restoration Centre has been involved in research for many years, and has established demonstration projects to show local authorities what can be done to restore rivers to their natural states.

Do some Internet research to find out more about one of the River Restoration Centre's projects, either the River Skerne (Figure 1.82) or the River Cole.

http://www.therrc.co.uk/case_studies/skerne_brochure.pdf

http://www.therrc.co.uk/case_studies/cole_brochure.pdf

- Aim to write about a side of A4 describing your chosen project.
- Consider why this stretch of river was restored and what was done.
- How successful has it been and who has benefited?
- Include simple maps and annotated photos in your report.

Flood forecasts and warnings

Forecasting a possible flood event and warning people is a behavioural response to flood risks. Throughout the UK, rivers and river basins are monitored remotely, using satellite communications and computers. Instruments are used to gauge river levels and precipitation amounts.

Authorities, such as the Environment Agency in England and Wales, monitor this information and issue warnings if flooding is likely. Three levels of warning can then be issued, using the Internet and emergency services like the police.

- **Flood Watch** means that flooding of low-lying land and roads is expected. People are urged to be prepared and to watch river levels.
- **Flood Warning** means that there is a threat to homes and businesses. People are urged to move items of value to higher floors and to turn off services, such as electricity and water. Evacuation is a possibility.
- **Severe Flood Warning** means that extreme danger to life and property is expected. People are urged to seek a high place in their home for safety, or to evacuate the property.

CASE STUDY

Flood management at Carlisle in Cumbria

In 2005, prolonged heavy rainfall caused the River Eden in Carlisle to burst its banks (Figure 1.75). Two people were killed and over 1700 properties were affected. Large parts of the city were evacuated and several main roads were inundated. Damage costing an estimated £500 million was caused.

As a result of the 2005 flood, the Environment Agency embarked on a two-phase plan to defend Carlisle from future floods.

- **Phase 1.** In 2008, a 4.5-km stretch of raised flood defences (flood walls and earth embankments) was completed along the River Eden and River Petterll (Figure 1.83). These defences, costing about £13 million, protect 1500 properties.
- **Phase 2.** During 2010, a further 5-km stretch of raised flood defences, costing £25 million, was completed – focusing on the River Caldew and protecting another 1000+ properties. A new pumping station was also installed.

Both phases are designed to protect properties from a 1-in-200-year flood. This level of flood protection also takes into account predicted changes resulting from global warming.

In addition to the construction of raised bank defences, the rivers have also been cleared of obstructions, and some footbridges have been raised to prevent blockages occurring during high-flow conditions. Areas of floodplain below the city have also been opened up to allow some flooding to take place. This has created new wetland areas and wildlife habitats.

In November 2009, Carlisle's new defences (only partly completed at the time) were tested by the massive rainfall totals that, elsewhere in Cumbria, led to devastating floods in Cockermouth and Workington (see pages 48-50). By using temporary defences on the River Caldew, the Environment Agency was able to successfully prevent flooding in Carlisle, even though the water rose to just 5 cm below the top of the temporary defences!

SKILLS BOX

Study Figures 1.83 and 1.84. The area in the photograph shows part of eastern Carlisle. Use the key on the photograph to identify the colours used to show the location of floodwalls and floodbanks.

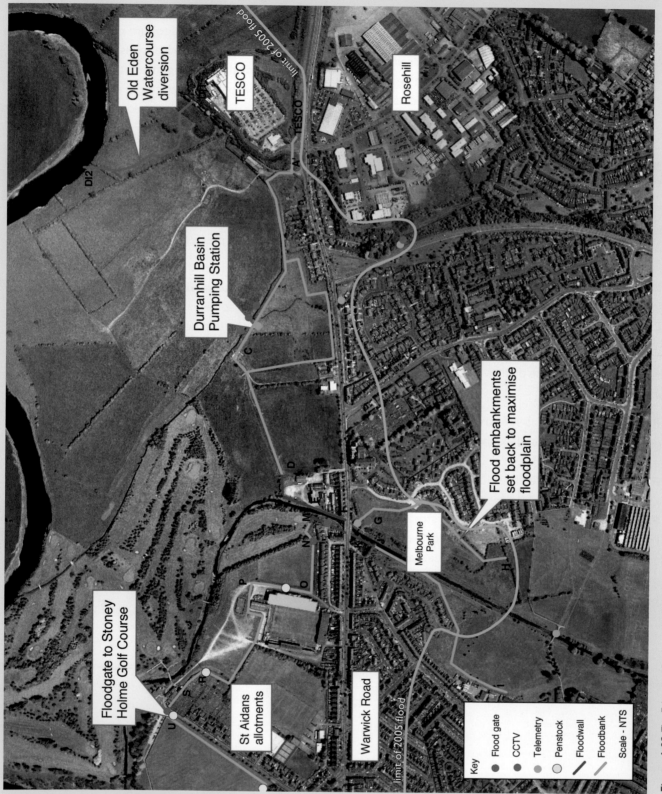

Figure 1.83 *The Eden and Petteril Flood Alleviation Scheme in Carlisle* ▲

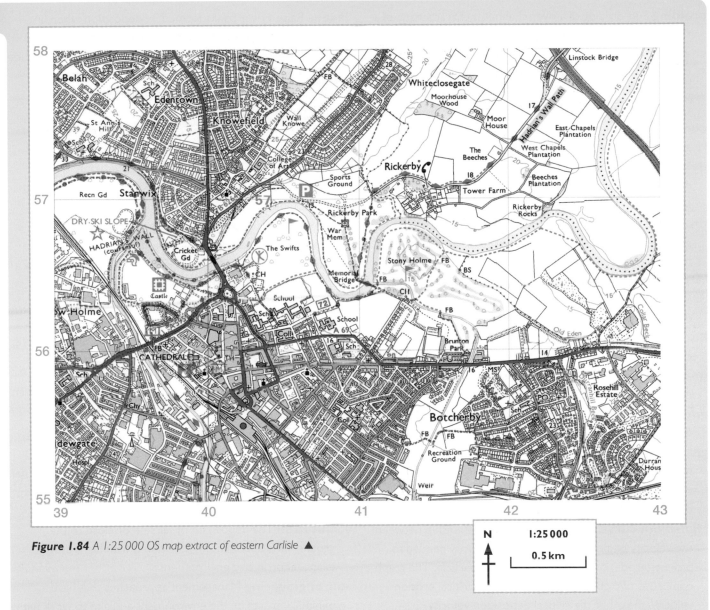

Figure 1.84 A 1:25 000 OS map extract of eastern Carlisle ▲

N 1:25 000
0.5 km

1 Describe and suggest reasons for the location of the raised flood defences (floodwalls and floodbanks) on the photograph. Bear in mind the limit of the 2005 flood (also on the photograph). Refer to specific locations using the photograph and the map.

2 Locate Melbourne Park on the floodplain of the River Petteril. Why do you think the raised flood defences have been set back at the edge of the floodplains?

3 What are the land uses on the floodplains of the River Eden and River Petteril that now lie between the rivers and the new raised flood defences? Do you think they are appropriate for such a high-risk flood area? Why?

4 Locate Durranhill Basin Pumping Station. Notice that a flood storage basin has been created here (enclosed on all sides by a floodbank). What is its purpose and how will the pumping station be used?

5 Comment on the location of the large Tesco store. Do you think this is a wise choice of location?

braiding Sub-dividing of a river channel, a characteristic of a river with a variable discharge

channel roughness Measurement of the smoothness of the wetted perimeter and the extent to which rocks protrude into the flow of a river. A smooth profile will result in little turbulence and a high velocity

corrosion River erosion whereby transported particles grind away at the river bed and banks

deltas An area of deposition (usually mud and sand) at the mouth of a river. Two common types of delta are arcuate (e.g. Nile) and bird's foot (e.g. Mississippi)

discharge The volume of water passing a given point in a given period of time, usually expressed in 'cumecs' (cubic metres per second) and calculated as: cross sectional area x velocity

drainage basin Area of land, bordered by a watershed, drained by a river and its tributaries

flooding Inundation of land not normally under water, e.g. river floodplain

floodplains Extensive, flat land immediately adjacent to a river that is prone to periodic inundation

gorge Steep-sided narrow valley often found immediately downstream from a waterfall

graded profile Smooth concave long profile of a river that has achieved a state of equilibrium with its environment

hard engineering Commonly, built structures such as concrete walls designed to resist natural processes

Hjulstrom curve Graph showing the relationship between the velocity of a river and the critical velocity needed to pick up (erode) or deposit an individual particle. The graph comprises two curves, one showing the critical erosion velocity, the other the critical deposition velocity

hydraulic action River erosion involving the sheer force of flowing water

hydraulic radius Measurement of channel efficiency calculated as cross-sectional area / wetted perimeter. The higher the value, the greater the efficiency (less energy lost by friction with bed and banks) of the channel

incised meander Steep-sided meander formed during a period of significant down-cutting following rejuvenation

kinetic energy When water moves downhill, potential energy is converted to kinetic energy and used to do 'work'

knick point Point in the river's long profile, often marked by a waterfall, that represents the extent to which a river has re-graded its profile following rejuvenation. Over time a knick point will gradually migrate upstream due to waterfall retreat

levees Raised river banks formed by deposition when, in flood, a river spills over the sides of its channel

magnitude-frequency analysis Analysis of the magnitude and frequency of events such as floods, which are often characterised by 'high magnitude, low frequency' events (negative association)

meander Sweeping bend or curve in a river, characterised by having a depositional slip-off slope on the inside bend and an erosional river cliff on the outside bend

potential energy Stored energy in a still body of water that has potential to do 'work'

pothole Circular hollow or depression formed by erosion in bedrock on a river bed

rapids Series of relatively small 'steps' in the long profile of a river forming a stretch of turbulent white water

rejuvenation Renewed period of river erosion resulting from a fall in base level (often sea level). As the river cuts down into its channel characteristic features form including incised meanders and river terraces

river restoration Restoring a river channel to its natural course following earlier intervention, e.g. channel straightening. River restoration usually involves soft engineering and aims to improve natural habitats and the amenity value of the river.

river terrace Abandoned floodplain left perched above the current floodplain following a renewed period of erosion (rejuvenation)

salinisation The deposition of salts on or close to the ground surface commonly associated with high rates of evaporation in arid environments

soft engineering Management approaches that have minimal impacts on the environment and aim to work with natural processes (e.g. planting trees in a river basin to reduce the risk of flooding)

solution Dissolving of chemicals as water flows over soluble rocks

storm hydrograph Line graph showing the discharge of a river over a period of time as it responds to an individual storm event

velocity Speed of flow of a river. The line of fastest flow in a river is called the thalweg

water balance Audit of water based on the equation: precipitation = runoff + evapotranspiration +/- soil moisture

waterfall 'Step' in the long profile of a river characterised by water cascading over a rock lip into a deep plunge pool

wetted perimeter Length (in metres) of river channel in contact with water at a river cross-section

Exam-style questions

1 (a) How does a river transport its sediment? *(4 marks)*

(b) Study Figure 1.2 (page 9). What effect would deforestation have on the drainage basin hydrological system? *(4 marks)*

(c) What are the causes of river rejuvenation? *(7 marks)*

(d) Discuss the advantages and disadvantages of soft engineering as a flood management strategy. *(15 marks)*

2 (a) Study Figure 1.42 (page 34). How has this pothole been formed? *(4 marks)*

(b) Study Figure 1.78 (page 55). What is the purpose of this form of hard engineering? *(4 marks)*

(c) What is a river terrace and how is it formed? *(7 marks)*

(d) Discuss the importance of drainage basin characteristics on the shape of the flood hydrograph. *(15 marks)*

3 (a) Study Figure 1.62 (page 43). Describe the characteristics of this incised meander. *(4 marks)*

(b) Suggest why the long profile of a river may not always be a smooth curve. *(4 marks)*

(c) Why is an understanding of the Hjulstrom Curve (Figure 1.22, page 21) essential in explaining the processes operating in a river? *(7 marks)*

(d) To what extent are human factors more important than physical factors in causing river floods? *(15 marks)*

EXAMINER'S TIPS

(a) What are the different methods of river transport? Consider using an annotated diagram.

(b) What happens when the trees are chopped down? Consider both the water pathways and the relative speeds of flow.

(c) Write about the two main causes, giving examples. Use the appropriate geographical terminology.

(d) Define soft engineering. Consider the advantages and disadvantages. Make comparative comments (cost, effectiveness, and so on). You must have some discussion. Use case studies to support your points.

(a) Refer to the evidence in the photo. Try to use the correct terminology.

(b) Identify the type of hard engineering. Focus on the purpose (function). Refer to the evidence in the photo.

(c) Give a clear definition. Write a detailed account of its formation using the correct geographical terminology. Consider using simple diagrams.

(d) What basin characteristics affect the shape of the flood hydrograph? Use the correct geographical terminology (such as peak discharge). Include some discussion. Draw simple hydrographs to illustrate your points.

2 Cold environments

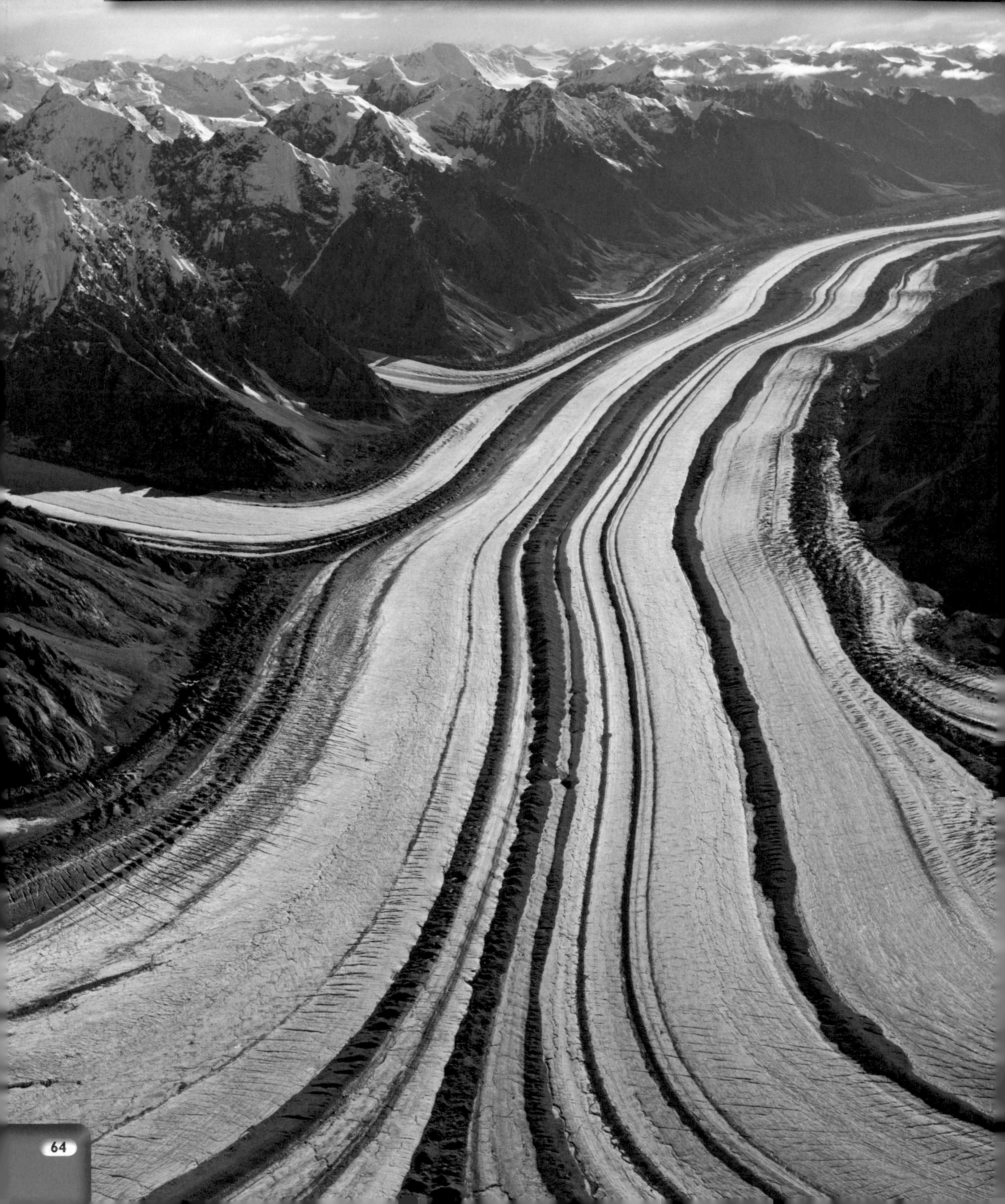

Introduction

Today, about 10% of the earth's land surface is covered by ice – Antarctica (85%) and Greenland (10%) account for most of this. In addition, vast swathes of Russia and Canada experience periglacial conditions, with low-growing tundra vegetation and permanently frozen soils. Mountain regions, such as the Himalayas, also experience extreme cold and are actively eroded by glaciers.

In the past, just 18 000 years ago, much of the USA and northern Europe were in the grip of an ice age. In the British Isles, only southern England remained ice-free.

In this chapter you will learn about the processes and landforms associated with cold environments. You will investigate some of the management challenges affecting these fragile environments, such as the need to balance economic exploitation and environmental conservation.

Books, music, and films

Books to read
Northern Lights by Philip Pullman

Music to listen to
'Snow (Hey oh)' by The Red Hot Chilli Peppers

Films to see
The Day After Tomorrow
March of the Penguins
Ice (1998, tv movie)
Ice Age 2

◀ *The Barnard Glacier in the St Elias Mountains is 53 km long*

Which country is the Barnard Glacier in?
What are the dark grey lines on the glacier?
Why do they form a symmetrical pattern?
Which way is the glacier flowing?
Has the glacier shrunk in size?

About the specification

'Cold environments' is one of the three Physical Geography option topics in Unit 1 – you have to study at least one.

This is what you have to study:
- The global distribution of cold environments – polar (land and marine-based), alpine, glacial, and periglacial.
- Glaciers as systems. Glacial budgets.
- Ice movement – types of flow: internal deformation, rotational, compressional, extensional, and basal sliding. Warm- and cold-based glaciers.
- Glacial processes and landscape development.
- Weathering in cold environments – specifically, frost shattering.
- Erosional landforms: corries, arêtes, pyramidal peaks, glacial troughs and associated features. Depositional landforms: types of moraine, and drumlins.
- Fluvioglacial processes: the role of meltwater erosion and deposition. Fluvioglacial landforms: meltwater channels, kames, eskers, and outwash plains.
- Periglacial processes: nivation, permafrost formation, frost heave, and solifluction. Periglacial landforms: nivation hollows, ice wedges, patterned ground, pingos, and solifluction lobes.
- The exploitation and development of tundra areas and the Southern Ocean:
  - the traditional economies of an indigenous population and recent changes and adaptations;
  - early resource exploitation by newcomers – whaling and/or sealing;
  - more recent development – oil in Alaska, fishing, tourism.
  The concept of fragile environments. The potential for sustainable development.
- The future of Antarctica, including the contemporary issues of conservation, protection, development, and sustainability in a wilderness area.

In this section you will learn about:
- the distribution of ice in the UK 18000 years ago
- the different types of cold environment
- the present-day distribution of cold environments

The UK's 'big freeze' – the winter of 2009/10

Do you remember the winter of 2009/10 (Figure 2.1)? It was the coldest winter for 30 years – and one of the four coldest winters in the last century. In parts of Scotland, the temperature fell below that of a domestic freezer! Schools were closed, many people were injured slipping on ice, and gas supplies became so low (due to high domestic demand to heat homes) that supplies to some industries had to be rationed. The UK also suffered widespread travel chaos, including severe shortages of road grit, which led to the closure of many roads. Newspapers reported the extreme conditions with headlines such as 'The Big Freeze' and 'Britain's Arctic Plunge'. But how typical were these conditions, compared with cold environments globally?

Figure 2.1 The Big Freeze of 2009/10 in the UK ▲

Cold environments of the past

About 18000 years ago, much of the Northern Hemisphere was plunged into an ice age. In the UK, only southern Britain escaped being covered by ice (Figure 2.2). Scientists believe that this was one of as many as 20 **glacial periods** – each one separated by a warmer **inter-glacial**. Collectively, this period of time – lasting from 2 million years ago until about 8000 years ago – is known as the **Ice Age**. Some people think that we may still be in the 'Ice Age', and that the current warm period of a few thousand years is just another inter-glacial!

The weathering and erosion processes that operated during the glacial periods were responsible for creating many of today's landscape features, including spectacular glacial valleys, jagged pyramidal peaks, and deep ribbon lakes.

Figure 2.2 The extent of ice cover over the UK 18000 years ago ▶

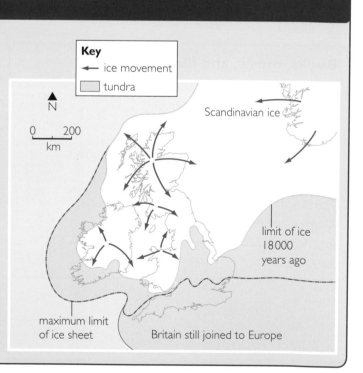

Key
- ← ice movement
- tundra

N

0   200
km

Scandinavian ice

limit of ice 18000 years ago

maximum limit of ice sheet

Britain still joined to Europe

Global cold environments today

Cold environments experience significant periods of time when the temperature is close to or below 0°C. In the most extreme cold environments, such as parts of Antarctica, the temperature can stay well below 0°C throughout the year – sometimes plummeting to as low as −60°C! Other, less-extreme, cold environments simply experience cold winters, such as the Alps in Europe or the Rockies in Canada.

There are four main types of cold environment: polar, alpine, glacial, and periglacial. The present-day global distribution of these four environments is shown on Figure 2.3.

Polar – These are the most extreme cold environments. In winter, temperatures often drop to –50°C! They include Antarctica, Greenland and some of the islands inside the Arctic and Antarctic Circles, such as Spitzbergen. Despite the snowy image of these polar environments, they are very dry – with relatively low amounts of precipitation (snow). There are also extensive areas of sea ice, particularly in the Arctic.

Alpine – Mountain areas, such as the Rockies and the Alps, experience very cold winters with heavy snow. Because of the high altitude, the temperature can drop to –10°C or less. The extreme winter cold is replaced in the summer with warmer weather, where the temperature can even exceed 20°C.

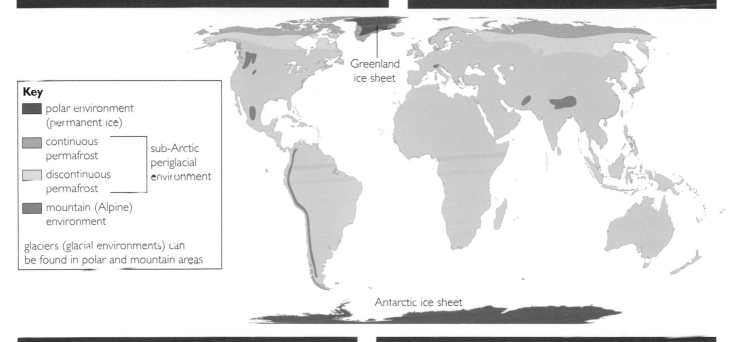

Key
- polar environment (permanent ice)
- continuous permafrost
- discontinuous permafrost
  } sub-Arctic periglacial environment
- mountain (Alpine) environment

glaciers (glacial environments) can be found in polar and mountain areas

Greenland ice sheet

Antarctic ice sheet

Periglacial – Periglacial literally means 'edge' of glacial. Periglacial environments are found on the fringes of polar or glacial environments, e.g. in parts of Siberia, Canada and Greenland. Periglacial areas experience permanently frozen ground (permafrost). During their brief warmer summers, the ground surface layer thaws – enabling hardy plants to grow. Periglacial environments are not permanently covered by ice.

Glacial – These environments are specifically associated with glaciers. While some enormous glaciers are found in polar environments, most of the world's actively moving glaciers are found high up in alpine mountain regions. The heavy winter snowfall in those areas provides the ice to feed the glaciers. Then, in the summer, meltwater lubricates the glaciers – helping them to move like giant conveyor belts down the alpine valleys.

Figure 2.3 *The global distribution of cold environments today* ▲

ACTIVITIES

1 Study Figure 2.2.
   a Locate your home town or city. Describe the conditions that you would have experienced there 18 000 years ago.
   b Which parts of the UK never experienced a covering of ice?
   c Use an atlas to identify the sources of ice flow in the British Isles.
   d Why is it important to understand past climates when trying to explain the formation of present-day landscapes?

2 Study Figure 2.3. For this activity you will need a blank world outline map and an atlas.
   a Make a careful copy of the map showing the present-day distribution of global cold environments.
   b Use the Internet to find a photograph to illustrate each of the environments.
   c Label the mountain ranges (alpine environments).
   d Identify and label a selection of regions/countries experiencing polar and periglacial conditions.

The glacial system

In this section you will learn about:

- the inputs and outputs that comprise the glacial system
- accumulation, ablation and the glacial budget (mass balance studies)
- changes in the mass balance of the Gulkana Glacier, Alaska

A glacier can be viewed as a system – with inputs, processes and outputs (Figure 2.4).

- The main input is snow. As it becomes increasingly compacted over many years, it gradually turns from low-density white ice crystals (snowflakes) to high-density clear glacial ice. Avalanches from surrounding mountainsides also provide inputs to the glacial system.

- The weight of the compacted ice then combines with gravity to make the glacier move slowly downhill. As it does so, it transports sediment that has dropped onto it – and also erodes its valley (processes in the system).
- The main outputs from the system include ice and meltwater – together with vast amounts of sediment.

Figure 2.4 *The glacial system* ▼

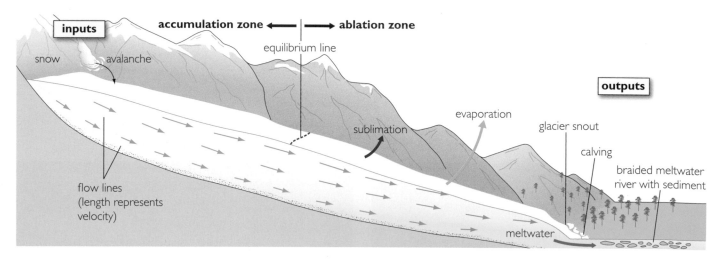

The glacial budget

Just as a financial budget involves credits (payments in) and debits (withdrawals), a glacier also has a budget. Look at Figure 2.4 again. As you can see, the glacier is divided into two zones:

- The **accumulation zone** is where there is a net gain of ice over the course of a year. Here the inputs (snow and avalanches) exceed the outputs.
- The **ablation zone** is where there is a net loss of ice during a year – the losses (melting, evaporation, etc.) exceed the gains.

The boundary where gains and losses are balanced is called the **equilibrium line**. Over a period of several years, variations in the glacial budget may result in the equilibrium line moving either up or down the glacier, causing the snout to advance or retreat. The glacial budget (usually

called **mass balance**) varies during the course of a year (Figure 2.5). In the summer, ablation will be at its highest, because of rapid melting of the ice. During the winter, higher amounts of snowfall and limited melting will result in accumulation being greater than ablation.

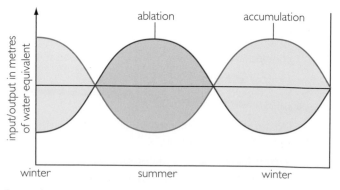

Figure 2.5 *Seasonal variations in the glacial budget* ▲

CASE STUDY

Glacial budget: Gulkana Glacier, Alaska

The United States Geological Survey has extensive mass-balance data for a number of North American glaciers. This case study examines data for the Gulkana Glacier in Alaska. Start by using the Internet to locate this glacier.

1 Study Figure 2.6.

    a How can you explain the positive mass balance values in the winter and negative mass balance values in the summer?

    b Glance down the winter values. Does there appear to be any trend between 1980 and 2005?

    c Now glance down the summer values. Can you detect any trend?

    d Make a prediction for the two empty columns. How do you think the net balance and the cumulative balance will change between 1986 and 2005?

2 Complete the two blank columns. The first few calculations have been done for you. Work with a partner to double-check the accuracy of your calculations.

3 Represent the **net balance** data in the form of a bar chart.

    a Draw a horizontal line representing a 0 net balance (where there is a balance between winter gain and summer loss). Time runs along this line from 1980 on the left to 2005 on the right.

    b Now draw bars above (for positive values) and below (for negative values). Use a blue colour for the positive bars and red colour for the negative values.

    c Comment on the pattern of the bars.

4 Represent the cumulative balance data in the form of a line graph. Be careful to choose an appropriate vertical scale. Describe the trends and contrast the patterns with the bar graph in question 3.

5 Write a summary few sentences describing the mass-balance trends from 1980-2005.

    a Is there any evidence that the glacier is advancing or retreating?

    b Might this be connected to climate change?

    c Access the website to see if any recent data has been added:
http://ak.water.usgs.gov/glaciology/gulkana/balance/index.html

Figure 2.6 Mass-balance data for the Gulkana Glacier, Alaska (1980-2005). The values are in metres of water equivalent. ▼

Date	Winter balance (m)	Summer balance (m)	Net balance (m)	Cumulative balance (m)	Equilibrium line altitude (m)
1980	1.13	-1.20	-0.07	-0.07	1738
1981	0.96	-0.93	+0.03	-0.04	1687
1982	1.55	1.67	0.12	-0.16	1746
1983	1.14	-1.11	+0.03	-0.13	1751
1984	1.30	-1.61	-0.31	-0.44	1768
1985	1.41	-0.73	+0.68	+0.24	1650
1986	1.09	-1.03			1682
1987	1.24	-1.37			1737
1988	1.26	-1.48			1759
1989	No data	No data	-0.70		1791
1990	1.36	-2.04			1794
1991	1.31	-1.37			1708
1992	0.98	-1.22			1758
1993	0.82	-2.49			1880
1994	1.37	-1.96			1777
1995	0.94	1.65			1806
1996	0.87	1.39			1768
1997	0.99	-2.68			1865
1998	0.79	-1.43			1793
1999	1.04	-2.15			1842
2000	1.44	-1.49			1704
2001	1.40	-2.08			1790
2002	0.76	-1.83			1833
2003	1.79	-1.80			1718
2004	0.93	-3.22			1851
2005	1.73	-1.99			1758

In this section you will learn about:

- the factors affecting ice movement
- the main types of ice movement – basal sliding and internal deformation
- variations in rates of flow along the length of a glacier

Types of ice movement

There are basically two broad types of ice movement – basal sliding and internal deformation (Figure 2.7).

Basal sliding involves the movement of a large block of ice – usually in a series of short jerks. It occurs in warm glaciers, where meltwater is present to help lubricate the base of the ice.

As you can see in Figure 2.7, meltwater forms on the upslope side of obstacles on the valley floor. Here the additional resistance causes an increase in pressure, which leads to localised melting of the ice – called **pressure melting**. The meltwater enables the ice to flow up and over the obstacle, although it often refreezes on the downslope side, where the pressure is reduced.

Internal deformation involves ice crystals slipping and sliding over each other (similar to the way in which individual grains of sugar move when heaped in a pile). The ice crystals may also become deformed or fractured, and gradually move downhill in response to gravity. Internal deformation occurs in both cold and warm glaciers – often at the same time as basal sliding.

Controlling factors

Ice moves in response to a number of controlling factors:

- *Gravity.* This is the downhill force that encourages ice to move. The steeper the gradient, the greater the pull of gravity.
- *Friction.* If the ice as a whole is to move forward, the friction exerted by the ground on the ice has to be overcome.
- *The mass of the ice.* The heavier the ice is, the more potential energy it has to move. However, more force will also be needed to overcome the increased friction caused by the extra weight.
- *Meltwater.* The presence of meltwater lubricates the base of the ice, allowing it to slip downhill. An increase in water pressure beneath the ice may also help to overcome friction by creating a buoyancy effect.
- *The temperature of the ice.* In some environments, like Greenland and Antarctica, the ice is so cold that it's frozen to the bedrock. These glaciers are called **cold glaciers**. In more temperate environments, such as Alaska and the Alps, temperatures are higher and there is more meltwater. These are called **warm glaciers**.

Figure 2.7 *Basal sliding and internal deformation* ▼

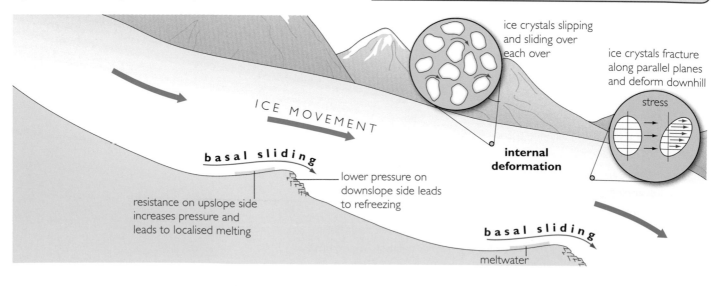

ice crystals slipping and sliding over each over

ice crystals fracture along parallel planes and deform downhill

stress

ICE MOVEMENT

basal sliding

internal deformation

lower pressure on downslope side leads to refreezing

resistance on upslope side increases pressure and leads to localised melting

basal sliding

meltwater

Variations in the rate of ice flow

With so many changing variables along the course of a glacier – such as glacier mass, steepness of the gradient, and the presence of meltwater – it's hardly surprising that there can be big variations in the rate of ice flow. For example, when there is a sudden increase in the gradient, the ice will flow faster and may become thinner as it's stretched. This is called **extensional flow** (Figure 2.8). This stretching often results in the ice cracking to form **crevasses**.

A reduction in the gradient further down the glacier will slow the ice flow, causing it to pile up and become thicker. This is called **compressional flow** (Figure 2.8). Here any crevasses that were opened up by the earlier extensional flow will now be closed.

Between the two zones of extensional and compressional flow, the ice moves in a curved or **rotational** manner. This type of movement is very important in the formation of corries (see page 75).

In common with a river, the fastest flow of ice is towards the centre. You can see the effect of this by looking at the shape of the crevasses in Figure 2.9. Notice how they curve downhill in the centre of both glaciers. This is because the friction exerted by the valley sides slows down movement at the edges of the ice. Ice will also tend to move faster at the surface than at depth, because of the increased friction of the bedrock.

Did you know?

Unlike rivers, glaciers can actually flow uphill! This occurs when there is sufficient weight of ice towards the back of a glacier to force the snout up and over a rise in the ground.

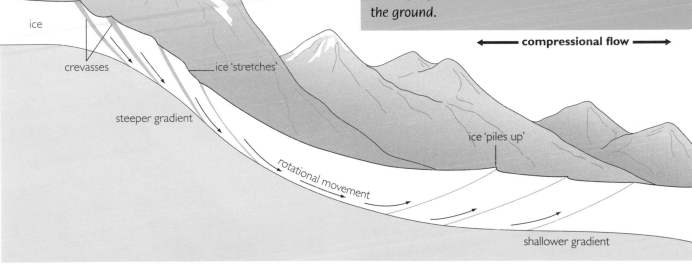

Figure 2.8 *Extensional and compressional flow* ▲

Figure 2.9 *Ice in the middle of a glacier moves faster than at the sides* ▼

ACTIVITIES

Study Figures 2.7 and 2.8.
1. How does basal sliding differ from internal deformation?
2. Explain why meltwater is formed on the upslope side of obstacles.
3. What type(s) of glacial movement would you expect to occur in cold glaciers? Explain your answer.
4. Under what circumstances would you expect glaciers to move rapidly?
5. Why are open crevasses evidence of extensional flow?
6. Give a reasoned description of what happens to these open crevasses when the gradient is reduced.

In this section you will learn about:
- weathering processes in a glacial environment
- the processes of erosion, transportation and deposition

Weathering in a cold environment

Weathering is defined as the breakdown or disintegration of rock in its original position (in situ), at or just below the ground surface. Two major weathering processes operate in glacial environments – frost shattering and carbonation.

Frost shattering

Frost shattering (also known as freeze-thaw) commonly affects bare rocky outcrops high up on a mountainside (Figure 2.10). The process begins when water (rainwater or meltwater) seeps into cracks and pores in the rock. Then, when the temperature drops to at least 0°C (freezing point), the water turns to ice – expanding in volume by about 9%. In a confined space, this expansion exerts stresses within the rock – enlarging the cracks and pores. When the temperature rises above 0°C, the ice thaws and more water seeps into the enlarged cracks. When this process of freezing and thawing is repeated many times, the cracks become so enlarged that chunks of rock begin to break away and either pile up as scree at the foot of the slope (Figure 2.10), or fall onto the glacier to be transported as **moraine**.

Frost-shattered rocks are very sharp and angular. As they become trapped under the ice, they form extremely abrasive tools – like the sand on sandpaper. Imagine how ineffective sandpaper would be if it had no sand! Frost shattering often prepares a landscape for glacial erosion by breaking up the surface – making it easier for a glacier to erode.

Glacial processes

Glacial erosion

There are two main forms of glacial erosion – abrasion and plucking.

Abrasion is the sandpapering effect of glacial ice as it grinds over and scours a landscape. The ice can do this because of the presence of the angular, frost-shattered rock fragments described above. Large rocks carried beneath the ice often scratch the bedrock to form **striations**. These scratches form useful clues for scientists studying the direction of ice flow in a post-glacial environment. Over time, the big rocks become pulverised by the weight of the glacial ice to become fine **rock flour**. This finer material tends to smooth and polish the underlying bedrock.

Figure 2.10 The effects of frost shattering in The Alps ▲

Carbonation

Carbonation is a process of chemical weathering that involves the slow dissolving of calcium carbonate in some rocks – particularly limestone. Carbon dioxide dissolved in water forms a weak carbonic acid. This reacts with calcium carbonate in the rock to form calcium bicarbonate, which then dissolves. The lower the temperature, the more effective the process of carbonation will be. This is because cold water can hold more dissolved carbon dioxide. This explains why carbonation is an important process in glacial environments.

Plucking or **quarrying** occurs when the refreezing of meltwater freezes part of the underlying bedrock to the base of the glacier. Any loosened rock fragments are then 'plucked' away when the glacier moves forward. Put very simply, it's like pulling out a loose tooth! This process is particularly common when a reduction in pressure under the ice on the downslope side of an obstacle leads to the refreezing of meltwater (see Figure 2.7 on page 70).

Figure 2.11 is a feature called a **roche moutonnee**. This resistant rock outcrop would have acted as an obstacle to a glacier moving down the valley. As you can see, it has one smooth side and one jagged side. The process of abrasion eroded the smooth side. The jagged downslope side reflects the process of plucking (Figure 2.12). Pressure melting (see page 70) was responsible for these two processes operating on either side of the obstacle.

Glacial transportation and deposition

Glaciers act like giant conveyor belts – transporting material on, in and below the ice. This material, called moraine, is deposited when the glacial ice starts to melt – just as the extra items used to decorate a snowman are left behind on the ground when it melts.

Figure 2.11 *A huge roche moutonnee in the USA* ▲

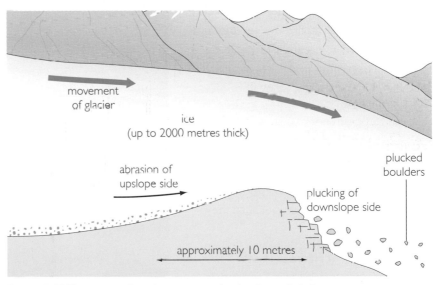

movement
of glacier

ice
(up to 2000 metres thick)

abrasion of
upslope side

plucked
boulders

plucking of
downslope side

approximately 10 metres

Figure 2.12 *The erosion of a roche moutonnee by abrasion and plucking* ▲

ACTIVITIES

Study Figures 2.11 and 2.12.

1 Draw a simple sketch of Figure 2.11 and add labels to describe the surface of a roche moutonnee.

2 What evidence is there that the process of abrasion has eroded the long side of Figure 2.11?

3 Which direction was the ice moving in to form the roche moutonnee in the photo? Explain your answer by adding an extended label to your sketch.

4 Explain, with the aid of a simple annotated diagram, how and why the processes of abrasion and plucking operate to form a roche moutonnee. Look back at page 70 to include references to pressure melting and refreezing.

In this section you will learn about:

- the characteristics and formation of landforms associated with glacial erosion (corries, arêtes, pyramidal peaks and glacial troughs)
- the characteristics and formation of landforms associated with glacial deposition (moraines and drumlins)

Skills

In this section you will:

▸ draw a sketch from a photograph
▸ recognise and describe glacial features using an OS map
▸ draw a sketch and a cross profile using an OS map

Valley glaciers form some of the most awe-inspiring landscapes in the world. Look at Figure 2.13, which shows part of the Mer de Glace in the French Alps. Even though this glacier is shrinking in size, it has gouged out a huge valley. Smaller glaciers – high up on the surrounding mountainsides – have also scoured out deep hollows, called cirques or corries. One of them is marked on Figure 2.13, along with two sharp-edged arêtes and a pointed pyramidal peak.

As well as eroding, glaciers are also capable of transporting huge quantities of sediment, which explains why the glacier in the photo looks so 'dirty'. Eventually, this sediment will be deposited to create a range of landforms – often hummocky in nature, and mostly left behind on the valley floor after the ice has retreated.

Erosional landforms

The glacial landscape in Figure 2.13 is very much 'work in progress'. It is impossible see most of the glacial features, because they are under the ice!

By contrast, look at Figure 2.14 – a spectacular glaciated valley in Scotland. It shows many of the features currently being formed in the French Alps (Figure 2.13). Notice the wide, steep-sided glacial trough, the truncated spurs on the mountainsides, and the sharp-edged arêtes. The processes of weathering and erosion formed these features when Scotland was in the grip of an ice age.

Figure 2.13 *The Mer de Glace (Sea of Ice) in the French Alps* ▲

Figure 2.14 *The glaciated valley of Glencoe in Scotland* ▶

Corries, arêtes and pyramidal peaks

A **corrie**, also known as a cirque (France) and a cwm (Wales), is an enlarged hollow on a mountainside (Figure 2.15). Its characteristic features include a steep back wall (often with a pile of scree at its base). Generally, the hollow is very deep and may contain a small lake – called a **tarn**. The front of a corrie often has a raised rock lip, which helps to explain the presence of the tarn.

Look at Figure 2.16 to see how corries are formed by a combination of snow-related processes and glacial erosion.

When two neighbouring glaciers cut back into a mountainside – each one eroding a corrie – the ridge between the two corries becomes narrower. This leads to the knife-edge ridges called **arêtes** that are common in both present-day glacial landscapes like the Alps (Figure 2.13), and post-glacial landscapes like Scotland (Figure 2.14). When three or more corries erode back-to-back, the ridge then becomes an isolated peak, called a **pyramidal peak** (Figure 2.17).

Figure 2.15 A typical corrie in the Scottish Highlands ▲

Figure 2.16 How a corrie is formed ▼

A Periglacial conditions

combined snow-related (nivation) processes, such as weathering and slumping, slowly enlarge a hollow on the mountainside

snow

B Glacial conditions

abrasion 'scoops' out the hollow

rotational slip of glacier leads to intense abrasion

plucking forms craggy back wall

glacier

rock lip where thinner ice is less erosive

C Post-glacial conditions

steep back wall

arête

scree

rock lip

tarn

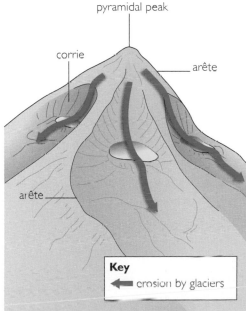

pyramidal peak

corrie

arête

arête

Key

◀ erosion by glaciers

Figure 2.17 How arêtes and pyramidal peaks are formed ▲

Glacial trough

When the Mer de Glace (Figure 2.13) finally melts away, it will reveal a spectacular U-shaped valley – or glacial trough – similar to the valley in Figure 2.14. **Glacial troughs** tend to be steep-sided and flat-bottomed. They are also generally straight, because of the immense power and inflexibility of the large glaciers that gouge them out. For instance, a valley glacier will just cut through any previously existing interlocking spurs to form **truncated spurs** up the valley side (see Figure 2.14).

Glacial troughs are often broadly symmetrical in cross-profile. However, they can have quite a variable long-profile. Look at Figure 2.18:

- If a glacier encounters weaker bedrock, or if a tributary glacier joins the main valley glacier to increase the volume of ice, there might be increased or enhanced erosion in parts of the trough (called **overdeepening**). Localised enhanced erosion can form a narrow – but deep – **ribbon lake**.
- A smaller tributary glacier may increase the overall volume of ice, but it has less erosive power than the main valley glacier. This means that when the ice finally melts, the tributary valley is left perched up above the main valley. A feature like this is called a **hanging valley**, and is often marked by a waterfall plunging down into the main valley.
- If bands of resistant rock cut across the valley, the glacial erosion will be reduced at those points. The same outcome might result from a sudden broadening of the valley – causing the ice to become thinner and less erosive.

Figure 2.18 Variations in the long profile of a glacial trough ▼

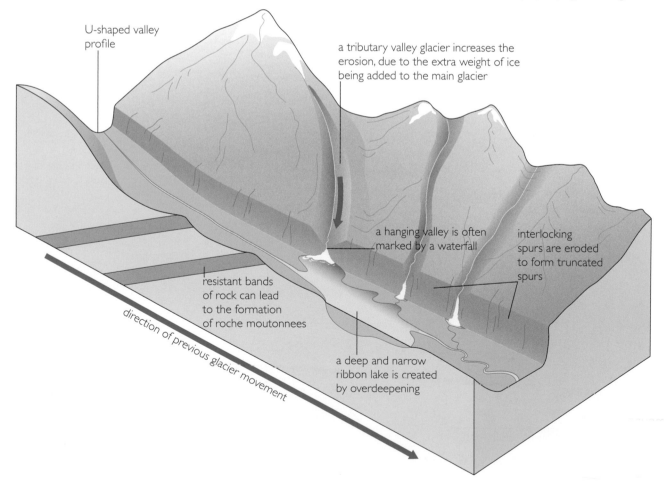

U-shaped valley profile

a tributary valley glacier increases the erosion, due to the extra weight of ice being added to the main glacier

a hanging valley is often marked by a waterfall

interlocking spurs are eroded to form truncated spurs

resistant bands of rock can lead to the formation of roche moutonnees

direction of previous glacier movement

a deep and narrow ribbon lake is created by overdeepening

Depositional landforms

Glaciers act like giant conveyor belts, transporting huge quantities of rock material – called moraine – down-valley. This material includes frost-shattered rocks from the valley sides, plus rocks freshly eroded by the ice through plucking.

When the ice eventually retreats from a glacial landscape, it leaves behind a variety of depositional landforms. Figure 2.19 shows a range of landforms deposited by the ice. You are expected to understand the characteristics and formation of two of these landforms – moraines and drumlins.

Moraines

Figure 2.20 shows the different types of moraine transported by a valley glacier:

- **Lateral moraines** consist largely of frost-shattered rocks that have fallen onto the glacier from the valley sides.
- When a tributary glacier joins the main glacier, two lateral moraines join up to form a **medial moraine**. This line of debris then continues its journey towards the centre of the main glacier.
- Rock material trapped within the ice – having been buried over the years by fresh layers of snow and ice – is called **englacial moraine**.
- The rock material that grinds along at the base of the glacier is called **ground moraine**. This moraine can be several metres thick.
- **Terminal moraine** marks the furthest extent of a glacier (Figure 2.19).

Beyond the terminal moraine lies a vast expanse of sand and gravel – deposited by braided meltwater rivers from the glacier. This is the **outwash plain** (Figure 2.19).

When a glacier finally melts or retreats, the material carried on or in the ice is dumped on the ground – on top of the ground moraine. Collectively, this jumble of poorly sorted rock debris is called **till**. As meltwater from the retreating glacier erodes this material, much of it is removed or re-shaped by the flowing water. This explains why it's often hard to distinguish between the different types of moraine in a post-glacial landscape.

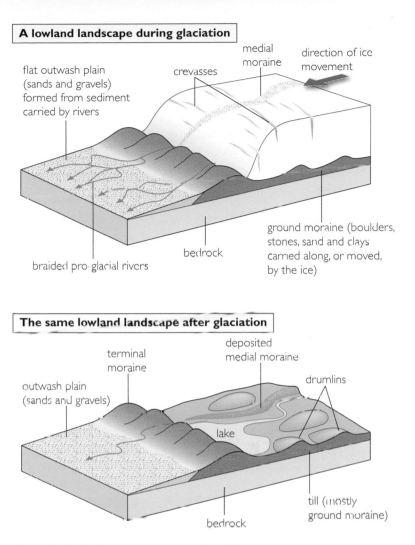

A lowland landscape during glaciation

flat outwash plain (sands and gravels) formed from sediment carried by rivers

crevasses

medial moraine

direction of ice movement

braided pro-glacial rivers

bedrock

ground moraine (boulders, stones, sand and clays carried along, or moved, by the ice)

The same lowland landscape after glaciation

terminal moraine

deposited medial moraine

drumlins

outwash plain (sands and gravels)

lake

bedrock

till (mostly ground moraine)

Figure 2.19 A landscape associated with glacial deposition ▲

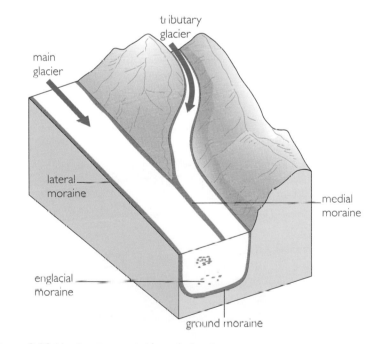

tributary glacier

main glacier

lateral moraine

medial moraine

englacial moraine

ground moraine

Figure 2.20 Moraines transported by a glacier ▲

Drumlins

Drumlins are egg-shaped hillocks that usually occur in clusters (called swarms) in low-lying valley bottoms. They are entirely depositional – consisting of a mixture of boulders and clay (ground moraine).

While it's a little uncertain how drumlins are formed (obviously, it's impossible to watch them forming under the ice!) they are thought to result from the moulding of rock debris on the valley floor by ice moving over it. The moving ice results in the features becoming streamlined, with a blunt end up-valley and a tapered end pointing down-valley (Figure 2.21). This characteristic shape allows scientists to interpret the direction of past ice movements.

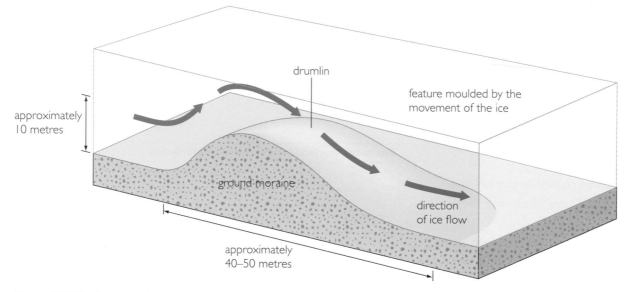

Figure 2.21 *The formation of a drumlin* ▲

ACTIVITIES

1 Study Figure 2.13.
   a What processes do you think are currently operating in this landscape? Justify your answer.
   b What evidence is there that the Mer de Glace is shrinking?
   c Imagine that you could look beneath the Mer de Glace. Describe what you would see.
   d When the Mer de Glace eventually retreats from the area, what do you think the valley will look like? Look at Figure 2.14 to help you with this question.
2 Study Figure 2.15.
   a Draw a sketch of the corrie.
   b Use Figure 2.16 to help you add some labels to your sketch.
   c Describe in your own words how a corrie is formed.
3 Study Figure 2.17.
   a Make a copy of the diagram.
   b Add detailed labels describing the formation of an arête and a pyramidal peak.
4 Study Figure 2.18. Suggest reasons why a glacial trough may vary in its long profile and its cross profile.

Internet research

Complete a short project describing the characteristics and formation of glacial depositional landforms. Include hand-drawn diagrams and annotated photos to support your written points. Concentrate on glacial deposition features only. Do not include fluvioglacial features.

You will find a great deal of material on the Internet. A couple of good sites are:

The Geography Site http://www.geography-site.co.uk/pages/physical/glaciers/deposit.html

About Geology has some good photos http://geology.about.com/library/bl/images/bliceindex.htm

CASE STUDY

The Cairngorms, Scotland – an OS map study

Study Figure 2.22, which shows part of the Cairngorms mountain range in Scotland. The Cairngorms now forms a National Park – one of only two in Scotland. Its landscape is extremely rugged, and reflects the action of ice over several glacial periods.

Locate Coire Bhrochain, which is a corrie. A semi-circle of bold black cliff symbols marks its back edge. This is an arête. The small black dots indicate deposits of scree.

1 Now draw a sketch of the corrie that contains Lochan Uaine. Add as many labels as you can to identify the main features of the corrie. Don't forget to include a scale, a north point and a title.

2 Why do you think a lake or tarn has formed in this corrie, but not in Coire Bhrochain?

3 Locate Sgor an Lochain Uaine (The Angel's Peak).
   a State its six figure grid reference.
   b What is its height above sea level?
   c Do you think this is an example of a pyramidal peak? Justify your answer.

4 What is the map evidence that suggests that Lairig Ghru is a glacial trough?

5 Locate Loch Einich.
   a Of what type of feature is Loch Einich an example?
   b Suggest why the valley glacier overdeepened its valley at this point.
   c Describe the shape of the valley to the north of the loch.
   d Use the bold contours (at 50-metre intervals) to draw a sketch cross profile from E-W across the valley to include Loch Einich. Remember that the blue contours in the lake are depths from the surface. Don't forget to include a scale and add as many labels as you can.

6 Using no more than 200 words, what evidence can you provide that the area in the map extract has been affected by glaciation?

Figure 2.22 *An OS map extract of part of the Cairngorms in Scotland* ▼

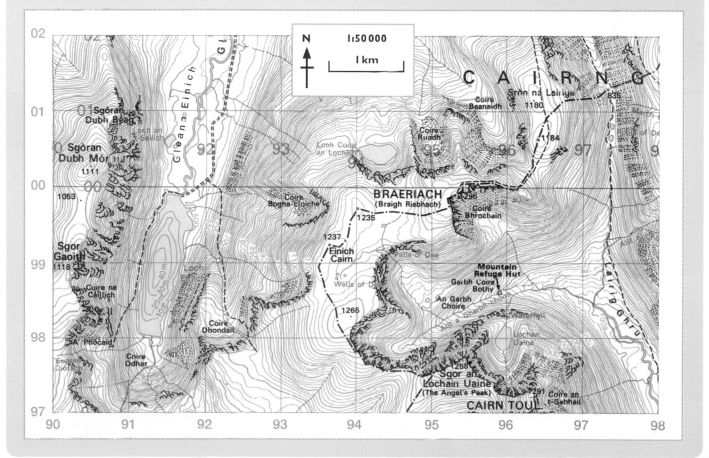

In this section you will learn about:
- the importance of meltwater as an agent in glacial environments
- the characteristics and formation of fluvioglacial features (meltwater channels, eskers, kames, and outwash plains)

If you stand on a temperate glacier (for example in Iceland or the Alps), you are likely to hear two things. One is a disturbing creaking sound – caused by the ice as it constantly adjusts its position. The other is the sound of rushing water.

The importance of meltwater

Meltwater is an extremely important element in many glacial environments:

- High up in the mountains, melting snow helps to enlarge the shallow nivation hollows that eventually become corries when the snow turns to ice.
- Meltwater helps to lubricate the base of a glacier and overcome friction – allowing the glacier to slip forward periodically (see right).
- Meltwater also transports moraine beneath the ice, where it provides tools for glacial erosion.
- If it refreezes, meltwater can bind the base of a glacier to broken rock fragments – allowing them to be 'plucked' away from the solid bedrock.
- Meltwater often forms rivers – both on top of and beneath the ice (Figure 2.23). These rivers often meander and exhibit many of the characteristics associated with 'normal' rivers.
- Meltwater will also erode channels and form distinctive depositional features – both in front of and beneath the ice.

Skills

In this section you will:
- discuss field evidence in the identification of fluvioglacial landforms
- interpret a radial graph

A glacial surge

It may be hard to believe, but huge – apparently static – glaciers are capable of sudden and rapid downhill movement. This is called a **glacial surge**. A glacial surge is an exceptional event, which occurs in cold environments when there is an unusual amount of summer melting. The meltwater lubricates the glacier, which allows a sudden forward movement to occur when friction with the underlying bedrock is overcome. In the 1980s, the Variegated Glacier in Alaska surged several times, reaching speeds of up to 65 metres a day. That's very fast for a glacier!

Figure 2.23 *A meltwater river on the Lake Gray Glacier in Chile* ▼

Fluvioglacial landforms

Meltwater-related features are called **fluvioglacial landforms** ('fluvio' = river/water). You will be expected to understand the characteristics of four of these landforms – meltwater channels, kames, eskers, and outwash plains.

Meltwater channels

A **meltwater channel** usually takes the form of a steep-sided – often dry – valley, carved into the landscape. Many possible scenarios can lead to the formation of a meltwater channel. Most commonly, it results from the overspill of a lake that builds up next to or in front of a glacier.

One of the best examples of a meltwater channel in the UK is Newtondale in North Yorkshire (Figure 2.24):

- During the last ice advance (70 000 – 10 000 years ago), the higher North York Moors remained largely unglaciated – forming a series of ice-free islands, surrounded by a vast ice sheet.
- Meltwater then began to collect between the islands and the ice – to form lakes in the lower valleys of the North York Moors.
- As the water levels in these lakes rose, they began to overflow into adjacent valleys – eroding deep meltwater channels.
- Newtondale was formed when water overflowed from Lake Wheeldale in a southerly direction towards Lake Pickering (Figure 2.24).
- When the ice finally retreated, these lakes emptied.
- Today, Newtondale forms a narrow gorge, which is about 80 metres deep and 5 km in length (Figure 2.25).

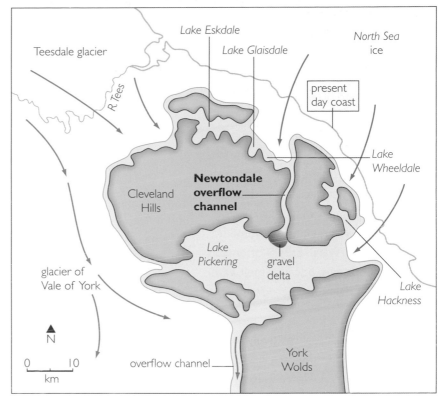

Figure 2.24 *The formation of Newtondale meltwater channel* ▲

Figure 2.25 *Newtondale, a meltwater channel in North Yorkshire* ▲

Kames and eskers

Kames and eskers are depositional features that form in contact with ice (Figure 2.26).

Eskers are long ridges of sand and gravel. They can be up to 30 metres high, and usually take the form of meandering hills – often stretching for several kilometres – that run roughly parallel to the valley sides. Their meandering shape suggests that they were formed by subglacial river deposition during the final stages of a glacial period – when the ice was melting away and no longer moving forward. Eskers often appear as discontinuous hills, because meltwater and postglacial rivers have eroded them away in places.

Kames form a group of relatively minor features. Like eskers, they are largely made up of sand and gravel. Figure 2.26 shows three different types of kame:

- *Kame terrace.* This is the most extensive type of kame. It results from the infilling of a marginal glacial lake. When the ice finally melts, the kame terrace is abandoned as a ridge on the valley side.
- *Kame delta.* A kame delta is a smaller feature that forms when a stream deposits material on entering a marginal lake. Kame deltas form small mound-like hills on the valley floor, and can be identified by their deltaic sedimentation characteristics.
- *Crevasse kame.* Some kames arise from the fluvial deposition of sediments in surface crevasses. When the ice melts, they are deposited on the valley floor to form small hummocks.

The key to identifying eskers and kames is that they are made up of sand and gravel – rather than clay-rich till like the surrounding countryside. So they tend to have different types of vegetation growing on them, when compared to the surrounding area (Figure 2.27).

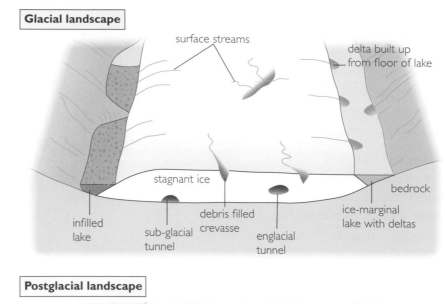

Glacial landscape

surface streams

delta built up from floor of lake

stagnant ice

bedrock

infilled lake

sub-glacial tunnel

debris filled crevasse

englacial tunnel

ice-marginal lake with deltas

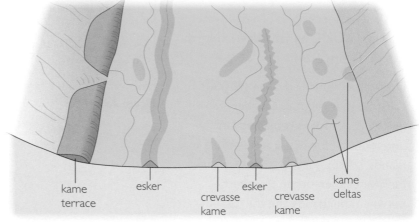

Postglacial landscape

kame terrace

esker

crevasse kame

esker

crevasse kame

kame deltas

Figure 2.26 *How kames and eskers are formed* ▲

Figure 2.27 *An esker and kames deposited on a till plain in Alaska, USA* ▲

Outwash plain

An **outwash plain** is an extensive – gently sloping – area of sands and gravels that forms in front of a glacier (Figure 2.19 on page 77). As the name implies, it results from the 'outwash' of sediment carried by meltwater streams and rivers.

During a glacial period, meltwater and the deposition of outwash will be very seasonal – restricted to the warmer summer months only. However, at the end of a glacial period, huge quantities of sediment will be spread out over the outwash plain by great torrents of meltwater. Today, some of the most extensive outwash plains can be seen in Iceland and Alaska, where large braided rivers choked with sediment meander their way across vast floodplains (Figure 2.28).

Figure 2.28 An outwash plain in Canada with a braided river system ▲

ACTIVITIES

1 Discuss the following scenarios in pairs.

a You are about to visit Newtondale (Figure 2.25) on a field trip. What evidence should you look for to decide whether this valley is a meltwater channel or a normal river valley?

b You have come across a number of low hills on a field trip to a lowland glaciated valley in Scotland. What evidence should you look for to help you decide whether these features are kames or eskers? How would you distinguish between the different types of kame? Use Figures 2.26 and 2.27 to help you.

2 Study Figure 2.29. The orientation of a pebble is measured using a compass to examine the orientation of its long axis. This has to be done in situ. Flowing water will always orientate pebbles in the direction of their long axes. So, by studying glacial sediments, it is possible to suggest the main flow directions of fluvioglacial rivers. However, outwash plains have multi-directional flow patterns, so a sediment sample often has more than one preferred orientation.

a The main orientation trend of the pebbles studied in Figure 2.29 is NE-SW. How can you tell this from the graph?

b What does this suggest about the dominant direction of river flow during the deposition of the sediment?

c What is the secondary orientation trend shown on the graph?

d Explain the presence of this second orientation trend.

e Suggest how and why a radial graph for a sample of till would differ from Figure 2.29.

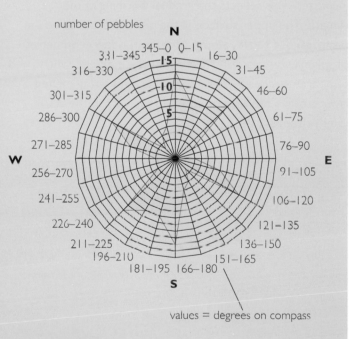

Figure 2.29 A radial graph showing the orientations of pebbles collected from fluvioglacial sediments in the Nant-y-Llyn Valley in North Wales ▲

In this section you will learn about:
- the global distribution of permafrost
- processes operating in periglacial environments
- the characteristics and formation of periglacial landforms

The periglacial environment

Today, vast swathes of Russia and Canada – together with Alaska and parts of northern Europe – are affected by periglacial conditions. Mountain ranges like the Alps also experience periglacial conditions, due to their high altitude. In the distant past – when ice sheets spread south over large parts of northern Europe – the periglacial zone also migrated to cover a huge swathe of North America and Europe.

Permafrost

Perhaps the most important characteristic of a periglacial environment is a persistently low temperature, which results in a deeply penetrating frost. With temperatures that rarely rise above 0°C, the ground can become permanently frozen to depths of over 100 metres. This is called **permafrost**. Under really extreme conditions – for instance in Siberia, where winter temperatures readily fall to −40°C – the permafrost can reach depths of over 1400 metres!

In Figure 2.30, you can see that the thickness of the permafrost decreases as the distance from the Arctic increases. This reflects the gradual warming of the climate. The upper surface of the permafrost is called the **permafrost table**. Above this, the top few centimetres of soil may temporarily melt during the summer, when temperatures briefly rise above 0°C. This soggy melted layer is called the **active layer**. In the winter, the active layer refreezes.

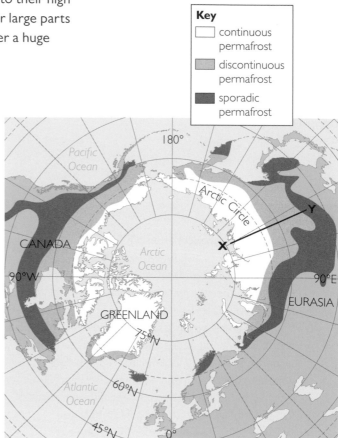

Key
- ☐ continuous permafrost
- ☐ discontinuous permafrost
- ■ sporadic permafrost

Figure 2.30 *The distribution of permafrost in the Arctic region* ▲▼

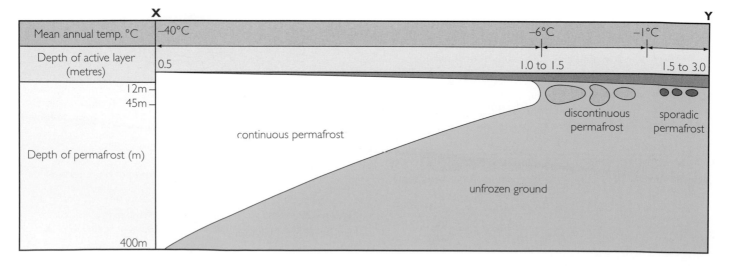

	X			Y
Mean annual temp. °C	−40°C		−6°C	−1°C
Depth of active layer (metres)	0.5		1.0 to 1.5	1.5 to 3.0
	12m — 45m —	continuous permafrost	discontinuous permafrost	sporadic permafrost
Depth of permafrost (m)		unfrozen ground		
	400m			

Periglacial processes

Processes related to frost

Frost shattering is one of the most important processes in a periglacial environment. Water expands when it turns to ice – breaking apart rocks and sediments and forming a rock-strewn landscape called **felsenmeer** (Figure 2.32). It also leads to the accumulation of scree at the base of cliffs (page 75). Frost shattering is most effective when there are diurnal (daily) fluctuations in temperature. However, it can also operate over a much longer time period – if freezing and thawing only occurs seasonally.

During a period of hard frosts in winter, you might have noticed that bare soil and lawns become very bumpy and irregular. This is the result of a process called **frost heave**, where freezing soil water just below the surface expands and pushes up the ground above. The growth of individual ice crystals can raise individual soil particles to create a spiky surface.

Nivation

The term **nivation** covers a range of processes associated with snow. You have already come across the term in the context of corrie formation (page 75). It includes the effects of frost shattering, which operates around the edges of the snow – gradually causing the underlying rock to disintegrate. Then, in the summer, meltwater removes any weathered rock debris to reveal an ever-enlarging **nivation hollow**. Slumping may also take place during the summer, when saturated debris collapses due to the force of gravity when it is no longer frozen into place. As long as the freshly weathered material is removed by either meltwater or slumping, the nivation hollow will continue to grow larger.

Periglacial landforms

Look at Figure 2.32. It shows some of the features associated with a periglacial environment. You will be expected to understand the characteristics and formation of nivation hollows, solifluction lobes, ice wedges, patterned ground, and pingos.

Solifluction

Solifluction is the term used in a periglacial environment to describe the downslope movement of rock and soil material in response to gravity. It's most likely to occur in areas which experience significant summer melting, and where there is a reasonably thick and saturated active layer. If this layer of saturated soil and rock lies on a steep slope, it will gradually slump downhill to form features called **solifluction lobes** (Figure 2.31).

Figure 2.31 *Solifluction lobes in the Yukon, Canada* ▼

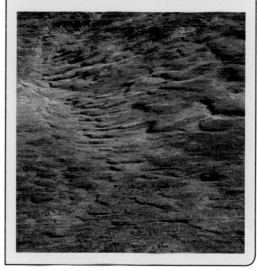

Figure 2.32 *Periglacial landforms* ▼

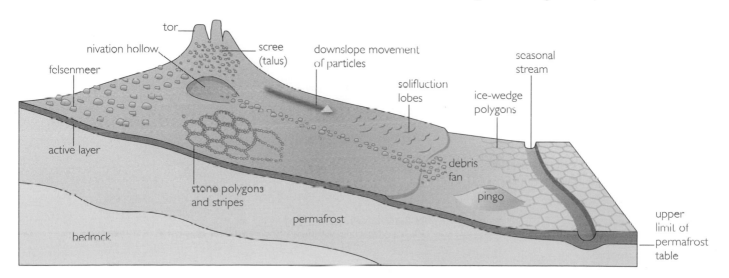

Ice wedges

When permafrost contracts in extremely low temperatures, cracks develop. During the summer, meltwater flows into these cracks and fills them up. The meltwater refreezes in the winter to form **ice wedges**, which expand and force the cracks to widen. Over a long period of repeated cycles of freezing and thawing, the ice wedges noticeably increase their size (Figure 2.33). They also start to influence the ground surface by forming narrow ridges – due to frost heave. In the summer, ponds of meltwater can form in the hollows between these ridges (Figure 2.34).

Patterned ground

As ice wedges become more extensive, a polygonal pattern may be formed on the ground surface – with the ice wedges and their ridges marking the sides of the polygons (Figure 2.34). Other features of **patterned ground** can be seen on slopes, where accumulations of stones produce stone stripes or stone polygons (Figure 2.35).

Stone polygons tend to form on shallower slopes, and are directly associated with ice wedges. Frost heave causes expansion of the ground and lifts soil particles upwards. This process is particularly effective for larger particles that can be thrust upwards by lenses of ice in the soil. Any smaller particles will probably be removed by wind or meltwater – leaving a concentration of larger stones lying on top of the ice wedges, marking out the polygonal pattern.

If the ground is more-steeply sloping, the polygons may become distorted as the stones gradually slide or roll downslope. Figure 2.35 shows how this distortion can lead to the formation of **stone stripes**, rather than polygons.

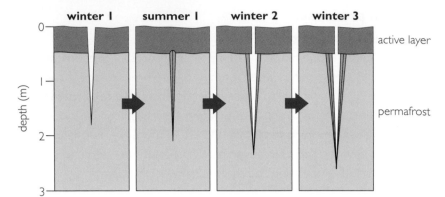

Figure 2.33 *How an ice wedge forms* ▲

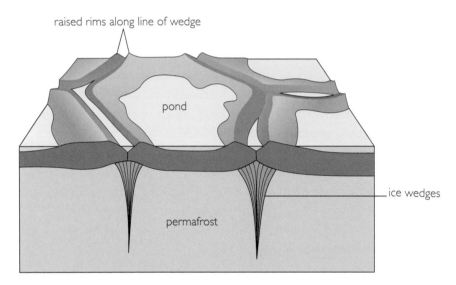

Figure 2.34 *Surface features associated with ice wedges* ▲

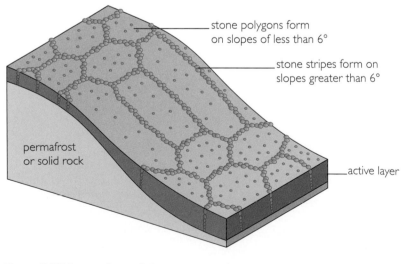

Figure 2.35 *Patterned ground* ▲

Figure 2.36 *A pingo near Tuktoyaktuk, Canada* ▲

Pingos

Look at Figure 2.36. The rounded hill in the photo is called a pingo. It is an ice-cored periglacial feature common in this part of northern Canada.

Canadian pingos commonly form on the site of a lake that has been infilled with sediment. The sediment insulates the ground and allows liquid water to collect underneath it. As the sediment gradually freezes during early winter, the water that has collected underneath it becomes increasingly confined – and its pressure increases. Eventually, the trapped water also freezes. Then, as it freezes, it expands – forcing the sediment lying on top of it upwards to form the characteristic pingo mounds. In the summer, when the pingo's ice core melts, its centre collapses to form a central dip or depression. This can sometimes become filled with water.

Pingos can reach up to 60 metres in height and 600 metres in diameter, although most of them are a lot smaller than that. There are thought to be over 1400 pingos in the Mackenzie delta region of Canada.

ACTIVITIES

1   Study Figure 2.30.
    a   Describe the distribution of the different types of permafrost in the area shown on the map.
    b   Under what temperature conditions is continuous permafrost likely to exist?
    c   How and why does the depth of the active layer change with distance from the Arctic?
    d   Suggest why periglacial conditions do not currently exist in the UK, or in most of Scandinavia, whereas they do exist at the same latitudes elsewhere in the world.

2   Study Figures 2.33 and 2.34.
    a   Describe how ice wedges form.
    b   How and why do ice wedges change over time?

    c   What effect do ice wedges have on ground surface features?
    d   What impact would surface ponds have on the development of ground surface features? Explain your answer.

3   Study Figure 2.35. Using simple sketches, explain in your own words the formation of patterned ground.

4   Study Figure 2.36.
    a   Make a careful sketch of the pingo in the photograph.
    b   Add labels to describe its characteristics.
    c   Alongside your sketch, write a short description of how it has formed.
    d   Do you think the ice core is still intact? Justify your answer.

In this section you will learn about:

- the traditional economy of the Arctic's indigenous peoples
- the early exploitation of resources
- recent developments in Alaska
- sustainable management in the White Mountains of Alaska

Skills

In this section you will:

▶ use the Internet to conduct research

Look back at Figure 2.3 (page 67) to remind yourself about the global distribution of cold environments. The most extreme cold environment is the polar zone. Typically, this is land or sea that is covered by ice – often really thickly. Adjacent to this zone is the less-severe periglacial environment. The natural vegetation of the periglacial zone is **tundra** (Figure 2.37).

Despite the extremely harsh conditions, the indigenous peoples of the polar and periglacial environments have long exploited their surroundings in a sustainable way. However, the recent discovery of valuable natural resources there, such as oil and gas, is now posing a considerable threat to the often-pristine wilderness. There is now a great deal of debate about the future sustainable exploitation of these environments.

Figure 2.37 *Tundra vegetation in the Yukon, Canada* ▲

Exploitation by indigenous peoples

An estimated 1.3 million indigenous people (such as the Inuit in Canada and the Sami in Scandinavia) live in the Arctic and tundra environments. They have lived a sustainable, largely subsistence, way of life there for thousands of years – depending on hunting, gathering and herding for their survival (Figure 2.38).

Of the 370 indigenous settlements in the tundra region, about 80% are coastal. In Greenland, for example, the Inuit spend their winters hunting seals and other animals on the frozen land and surrounding sea ice, while in the summer – when the sea ice thaws – they use small boats to fish in the open coastal waters. In the past, traditional hunting techniques were used, but today many indigenous people have access to modern high-velocity rifles and snowmobiles. Most of the remaining indigenous settlements are found along river valleys in Russia and Scandinavia, where the people concentrate on herding animals such as reindeer.

Figure 2.38 *An Inuit hunter in Nunavut, Canada, with a Caribou he has just shot* ▲

Early resource exploitation

For thousands of years, the Arctic Ocean provided rich fishing grounds for both indigenous peoples and also fishermen from countries further south. Early visitors hunted seals and whales, as well as catching marine and freshwater fish. Northern European nations, such as the UK, Norway and Iceland, made huge profits from whaling and from the fishing industry, particularly for cod.

Recent developments

However, in recent years, oil and gas exploration has led to significant industrial developments in the Arctic region (Figure 2.39). New roads have been constructed, pipelines laid, and settlements established to service the new energy industries.

These new industrial developments have already had some negative impacts on the Arctic's indigenous peoples – by interfering with the migration and hunting of caribou, and by threatening fragile marine ecosystems and fisheries (e.g. through oil spills like the *Exxon Valdez* in Alaska).

Other human developments in the Arctic region include: mining, hydroelectric power, and even bombing ranges – the former Soviet Union used to use the Arctic as a nuclear testing ground! Nowadays, an ever-increasing number of tourists are visiting the Arctic region, often for adventure tourism. Even though fishing remains an extremely important industry in the Arctic, poor fishery management in recent decades has led to over-exploitation and a collapse of fish stocks, particularly for cod.

The greatest development pressure is in Northern Scandinavia, where holiday cabin resorts, road construction, mining (copper and iron in northern Sweden), HEP and wind farms all threaten the access of Sami reindeer herders to their traditional grazing areas. An estimated 35% of their traditional grazing land has already been affected, and – by 2050 – as much as 78% of the coastal ranges may be unavailable for them to use.

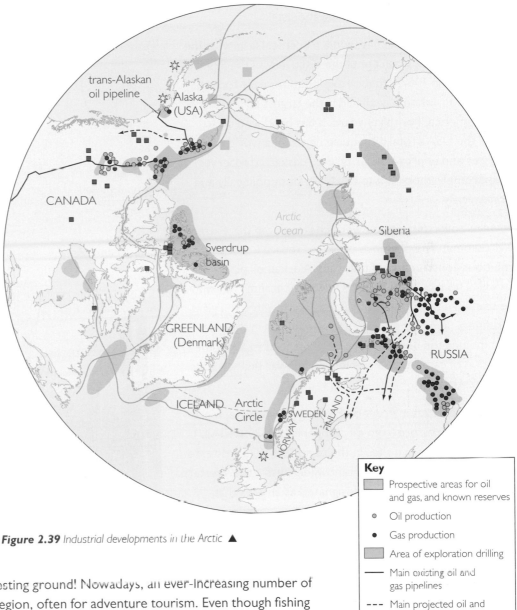

Figure 2.39 *Industrial developments in the Arctic* ▲

Key

▨	Prospective areas for oil and gas, and known reserves
○	Oil production
●	Gas production
▨	Area of exploration drilling
—	Main existing oil and gas pipelines
- - -	Main projected oil and gas pipelines
—	Major navigation routes
■	Mining sites
✸	Major oil spills

Resources and sustainable management in Alaska

Northern Alaska has large reserves of oil and gas, in a coastal region called the North Slope. This part of Alaska is an extensive Arctic wilderness, and there are two designated protected areas there (Figure 2.40). As you can imagine, there has been considerable conflict between the oil and gas companies and conservationists!

Drilling for oil in Alaska began in a piecemeal fashion in the early 1900s, but it wasn't until the discovery of the Prudhoe Bay oilfield – in 1967 – that oil exploitation really took off there. The Prudhoe Bay oilfield is the largest oilfield in North America. For thirty years, it has produced up to 20% of the USA's total oil requirements. In a world of rising oil prices and energy insecurity, the Alaskan oil reserves are extremely important to the USA, both politically and economically.

However, oil exploitation is very challenging in this extreme environment. The presence of winter sea ice led to the ambitious construction of the trans-Alaskan oil pipeline in 1974. This pipeline snakes its way right across Alaska to the southern tanker port of Valdez, where there is a giant oil refinery. In order to prevent the warm oil in the pipe from melting the underlying permafrost, and to allow the migration of caribou, the whole pipeline was built on stilts (Figure 2.41).

As the oil reserves in Prudhoe Bay begin to decline, there is increasing pressure from oil companies and politicians to develop alternative sites in the region. In 1923 – despite being recognised as an important wilderness – the Western Arctic Reserve (WAR) was designated as an area that could be developed for oil and gas, 'in an emergency'. Conservationists argue that – with its extensive wetlands and large population of threatened species – the WAR should continue to be protected from development. It is also home to groups of indigenous people who depend on the wildlife for their food, clothing and shelter. Almost 500 000 caribou – the largest herd in Alaska – live in and migrate through the reserve.

So, if the oil and gas reserves in the Western Arctic Reserve are exploited in the future, great care will have to be taken to minimise the environmental damage and the impact on indigenous people.

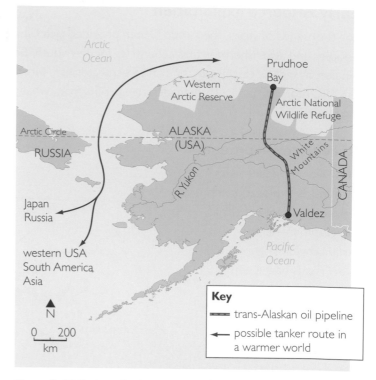

Figure 2.40 *Alaska – developments in an Arctic environment* ▲

Figure 2.41 *The trans-Alaskan oil pipeline* ▲

Wildlife and adventure tourism is becoming increasingly popular in northern Alaska. If the natural environment is damaged further by industrial development, fewer tourists will want to visit the area.

Sustainable management in Alaska's White Mountains

The White Mountains (Figure 2.40) is an alpine tundra landscape that provides many attractions and opportunities for tourism. In the summer, there is camping, hunting, fishing and walking (Figure 2.42). In the winter, cross-country skiing and snowmobiling.

About 400 000 hectares of the White Mountains have been designated as a protected area, called the White Mountains National Recreation Area. The US Bureau of Land Management (BLM), whose primary concern is to encourage responsible tourism while preserving the fragile natural environment, manages this area.

Figure 2.42 *Kayaking in the White Mountains National Recreation Area* ▲

- A network of all-terrain vehicle routes has been established – making use of artificial soil protection sheets to prevent damage to the ground and melting of the permafrost. Drainage ditches dug alongside these routes prevent waterlogging.
- Panning for gold is permitted in a small area of Nome Creek, but the panners can only use hand tools.
- All hunters in the area need permits, which strictly controls the level of hunting. Rangers patrol the area to prevent illegal hunting.
- There is regular wildlife monitoring to ensure that stocks are thriving and not being adversely affected by visitor pressure.
- The BLM has established a number of hiking trails. Cabins with special gutters to capture rainwater have been built along these routes.

1 Study Figure 2.38. In what ways, and for what reasons, do indigenous people manage their environment sustainably?
2 Study Figure 2.39.
  a Describe the industrial developments in the Russian Arctic.
  b Where is the Sverdrup Basin and what are the developments taking place there?
  c Why is there a need to transport oil and gas by pipeline from the Arctic?
  d What environmental problems can you think of that could be associated with constructing these pipelines?
  e Describe the locations of the main concentrations of mining sites.
  f How could mining be a threat to fragile Arctic environments?
  g A number of major oil spills are shown on the map. Use the Internet to research the *Exxon Valdez* oil spill in 1989. What happened and what damage was done to the environment?
3 Do you think that the Western Arctic Reserve should be exploited for oil and gas in 'an emergency'? Explain your answer.
4 In what ways, and for what reasons, is the White Mountains National Recreation Area being managed sustainably? You will find more information at: http://www.earthjustice.org/library/background/the_western_arctic.html

In this section you will learn about:
- what makes Antarctica special
- the role of the Antarctic Treaty – an international agreement
- development pressures and management issues

Skills

In this section you will:
- conduct research using the Internet

What makes Antarctica special?

- Antarctica is the remotest continent on Earth.
- It's also the coldest, windiest, and least populated.
- And, with a land area of 7 million square kilometres, it's the smallest continent – apart from Australasia.
- However, the most distinctive feature about Antarctica is its gigantic ice sheet – with an average thickness of about 2000 metres. In fact, just 0.4% of Antarctica is not covered by ice. This Antarctic ice sheet stores about 70% of the Earth's fresh water!
- Perhaps surprisingly, given all that snow and ice, Antarctica is also one of the driest places on Earth – with average precipitation values below those of many of the world's deserts.
- However, despite its harsh environment, Antarctica is home to penguins and seals, as well as many forms of algae, and its surrounding seas possess abundant wildlife.

Figure 2.43 *The remote landscape of Antarctica* ▲

> **Did you know?**
>
> The coldest temperature ever recorded on Earth was −89.2°C. It was recorded at the Russian research station Vostok in Antarctica in 1983.

The Antarctic Treaty

Seven countries have made territorial claims to segments of Antarctica (Figure 2.44). However, since 1959, an international agreement – called the Antarctic Treaty – has bound them (plus many other nations) in an unprecedented agreement to preserve the continent. Despite the discovery of valuable minerals there, the Antarctic Treaty has held firm, and the main function of the continent remains as a base for scientific research into – amongst other things – climate change. Under the Antarctic Treaty, native fauna is protected and disturbance is kept to a minimum. Tourism is also controlled (Figure 2.45).

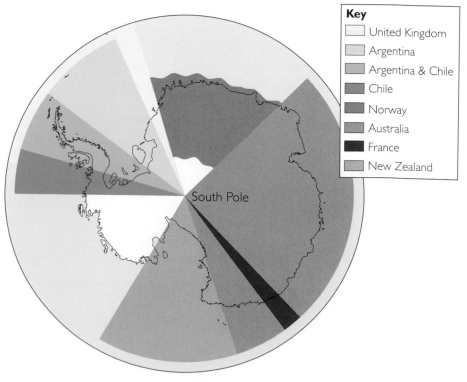

Key
- United Kingdom
- Argentina
- Argentina & Chile
- Chile
- Norway
- Australia
- France
- New Zealand

South Pole

Figure 2.44 *Territorial claims to Antarctica* ▲

Pressures and management issues

Looking into the future, Antarctica and the seas that surround it face a number of issues (Figure 2.45). However, perhaps the greatest threat of all is from climate change and global warming. As the seas around Antarctica become warmer, and the ice sheet continues to melt, the ecology of the entire region might well be harmed. And if the Antarctic ice sheet does melt, the whole world will feel the effects, not just Antarctica. So, Antarctica needs to be managed sustainably in order to preserve its pristine wilderness environment, both on land and at sea, for the future.

Figure 2.45 *Development pressures on Antarctica* ▼

Pressure	Impact on Antarctica
Falling numbers of krill	Krill are tiny crustaceans, near the bottom of the food chain, on which important Antarctic species depend. In recent decades, they have been extensively harvested by humans to provide fishmeal and animal feed.
	As well as being over-harvested by humans, continued global warming could adversely affect the phytoplankton on which krill feed. This, in turn, could have a devastating effect all the way up the Antarctic food chain.
General over-fishing	Several species of fish, such as the Antarctic Icefish have been over-fished.
Increasing levels of tourism	Since the early 1990s, the number of tourists visiting Antarctica has increased from 5000 to over 35 000 a year. Over 60% come from the USA, Germany and the UK. Most travel by ship, and increasing numbers are going ashore to take photographs (Figure 2.46) and enjoy activities like cross-country skiing and mountaineering.
	In 2009, the signatories of the Antarctic Treaty agreed to limit the number of tourists allowed to go ashore in Antarctica. They also agreed to a ratio limit of 20 tourists per guide. In addition, only one cruise ship at a time is allowed to land people on any given site.
	Despite these tighter regulations, scientists are concerned about the impacts on the Antarctic ecosystem of invasive species, such as the Mediterranean mussel, being brought into the region on the hulls of cruise ships. There is also an increased risk of pollution and oil leaks from ships.

Did you know?

In the past, whales were extensively hunted in Antarctic waters, but the establishment of the Southern Ocean Whale Sanctuary in 1994 has now stopped any commercial whaling in Antarctica. However, some whaling still occurs for the purposes of 'scientific research', particularly by Japan.

Internet research

Use the information in this section, together with research on the Internet, to help you find out more about the management issues facing Antarctica for the future. Consider the following questions:

- What are the development and environmental pressures on Antarctica? Find out more about the issues associated with over-fishing (krill and other fish) and tourism.
- Consider the need for sustainable management. Should fishing be banned? Should tourists be allowed to visit Antarctica? If so, how should this pressure be managed sustainably?
- Is the Antarctic Treaty likely to hold firm in the future? How might its functions change?

Here are a couple of websites to get you started:
http://www.doc.ic.ac.uk/%7Ekpt/terraquest/va
http://en.wikipedia.org/wiki/Antarctica

Figure 2.46 *Tourists photographing emperor penguins in Antarctica* ▼

arêtes Steep-sided knife-edge mountain ridge often marking the edges of a corrie or the watershed between two valleys (troughs)

basal sliding Large-scale and often quite sudden movement of a portion of ice in a glacier usually lubricated by sub-glacial meltwater

cold-based glacier Glacier where the base temperature is too low to enable liquid water to be present and where the glacier will probably be frozen to the ground

compressional flow 'Piling-up' or thickening of ice (glacier) due to a decrease in the long profile valley floor gradient

corries Enlarged armchair-shaped hollows on a mountainside characterised by a steep backwall and a hollowed-out bowl, occasionally containing a lake (tarn)

drumlins Egg-shaped depositional features, often in groups (swarms), resulting from the moulding of sub-glacial (ground) moraine by moving ice moving over the top

eskers Sinuous (winding) ridges found on the floor of a glacial trough, formed by fluvioglacial deposition in a meandering sub-glacial river

extensional flow Stretching or thinning of ice (glacier) in response to an increase in gradient (steep slope)

fragile environments Natural environments where processes operate slowly and where ecosystems can be easily harmed and take a long time to recover

frost heave Small-scale upwards displacement of soil particles resulting from the freezing and expansion of water just below the ground surface

frost shattering Also known as freeze-thaw, a physical weathering process involving alternating freezing and thawing of water in joints and pores within rocks

glacial surge Relatively rapid but usually short-term movement of a glacier

glacial system Inter-relationships between components in a glacial environment, often sub-divided into inputs, processes, and outputs

glacial troughs Glacially-enlarged river valley characterised by having a broad flat base and steep sides (U-shaped)

glacier budget Balance between inputs (accumulation), such as snowfall and avalanches, and outputs (ablation), such as calving and melting

hanging valleys Tributary glacial trough perched up on the side of a main glacial trough and often marked by a waterfall

ice wedges V-shaped ice-filled features formed by the enlargement of surface cracks by frost action. In time the cracks will become infilled with sediment

internal deformation Small scale inter- and intra-granular movement or deformation of ice crystals in response to gravity and mass

kames Mounds or hillocks found on the floor of a glacial trough formed by fluvioglacial deposition, e.g. kame terraces, formed by deposition in a marginal lake

meltwater channels Often narrow and steep-sided valleys formed by torrents of meltwater at the end of a glacial period

moraines Glacial deposits comprising largely angular and unsorted debris transported on (supraglacial), in (englacial), or under (sub-glacial) the ice. Many types of moraine can be identified, such as lateral (edge of glacier), medial (centre of glacier) and terminal (snout of glacier)

nivation hollows Shallow hillside hollow resulting from a concentration of snow-related processes such as frost shattering and slumping

nivation Snow-related processes, such as weathering and mass movement, that operate collectively to form shallow hollows in the landscape

outwash plains Often vast area of well-sorted and rounded sand and gravel deposits extending for some distance in front of a glacier, carried by meltwater

patterned ground Concentration of large stones on ground surface, usually associated with polygonal patterns of ice wedges, often forming stripes on slopes due to gravity

periglacial Environments experiencing long cold winters and short warm summers, typically with frozen ground (permafrost) but not covered by ice (glacial)

permafrost Permanently frozen soil and rock, a key characteristic of a periglacial environment

pingos Ice-cored mounds found in periglacial environments, formed by the freezing of sub-surface water bodies and subsequent swelling of the ground surface

pyramidal peaks Remnant of intense glacial erosion taking the form of a very steep-sided isolated peak

rotational flow Concave or arcuate flow typically experienced in a corrie and responsible for increased erosion (over-deepening)

solifluction lobes Extended lobes (tongues) of saturated soil formed by solifluction on a hillside

solifluction Gradual downhill slumping of saturated soil and rock, usually in summer when the upper surface zone (active layer) melts and becomes heavy and waterlogged

truncated spurs Former interlocking spurs that have been eroded by a glacier to form a steep valley side

warm-based glacier Glacier where the base temperature is high enough to enable meltwater to exist and therefore basal sliding to occur

Exam-style questions

1 (a) Study Figure 2.3 (page 67). Outline the global distribution of permafrost. *(4 marks)*

(a) Consider the overall distribution / patterns of continuous and discontinuous permafrost. Refer to regions on the map.

(b) Explain the formation of pingos. *(4 marks)*

(b) Use the correct geographical terminology. Focus on formation rather than just describing its features.

(c) How does a glacier erode its valley? *(7 marks)*

(c) Identify and describe the glacial processes involved. Make reference to the types of flow too, i.e. compressional, extensional. Use geographical terminology.

(d) Discuss the issues associated with the exploitation of cold environments for short-term gain. *(15 marks)*

(d) You must focus on short-term gains. Consider a range of reasons for exploitation (e.g. resources, tourism) and ensure that you refer to different types of cold environment. Make sure you discuss the issues (e.g. economic gain versus conservation).

2 (a) Study Figure 2.15 (page 75). Describe the characteristic features of this corrie. *(4 marks)*

(a) Refer only to the corrie in the photo. Use correct terminology.

(b) Study Figure 2.5 (page 68). Suggest reasons for the seasonal patterns of the glacial budget. *(4 marks)*

(b) Identify the pattern and suggest why ablation and accumulation fluctuated with the seasons.

(c) What are ice wedges and how does their development affect the periglacial landscape? *(7 marks)*

(c) Give a brief definition (with a diagram?). Make connections between ice wedge development and landscape features, such as stone polygons. Use geographical terminology.

(d) Discuss the advantages and disadvantages of sustainable management in cold environments. *(15 marks)*

(d) Define sustainable management. Refer to both advantages and disadvantages and include some discussion. Try to cover all cold environments and refer to case studies.

3 (a) Study Figure 2.14 (page 74). Describe the landforms resulting from glacial erosion. *(4 marks)*

(b) Study Figure 2.28 (page 83). Describe the typical characteristics of an outwash plain. *(4 marks)*

(c) What is a meltwater channel and how is it formed? *(7 marks)*

(d) Discuss the issues associated with the exploitation of Antarctica. *(15 marks)*

Coastal environments

A view along the south-eastern coast of Australia

Are the rocks in the photo easily eroded?

What are the coastal landforms in the photo?

Why has sand been deposited in the small bays?

Why are there no settlements here?

How might this view change in the future?

Introduction

For many of us, the coast is an important playground – it provides opportunities for walking, swimming, surfing, diving, sailing, and sun-bathing. Coastal environments provide some of the world's most spectacular scenery, including towering cliffs, jagged isolated rock pinnacles, and broad sandy beaches.

The vast majority of the world's population live within a few miles of the coastline. This emphasises the need for careful management. Cliff collapse, flooding, and resource management are major issues affecting coastal areas today.

In this chapter you will learn about the processes and landforms associated with coastal erosion and deposition. And you will investigate some of the management strategies that attempt to balance the various economic, social, and environmental demands at the coast.

Books, music, and films

Books to read

There is no such things as a natural disaster by Chester Hartman and Gregory D Squires, about the New Orleans flood of 2005 (2006)

Music to listen to

'After the Storm' by Mumford and Sons

Films to see

The Island
Too Hot To Handle (documentary, 2006)

About the specification

'Coastal environments' is one of the three Physical Geography option topics in Unit 1 – you have to study at least one.

This is what you have to study:
- The coastal system – which means constructive and destructive waves, tides, and sediment sources and cells.
- Coastal processes: marine erosion, transportation, and deposition; land-based sub-aerial weathering, mass movement, and run-off.
- Landforms of erosion: headlands and bays, blow holes, arches and stacks, cliffs, and wave-cut platforms. Landforms of deposition: beaches and associated features, such as berms, runnels and cusps, spits, bars, dunes, and salt marshes.
- A case study of coastal erosion, which should look at the specific physical and human cause(s) of the erosion and its physical and socio-economic consequences.
- Sea level change – eustatic and isostatic change. Coastlines of submergence and emergence and associated landforms. The impact of present and predicted sea level increase.
- A case study of coastal flooding, which should look at the specific physical and human cause(s) of the flooding and its physical and socio-economic consequences.
- Coastal protection objectives and management strategies:
  - hard engineering – sea walls, revetments, rip-rap, gabions, groynes, and barrages;
  - soft engineering – beach nourishment, dune regeneration, marsh creation, and land use and activity management.
- Case studies of two contrasting areas – one where hard engineering has been dominant and one where soft engineering has been dominant. The investigation of issues relating to the costs and benefits of such schemes, including the potential for sustainable management.

In this section you will learn about:
- the coastal system
- the different types of waves, and their formation
- the importance of tides on the coast

The **coastal environment** forms the boundary between the land and the sea (Figure 3.1). But where exactly does it begin and end? Well, it certainly stretches out into the sea for several hundred metres – to include the seabed offshore and the various processes affecting waves and the movement of sediment. In the other direction, it includes settlements (e.g. seaside resorts and industrial ports), and natural environments (e.g. sand dunes and salt marshes).

The factors that affect the coast extend well beyond the coastline into the open sea, where waves are generated. They also extend inland – to the rivers that transport over 90% of the sediment found at the coast. In addition, weather and climate can influence the processes of weathering and mass movement.

The coastal system

To help you understand the above processes and interactions more easily, think of the coast as a system – with inputs, processes and outputs (Figure 3.2). The components of the coastal system are all interlinked. If one aspect changes naturally – or is changed by people's management decisions – it will have a knock-on effect.

Waves

Waves are usually formed by wind moving over the ocean. As the wind drags over the surface of the water, friction causes a disturbance and forms waves. Even though individual waves appear to move across the ocean, this is just an illusion. They actually form part of a relatively static orbital motion, with little horizontal movement of the individual water molecules (Figure 3.3).

However, when waves approach a coast, the offshore profile of the seabed begins to distort their orbital motion. This makes them grow taller and steeper – and eventually break (Figure 3.3). At this point, water rushes up the beach as **swash**, before drawing back to the sea as **backwash**.

The energy of a wave depends on the strength (speed/velocity) and duration of the wind, plus the distance of open water over which the wind has blown (**the fetch**).

Residential and commercial properties

Transport infrastuctu

Tourisr

Sea, waves and sediment transfer

Recreational open space

Rock groynes and beach nourishment to protect the coast from flooding

Figure 3.1 *The coastal environment at Felixstowe in Suffolk* ▲

Figure 3.2 *The coastal system* ▼

INPUTS

Marine	**Land**
• Waves	• Rock type and structure
• Tides	• Tectonics

Atmospheric	**People**
• Weather/climate	• Human activities
• Climate change	• Economics
• Solar energy	• Recreation/tourism
	• Sea defences

PROCESSES

Weathering and mass movement	**Transportation**
• Freeze-thaw	• Sediment cells
• Salt weathering	• Longshore drift
• Carbonation	**Deposition**

Erosion	
• Hydraulic action	
• Abrasion/corrasion	
• Solution	

OUTPUTS

Erosional landforms	**Depositional landforms**
• Cliffs	• Beaches
• Stacks	• Spits
• Wave-cut platforms	• Salt marshes
	• Sand dunes
	• Mangrove swamps

Although waves vary enormously, it's possible to identify two broad types – **constructive** and **destructive** (Figure 3.4). These two wave types have significant impacts on the different processes and landforms found on a coast.

Occasionally, tectonic activity – such as earthquakes or volcanic eruptions –may form huge waves called tsunamis.

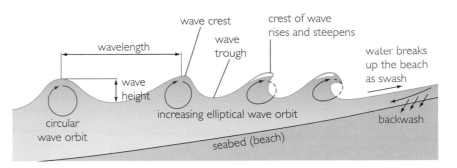

Figure 3.3 *Waves approaching and breaking on the shore* ▲

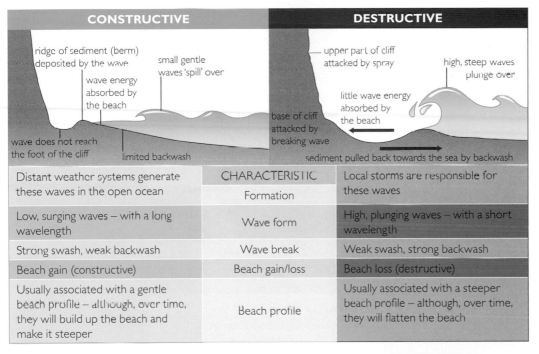

◄ **Figure 3.4** *Constructive and destructive waves*

CONSTRUCTIVE	CHARACTERISTIC	DESTRUCTIVE
Distant weather systems generate these waves in the open ocean	Formation	Local storms are responsible for these waves
Low, surging waves – with a long wavelength	Wave form	High, plunging waves – with a short wavelength
Strong swash, weak backwash	Wave break	Weak swash, strong backwash
Beach gain (constructive)	Beach gain/loss	Beach loss (destructive)
Usually associated with a gentle beach profile – although, over time, they will build up the beach and make it steeper	Beach profile	Usually associated with a steeper beach profile – although, over time, they will flatten the beach

Tides

Tides are variations in sea level, caused by the gravitational pull of the moon and the sun. In the UK, we have two high tides a day – 12 hours 25 minutes apart. The relative difference in height between high tide and low tide is called the **tidal range**. A high tidal range is associated with relatively powerful currents as the tides rise and fall.

Tides are important influences at the coast. For instance:

- If there is a small tidal range, the power of the waves (and therefore erosion) is concentrated on a relatively narrow section of cliff.
- If there is a large tidal range, vast tracts of sand can become exposed at low tide. When this sand dries out, onshore winds can transport the sand to form sand dunes.
- Tidal currents are important in moving sediment to and from a beach.

ACTIVITIES

1 Study Figure 3.1.
   a Work in pairs to suggest inputs, processes and outputs indicated by the photo.
   b Use the Internet to select a coastal photo of your own (similar to Figure 3.1). Add labels to indicate some of the features and characteristics of the coastal environment shown in your photo.
2 Study Figure 3.4.
   a Draw two large diagrams to show constructive and destructive waves. Add labels to identify some of the key characteristics of these waves.
   b Explain how – over time – constructive waves can alter the beach profile to encourage the formation of waves that are more destructive in their characteristics.
   c What affect do destructive waves have on the beach profile? How does this affect the waves approaching the beach?

In this section you will learn about:
- weathering at the coast
- mass movement processes at the coast
- marine processes (erosion, transportation and deposition)

The coast forms a dynamic boundary between the land and the sea. Its ever-changing shape and its many landforms result from the interaction of a range of physical processes (Figure 3.5).

Weathering at the coast

Weathering is the gradual breakdown of rocks in their original place (in situ) at – or close to – the ground surface. It is very common at the coast, where there is a plentiful supply of water and regular wetting and drying. The presence of vegetation, organic acids, and land and marine organisms also promotes different weathering processes there.

Freeze-thaw weathering

Freeze-thaw weathering (also known as frost shattering) involves water percolating into cracks and pores in a cliff and then freezing to form ice. When the water freezes, it expands in volume by about 9%. This expansion exerts stresses on the rock, which force the cracks to widen. With regular freezing and thawing, some fragments of rock eventually break away to collect at the base of the cliff as **scree**.

Coastal locations encourage freeze-thaw weathering. This is because cliffs provide extensive bare rocky outcrops that are often cracked and weathered. Also, coastal areas tend to have wetter climates than those inland – providing plenty of water to seep into the cracks. However, the slightly milder climate at many coasts means that freezing is not very common there. It's only during exceptionally cold winters that this weathering process is really effective.

However, despite fewer 'freeze' periods at the coast, freeze-thaw weathering is still important in coastal locations. For example, in 2001 – following a very wet autumn and a cold February – freeze-thaw weathering triggered several major rockfalls along the south coast of England (Figure 3.6). Chalk, a permeable and porous rock, was the main rock to be affected.

Freeze-thaw weathering generates angular rock fragments that can be picked up by the sea and used as tools for marine erosion. Without rocks like this, erosion by the sea would be far less effective.

Weathering
- Freeze-thaw
- Salt weathering
- Chemical weathering

Mass movements
- Rockfalls
- Landslips

Marine processes
- Erosion
- Transportation
- Deposition

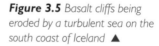

Figure 3.5 *Basalt cliffs being eroded by a turbulent sea on the south coast of Iceland* ▲

Figure 3.6 *A major rockfall at The White Cliffs of Dover in February 2001, as a result of freeze-thaw weathering* ▼

Salt weathering

Salt water is readily available in the tidal (or wave splash) zone. When it evaporates, it leaves behind salt crystals. These crystals can grow larger over time and exert stresses in the rock – in much the same way as ice – causing it to break apart. Salt is also capable of corroding rock (particularly if it contains traces of iron).

Wetting and drying

Frequent cycles of wetting and drying are common on the shore. Clay rich rocks, such as shale, expand when they get wet and contract when they dry out. This can cause them to crack, and contribute to their break up.

Carbonation

Carbonation involves the slow dissolving of calcium carbonate from rocks such as limestone and chalk. When it absorbs carbon dioxide from the air, water forms a weak carbonic acid. This acid reacts with the calcium carbonate in certain rocks to form calcium bicarbonate. This, in turn, is easily dissolved. The cooler the temperature of the water, the more carbon dioxide will be absorbed. This increases the effectiveness of the carbonation.

Biological weathering

Several types of biological weathering take place here:

- Plant roots prise apart rocks on cliffs when they grow into small cracks and then thicken and deepen with age.
- Water passing through decaying vegetation becomes acidic – leading to enhanced chemical weathering.
- Birds (e.g. puffins) and animals (e.g. rabbits) dig burrows in cliffs.
- Marine organisms are also capable of burrowing into rocks (e.g. piddocks) or secreting acids (e.g. limpets).

Mass movement at the coast

Mass movement is the downhill movement of rock and soil in response to the force of gravity. It includes processes such as rockfalls, landslides, mudflows, rotational slips, and soil creep (Figure 3.7).

Mass movement at the coast is quite common, because of the constant undercutting of the cliffs by the sea. This undercutting makes the cliffs unstable and prone to collapse. Weathering processes, such as freeze-thaw, also weaken cliffs and cause them to collapse (Figure 3.6).

Figure 3.7 *Mass movement processes at the coast* ▼

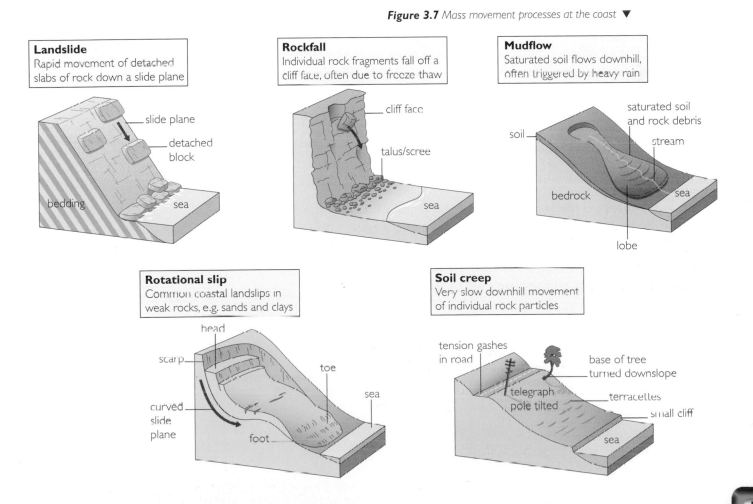

Look at Figure 3.8. It shows the dramatic collapse of the cliffs at Scarborough, North Yorkshire, in 1993. This huge landslip resulted in the demolition of the Holbeck Hall Hotel – and completely re-shaped the coastline. Less-dramatic mass movement events – such as rockfalls (Figure 3.6) and rotational slips – happen all along the coastline. They are important mechanisms for producing beach material for the sea to transport and redistribute.

Figure 3.8 *The dramatic landslip that destroyed the Holbeck Hall Hotel at Scarborough in 1993* ▲

Marine processes

Marine erosion

There are four main processes of marine erosion:

- **Hydraulic action**. Waves breaking at the foot of a cliff can carry out a significant amount of erosion. When air is blasted into cracks (a process called **cavitation**), loose rocks are dislodged and removed. Gradually, a wave-cut notch forms – close to the high-tide line – undercutting the cliff, and eventually leading to cliff collapse and retreat.
- **Corrasion**. Rocks caught up in surging seawater are hurled at a cliff face – causing it to be chipped and gouged.
- **Abrasion**. The sandpapering effect of loose rocks being constantly scraped across bare rock, e.g. a wave-cut platform. It often causes smoothing and polishing of the rock surface.
- **Attrition**. When rocks carried by seawater bash against each other – gradually making them smaller and smoother.

Marine transportation

There are also four main ways in which coastal sediment can be transported:

- **Traction** involves the rolling of relatively large and heavy rocks on the seabed.
- Slightly smaller and lighter rocks might adopt a bouncing motion, called **saltation**.
- A lot of lighter sediment is **suspended** within the water. This accounts for the often-murky appearance of the sea – particularly in river estuaries.
- Sediment that has dissolved completely will be transported in **solution**.

Marine deposition

Deposition occurs in low-energy environments, like bays and estuaries. When sand is deposited on a beach – and dries out – it might be whisked up by the wind and transported further inland, to form sand dunes at the back of the beach. In a river estuary, mud can build up on the sheltered side of a spit to eventually form a salt marsh (pages 108-109).

Sediment cells

A **sediment cell** is a stretch of coastline, usually bordered by two prominent headlands, where the movement of sediment tends to be contained. Eleven major sediment cells have been identified in England and Wales (Figure 3.36, page 119). These eleven cells form the basic units for coastal management.

Each major sediment cell can be divided into a number of sub-cells. One of these sub-cells lies in Christchurch Bay (Figure 3.9). This sub-cell is part of Major Sediment Cell 5 on Figure 3.37.

Within the Christchurch Bay sub-cell, the marine processes of erosion, transportation and deposition are all interlinked. Areas of deposition, like Dolphin Bank, also form potential 'sources' of sediment, as well as 'sinks'.

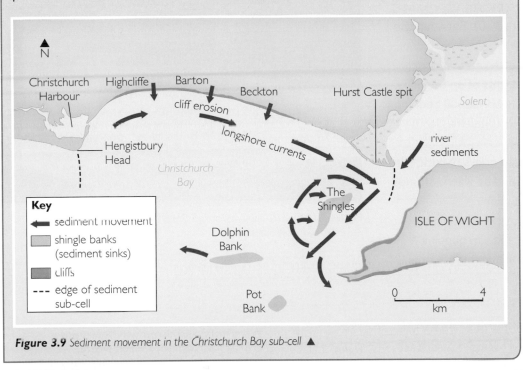

Figure 3.9 Sediment movement in the Christchurch Bay sub-cell ▲

Internet research

Use the Internet to find one or more photos of mass movement processes at a named stretch of coast. Add detailed labels to the photo(s) to describe the processes at work on that coastline, and identify evidence for them. Use Figure 3.7 to help you. Also include labels to describe any possible weathering and marine processes that might be at work there.

ACTIVITIES

1 Study Figure 3.5. Having read through this section, suggest the precise processes that are likely to be operating on this stretch of the Icelandic coast. Use evidence from the photo to support your suggestions.

2 Study Figure 3.8.
   a Draw a sketch of the photo.
   b Add detailed labels to your sketch to describe the main features of this landslip, plus its effects on the coast.
   c Suggest some possible reasons why this landslip might have occurred.
   d What effects would this landslip have on marine processes, do you think?

3 Study Figure 3.9.
   a Name the coastal features that form the edges of the Christchurch Bay sediment sub-cell.
   b Identify the inputs of sediment to this sub-cell.
   c Deposits of sediment are called sediment sinks. Identify some sediment sinks on the map?
   d Hurst Castle Spit needs some shingle to nourish its beach and help protect it from coastal erosion. Suggest where this shingle might come from and why?
   e Cliff erosion has generally ceased at Barton, following the construction of coastal defences there. What effects will this have on the sediment cell?

In this section you will learn about:
- landforms of coastal erosion (headlands and bays, blowholes, arches and stacks, cliffs and wave-cut platforms)
- landforms of coastal deposition (beaches, spits, bars, dunes and salt marshes)

Skills

In this section you will:
- interpret features of coastal erosion, using an OS map and a photo
- interpret features of coastal deposition, using an aerial photo
- use the Internet for research

Landforms of coastal erosion

Figure 3.10 shows one of the most famous stretches of coast in the world. Many people travel along the Great Ocean Road just to see spectacular examples of:

- **Headlands and bays**. Headlands are sections of coast that protrude out into the sea (there are several in the distance in Figure 3.10). Between the headlands are sandy bays.
- **Cliffs**. The cliffs along this stretch of coast rise vertically from the shore to heights of up to 45 metres. You can see clearly the horizontal bedding planes of the sedimentary sandstones that form the cliffs.
- **Stacks and stumps** – isolated pillars of rock, lying just off the coast and surrounded by water. Stumps are completely covered at high tide.
- **Wave-cut platforms** – flat rocky platforms extending out from the coast and surrounding the isolated stacks. Wave-cut platforms are only exposed at low tide.

Figure 3.10 *Coastal landforms on the Great Ocean Road in southern Australia, near Melbourne* ▲

Headlands and bays

Headlands and bays commonly form when rocks of different strengths are exposed together at a coast (Figure 3.11). Tougher, more-resistant rocks (like limestone and granite) tend to form headlands. Weaker rocks (like shale and clays) are eroded to form bays.

Once formed, headlands and bays interfere with the incoming waves. They make the waves bend, or refract. As Figure 3.11 shows, wave refraction focuses the waves' energy on the headlands. This explains the formation of eroded landforms like steep cliffs, arches and stacks at headlands. By contrast, the waves' energy is spread out and reduced in the bays. This results in the deposition of sand and shingle to form beaches.

Figure 3.11
Headlands and bays and wave refraction ▼

Wave energy is concentrated (converging orthogonals) at the headlands, where features of erosion are formed. In the bay, energy is spread out (diverging orthogonals) and sediment is deposited to form a beach.

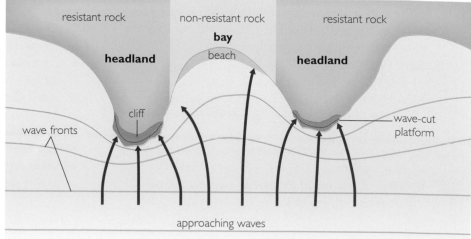

Cliffs and wave-cut platforms

When waves break against the foot of a **cliff**, erosion tends to be concentrated close to the high-tide line. This creates a distinctive **wave-cut notch**, and begins to undercut the cliff. Eventually, when it becomes unstable, the upper part of the cliff collapses. In this way – over hundreds of years – the cliff slowly retreats inland. It leaves behind it a smooth and rocky **wave-cut platform**.

Figure 3.12 shows part of the impressive chalk cliffs near Eastbourne in West Sussex. Constant wave action against the base of these cliffs maintains their steep profile. As you can see, a very extensive wave-cut platform is exposed at low tide.

A cliff's profile, and its rate of retreat inland, is affected by a number of factors (Figure 3.13).

Figure 3.12 *Cliffs and a wave-cut platform near Eastbourne* ▲

Figure 3.13 *Factors affecting a cliff's profile and its rate of retreat* ▼

Steep cliff

resistant rocks create steep cliffs

the lack of a beach results in undercutting and cliff collapse

rocks dipping away from the sea

long fetch and high-energy waves

Low-angle cliff

slumping

weak rocks, prone to collapse

collapsed cliff builds up

wide beach prevents waves reaching the cliff

rocks dipping towards the sea

low-energy waves are unable to remove material at the cliff base

Factors affecting the rate of retreat

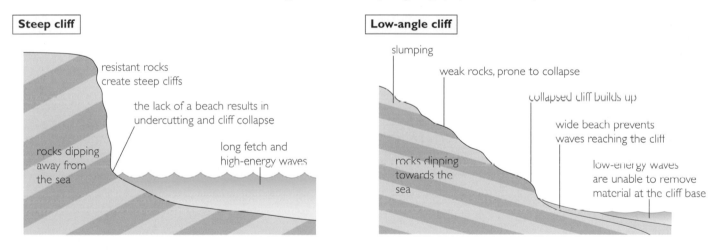

Weathering and mass movement
rocks prone to weathering and mass movement will be subject to severe erosion

Rock type
weak rocks result in rapid retreat: till (Holderness Coast) 1–10 m/year; sandstone (Devon) 1 cm–1 m/year; limestone (Dorset) 1 mm–1 cm/year; granite (Cornwall) 1 mm/year

Wave energy
high-energy waves, driven by strong prevailing winds and a long fetch, will increase the rate of retreat

(other factors leading to rapid rates of cliff retreat include the absence of a beach, rising sea levels, and human activities such as coastal defences elsewhere leading to increased erosion)

Caves, blowholes, arches, and stacks

Figure 3.14 shows a headland protruding into the sea. Several coastal landforms have been labelled on this diagram. However, even though these landforms have been labelled separately, they are all connected. They show a sequence of landform development.

- Joints or faults in the rock forming a headland create lines of weakness. These lines of weakness are then exploited by the sea. They might be enlarged at the base to form **caves**. Or they might simply be extended upwards to reach the surface as narrow, chimney-like **blowholes** – occasionally spouting fountains of water during stormy high tides.
- If two caves on either side of a headland join up, an **arch** will form. Slowly this will become enlarged by marine and weathering processes, and become wider at its base.
- Eventually, the roof of the arch will become unstable and collapse – leaving an isolated pillar of rock, called a **stack**. This feature will continue to be eroded by the sea. Eventually it will collapse so much that it only appears above the surface at low tide. At this point it's called a **stump**.

Figure 3.14 Coastal erosion features on a headland ▼

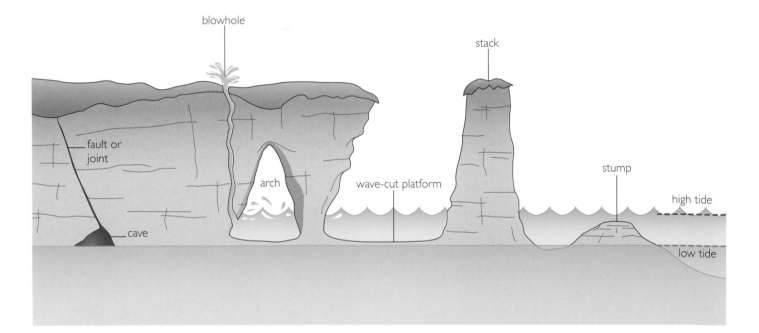

ACTIVITIES

1 Study Figure 3.10.
   a Describe the coastal erosion landforms shown in the photo.
   b How active is the erosion at this coast, do you think? Justify your opinion with evidence from the photo.
   c How do you think this coastal landscape might change over the next couple of hundred years?
2 Study Figure 3.11.
   a How does wave refraction account for the dramatic features of coastal erosion often found at headlands (e.g. arches and stacks)?
   b How does wave refraction explain the deposition of sediment in bays?
   c If sediment continues to build up in the bay, and the headlands continue to be eroded, do you think that the coast will eventually become straight? Explain your answer.
3 Study Figure 3.13. With reference to specific examples, describe the factors affecting cliff profiles. (Use the Internet and reference books from the library to help you with this essay question.)

CASE STUDY

Coastal erosion features at Trevose Head, Cornwall

Figure 3.15 shows part of the north Cornish coast. The main feature along this stretch of coastline is a headland called Trevose Head. This headland has been eroded by the sea to form a number of smaller headlands and bays.

Mother Ivey's Bay (Figure 3.16) is on the eastern side of the headland (in grid square 8676). The photo was taken looking northwest towards Merope Rocks. These are examples of stacks.

1 Study Figure 3.16. Describe the coastal landforms shown in the photo. You might find it easier to draw a simple sketch and use labels to identify the features.

2 Study Figure 3.15.

a Locate Trevose Head at 850766. Why do you think this location was chosen for a lighthouse?

b Merope Rocks are good examples of stacks. Give another example of a stack from the map.

c With the aid of simple sketch maps or diagrams, suggest how Merope Rocks may have become isolated from the main headland.

d Describe the location and the extent of wave-cut platforms on the map extract.

e Why do you think an extensive beach has formed in Harlyn Bay?

f What are the opportunities and attractions for tourism along this stretch of coast?

Figure 3.16 *Mother Ivey's Bay, with Merope Rocks in the distance* ▲

Figure 3.15 *A 1:25 000 OS map extract of Trevose Head in north Cornwall* ▼

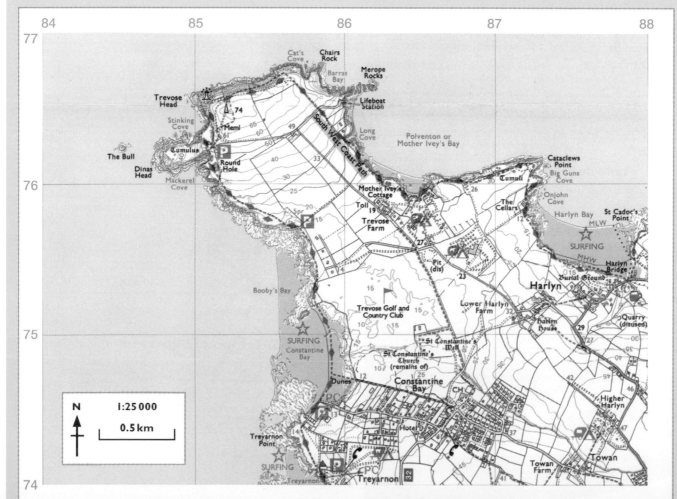

Landforms of coastal deposition

Coastal deposition takes place on sheltered stretches of coast. Sediment that can no longer be transported along the seabed, or suspended in the water, is deposited to form features such as beaches and spits.

Beaches

Beaches are commonly found in bays. There, wave refraction creates a low-energy environment that leads to deposition (Figure 3.11). A beach can be made of sand or shingle, depending on factors like the nature of the nearby sediment and the power of the waves. Figure 3.17 describes some of the typical features of a beach.

There are two broad types of beach – swash-aligned and drift-aligned. Figure 3.18 describes how each type is formed.

Spits and associated features

Spits form on stretches of coast where the beaches are drift-aligned. When sand or shingle is moved along a coastline by longshore drift (Figure 3.18), it may reach a point where the coast changes direction. At this point, sediment will build up and begin to protrude into the sea. This is the start of the formation of a **spit** (Figure 3.19). Occasionally, a spit may extend out from the coast to link an island to the shore. This feature is known as a **tombolo**.

Look at Figure 3.20. It shows Blakeney Point, a spit on the north coast of Norfolk. It has many of the features shown in part B of Figure 3.19. On the spit itself, sand has been deposited by the sea and then blown further inland by the wind to form **sand dunes** (Figure 3.21). Mud has also been deposited in the sheltered waters behind the spit. This has slowly become vegetated to form **salt marsh** (Figure 3.22).

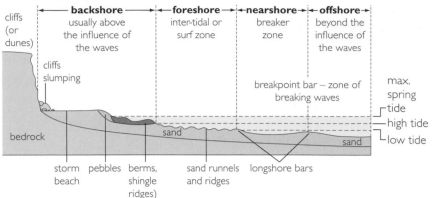

Figure 3.17 *The typical features of a beach* ▲

Figure 3.18 *How swash-aligned and drift-aligned beaches are formed* ▼

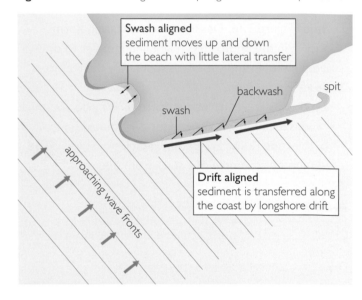

Figure 3.19 *How a spit is formed* ▼

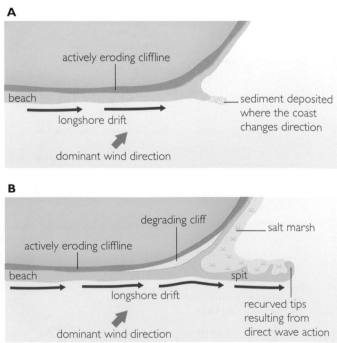

◀ **Figure 3.20** *Blakeney Point spit in Norfolk*

Sand or shingle can sometimes build up across a bay to form a **bar**. This feature is particularly obvious at low tide, when it becomes completely exposed. At high tide, it makes the water shallower at that point, which often causes waves to break early – ideal for surfing! In some places, sand and shingle bars have been driven onshore over thousands of years by rising sea levels to form **barrier beaches**.

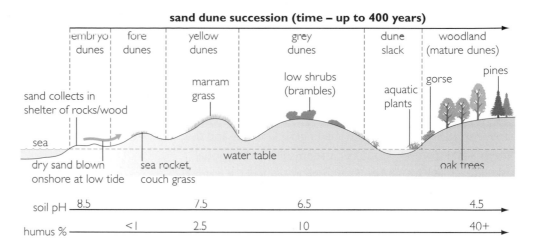

Figure 3.21 *How sand dunes are formed* ▲

ACTIVITIES

1 Study Figure 3.17. Compare the beach landforms associated with sand and shingle (pebbles) and suggest reasons why they differ.

2 Study Figures 3.18 and 3.19. The dominant (prevailing) wind direction is from the southwest. Do you think the beach in Constantine Bay (Figure 3.15) is swash-aligned or drift-aligned? Explain your answer with the aid of a simple sketch map.

3 Study Figure 3.21.
   a Explain why sand dunes will only form in certain coastal environments.
   b Are sand dunes formed by the actions of the sea or by the wind? Explain your answer.
   c Sea rocket and couch grass are **pioneer species**. This means that they can tolerate extreme environmental conditions. Suggest what those conditions might be here.
   d Marram grass is the tall rather sharp and spiky grass that most people associate with sand dunes. Its growth is stimulated by burial, and it has deep and very matted roots. Suggest why this makes it an ideal plant for these conditions.
   e The range of plants, insects and animals (biodiversity) increases with the distance from the sea. Use Figure 3.21 to suggest why this might occur.
   f Localised wind erosion can cause depressions, called blowouts. Sometimes a dune slack may form in a blowout. How and why are conditions in a dune slack quite different from those elsewhere in the sand dunes?
   g Suggest ways in which human activities could alter the natural sand dune ecosystem.

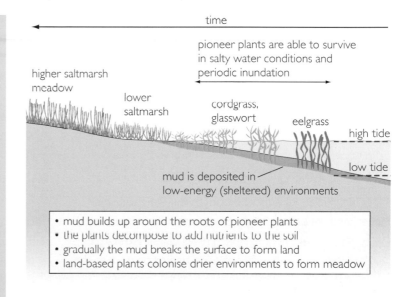

Figure 3.22 *How salt marshes are formed* ▲

Internet research

Conduct a research study of the depositional coastal landforms at Blakeney Point (Figure 3.20). With the aid of labelled maps and annotated photos, describe the formation and characteristics of the spit. Also describe the location, characteristics and formation of the sand dune and salt marsh areas.

- The OS 'Getamap' service is excellent for finding an OS map of the area http://www.ordnancesurvey.co.uk/oswebsite/getamap
- Wikipedia has useful information about Blakeney Point.
- Google Images also has plenty of photos (search 'Blakeney Point spit')

In this section you will learn about:
- the causes of sea level change
- landforms resulting from rising sea level
- landforms resulting from falling sea level

Skills

In this section you will:
- complete an annotated sketch from a photograph
- use the Internet for research

Sea level in the past was very different from today's level. During glacial periods, it was so low that it was possible for early man to walk across both the English Channel and the North Sea! However, in future, it seems likely that global warming could cause a significant rise in sea level. If it does, this could flood vast areas of low-lying coast, including whole countries like Bangladesh and The Maldives.

Sea level change – a balancing act

Sea level is the position of the sea relative to the land. If the amount of water in the oceans changes over time, so will sea level. This is called **eustatic** change. In cold glacial periods, precipitation falls as snow rather than rain. Water becomes confined to the land – in the form of snow and ice – and less liquid water flows into the oceans. As a result, sea level falls – which explains why, during previous glacial periods, the English Channel formed dry land.

During glacial periods, the enormous weight of the overlying ice (which might be several kilometres thick) actually makes the land sink. This is called **isostatic subsidence**. Then, when the ice begins to melt at the end of the glacial period, the reduced weight of ice on the land makes it more buoyant and it starts to rise. This is called **isostatic recovery**.

Eustatic changes occur relatively quickly, but isostatic changes take much longer and are often more complex. At the end of the last glacial period in Europe (about 8000 years ago), meltwater led to a relatively rapid rise in sea level – forming the English Channel and the North Sea. However, despite the removal of the huge weight of ice, the land only started to rise very slowly. That rise is still continuing today.

In the UK, isostatic recovery is complex. In fact, it's occurring at a slight tilt – with land in the north and west rising, but land in the south and east slightly sinking! This explains why engineers are concerned about the increased flood risk in the London area as land subsidence combines with the rise in global sea level caused by global warming.

Sea level rise associated with **global warming** is an important issue for the future, with increases of several centimetres or more widely forecast over the next few decades. The two main causes of sea level rise as a result of global warming are:

- the **thermal expansion** of water as it becomes warmer
- more water being added to the oceans following the melting of freshwater ice, e.g. the Greenland ice sheet. BUT the melting of sea ice (such as the Arctic ice cap) has no impact on sea level, because the sea ice is already displacing the ocean.

Did you know?

Sea level has risen by about 100 metres since the end of the last glacial period. During the main ice-melting phase, it rose at an average rate of about 10 mm a year. This compares with the estimated current rate of sea level rise – associated with global warming – of about 1.8 mm a year.

In some parts of the world, **tectonic uplift** can also have a direct impact on the coast – and therefore sea level. For example, in parts of Italy and Greece, tectonic uplift has raised land at the coast to create sheer cliffs (Figure 3.23).

Figure 3.23 *Cliffs formed by tectonic uplift at Sougia, Crete (Greece)* ▶

Landforms associated with rising sea level

The effect of a rise in sea level is to flood the coast – creating a **submergent coastline**. Low-lying coastal features (such as spits, beaches, and deltas) disappear beneath the waves. One of the most distinctive features associated with a rise in sea level is a **ria**. This is the permanently flooded lower course and estuary of a river as it approaches the sea. Rias are common in southwest England, where they create wide, sheltered moorings for leisure craft (Figure 3.24).

Figure 3.24 *The ria at Salcombe in Devon* ▼

In some mountainous coastal regions, such as western Scotland and along the coast of Norway, spectacular drowned glacial valleys – called **fjords** – are formed. Fjords are generally extremely deep – with dramatic, often sheer, rock walls forming their sides. The deep water allows large cruise ships to bring tourists to see this spectacular glacial landscape (Figure 3.25).

As sea level rises, it can cause seabed sediments to be bulldozed towards the coast. This has happened in parts of southern England, and has resulted in the formation of barrier beaches. Chesil Beach in Dorset is a shingle ridge, 29 km long, that was driven onshore by rising sea level at the end of the last glacial period. Today it forms an impressive coastal landform that links the coast to the Isle of Portland near Weymouth.

Figure 3.25 *Geirangerfjord in Norway* ▲

Landforms associated with falling sea level

The effect of a fall in sea level is to expose land normally covered by the sea – creating an **emergent coastline** (Figure 3.26). Cliffs that are no longer eroded and undercut by the waves become degraded – forming relic cliffs. Over time, scree begins to pile up at their foot and their slope angle becomes less steep. Vegetation also gradually begins to colonise the slope. The old beach becomes a **raised beach**, and may also become covered with vegetation eventually. Sometimes, dramatic coastal features – like arches and stacks – are also left high and dry (Figure 3.27).

Figure 3.26 *Landforms associated with an emergent coastline* ▼

Figure 3.27 *Landforms associated with falling sea level at Balintoy Harbour, Antrim, Northern Ireland* ▲

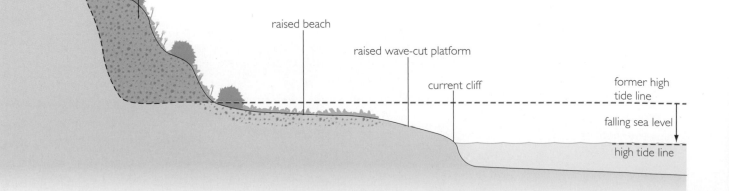

degraded (relic) cliff

raised beach

raised wave-cut platform

current cliff

former high tide line

falling sea level

high tide line

Sometimes, sea level change results in a landform called a **slope-over-wall cliff** (Figure 3.28). This feature is the result of a combination of sea level fall and sea level rise. It helps to show the complex causes of many coastal features. Examples of slope-over-wall cliffs can be found on the north Cornish coast and the Gower in south Wales.

Figure 3.28 How slope-over-wall cliffs are formed ▶

ACTIVITIES

1. Study Figure 3.24.
   a Describe the characteristics of the ria (drowned river valley) in the photo.
   b What would the valley have looked like before the coastline became submerged?
   c Assuming that global warming continues to cause sea level to rise, how might the view in the photograph be different in 100 years' time?
2. Study Figure 3.26. Imagine that sea level starts to rise again. Describe the effect of the rising sea level on the landforms in Figure 3.26. Use simple diagrams to illustrate how and why the landforms might be changed.
3. Study Figure 3.27. What evidence is there that sea level has fallen at Balintoy? Use an annotated sketch of the photo to describe the characteristics of the features you have identified.
4. Study Figure 3.28. With the aid of a diagram, suggest how the current slope-over-wall cliff profile might be modified if the UK was plunged into another glacial period.

Internet research

1. Discover how sea level has fluctuated in the last 25 000 years by accessing The Other Side website at http://www.theotherside.co.uk/tm-heritage/background/channelform.htm#map
2. With the aid of maps, describe the formation of the English Channel.
3. Use the Internet to create a bank of photos to show the landforms associated with sea level change. Add labels to describe the main features of the photos you decide to use. This will be helpful for your revision.

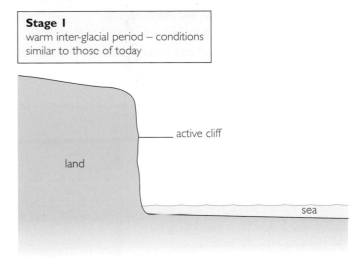

Stage 1
warm inter-glacial period – conditions similar to those of today

active cliff

land

sea

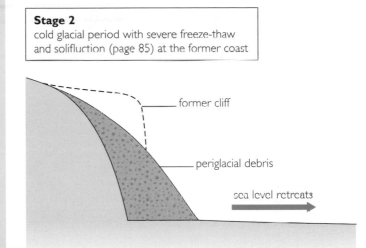

Stage 2
cold glacial period with severe freeze-thaw and solifluction (page 85) at the former coast

former cliff

periglacial debris

sea level retreats

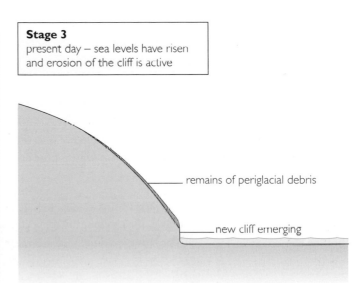

Stage 3
present day – sea levels have risen and erosion of the cliff is active

remains of periglacial debris

new cliff emerging

In this section you will learn about:

- the physical and human causes of cliff collapse
- economic, social and environmental impacts of cliff collapse
- two case studies of cliff collapses

CASE STUDY

Cliff collapse: Algarve, Portugal, 2009

On 21 August 2009, five holidaymakers were killed by a cliff collapse at Maria Luisa Beach in Albufeira, southern Portugal. Four of the five were members of the same Portuguese family. They were sunbathing at the foot of a sandstone cliff when part of it collapsed without warning – burying them (Figure 3.29).

A number of physical and human factors combined to cause this deadly cliff collapse:

- The sandstone cliff was already severely weathered and cracked (as you can see in Figure 3.29), and had been weakened.
- An earthquake measuring 4.2 on the Richter Scale struck the area a few days before the cliff collapse. This earthquake may well have further weakened the cliff and made it more likely to collapse.
- At high tide, the sea actively erodes the foot of this cliff (see the wave-cut notch on the left of Figure 3.29). This continuous undercutting would increase the likelihood of a cliff collapse.
- According to the Portuguese National Institute of Hydrology, recent very high tides might have added to the cliff's erosion just prior to the collapse.
- Rockfalls are common with these crumbling cliffs. Warning signs written in several languages were put up to alert people to the potential danger of cliff collapse. It seems likely that the victims chose to ignore these warning signs because it was cooler and shadier close to the foot of the cliff.

The civil authorities were well aware of the danger of cliff collapse, and inspected the cliff regularly. In fact, they inspected the cliff just a week before the collapse and found no abnormalities. So it could be that the earthquake was largely responsible for this cliff collapse.

Figure 3.29 *Rescue workers digging out the bodies of the five Portuguese victims after a section of sandstone cliff collapsed on top of them* ▲

Cliff collapse: Birling Gap, East Sussex, UK

Birling Gap is a coastal hamlet close to Eastbourne (on the famous Severn Sisters stretch of chalk coast in East Sussex). The cliff there has retreated steadily over the years. Several Victorian cottages have already collapsed onto the beach (Figure 3.30). Now just a few remain, and the government has decided that it's too expensive to build sea defences to protect them. Most of the cottages are empty and will soon be demolished. A nearby hotel and café will also be demolished eventually, as the cliff edge continues to retreat towards them.

EXTREME CAUTION

These cliffs are continuously eroding. There is a risk of rock falls at all times, which could cause death or injury

Figure 3.30 *Birling Gap, East Sussex – continuous cliff collapse* ▲

Several physical factors are combining to cause the continuous cliff collapse:

- The chalk cliffs are broken and shattered in this area, which makes them very vulnerable to freeze-thaw. This is helping to increase the rate of cliff retreat (currently an estimated 75 cm a year).
- The absence of a wide beach at Birling Gap means that the sea reaches the cliff foot at high tide and continuously undercuts and weakens it.
- With a long south-westerly fetch, powerful waves frequently pound this vulnerable coastline.

The National Trust, which owns some of the cottages at risk – together with English Nature and other conservation bodies – is in favour of allowing the coastal erosion to continue. This stretch of coast has naturally developing habitats that would be damaged if sea defences were constructed here. Also, with the sea level due to rise with global warming, the rate of erosion is likely to increase – making the costs of any coastal defences to prevent or reduce it even more expensive.

1   Study Figure 3.29.
    a   What clues are there in the photograph that these cliffs are potentially unstable?
    b   Present in the form of a table the physical and human factors that may have contributed to the fatal cliff collapse at Albufeira.
    c   Work in pairs to suggest the likely economic, social and environmental impacts of the cliff collapse.
2   Study Figure 3.30.
    a   What evidence is there in the photograph that these cliffs are being actively eroded?
    b   Why do you think there is a ladder down to the beach?
    c   Explain the physical factors that cause the rapid rate of cliff retreat at Birling Gap?
    d   Do you think the decision to let the cliffs continue to collapse and retreat was the correct one? Explain your answer.

Use the Internet to search for additional information about the Albufeira cliff collapse and the continuous erosion at Birling Gap. Remember that your Specification requires information about the physical and human causes and the economic, social and environmental impacts of cliff collapse.

There are some excellent photos of the Portuguese collapse at: http://www.iberiaimages.com/albums.asp?AID=1251

and some good photos of Birling Gap at: http://www.english-nature.co.uk/imagelibrary/searchresults.cfm?category=103&page=8&thumbnail=Y

Close to Birling Gap is Beachy Head, which has also been affected by some spectacular cliff collapses in the last ten years or so. There is plenty of information about these collapses on the Internet.

In this section you will learn about:

- physical and human causes of coastal flooding
- economic, social and environmental impacts of coastal flooding

CASE STUDY

Coastal flooding: Towyn, Wales, 1990

On 26 February 1990, sea defences in the Welsh coastal resort of Towyn were breached. Thousands of homes in the town were then flooded with seawater (Figure 3.31). It was one of the worst coastal floods in the UK in recent decades.

The main cause was:

- a storm surge of 1.5 metres
- driven onshore by gale-force westerly winds
- combined with a very high tide.

The result was a 400-metre-long breach of Towyn's sea wall – the effects of which were very severe:

- Seawater flooded four square miles of low-lying land and affected 2800 properties (Figure 3.32).
- Some areas of the town had to be evacuated.
- Many people were forced to live in temporary accommodation for up to a year after the event.
- The main railway line was severely damaged by the flooding.

To protect Towyn from future seawater flooding, rock revetments were built on the coast, as part of a £2 million flood-defence scheme. Then, in 2010 – to improve Towyn's protection – an additional £600 000 was committed to restore the flood banks on either side of the River Clwyd estuary, between Kinmel Bay and Rhyl (Figure 3.31).

However, despite the improvements to Towyn's flood defences, a recent report has suggested that likely future rises in sea level (as a result of global warming) could reduce the recurrence interval of a similar event to that of 1990 from 1 in 200 years to 1 in 75 years.

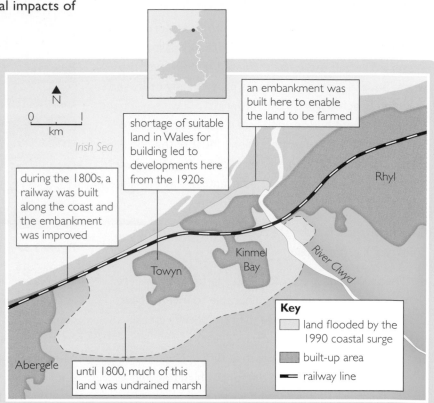

an embankment was built here to enable the land to be farmed

shortage of suitable land in Wales for building led to developments here from the 1920s

during the 1800s, a railway was built along the coast and the embankment was improved

until 1800, much of this land was undrained marsh

Irish Sea

Rhyl

River Clwyd

Kinmel Bay

Towyn

Abergele

Key
- land flooded by the 1990 coastal surge
- built-up area
- railway line

Figure 3.31 *The extent of the flooding in Towyn in 1990* ▲

Figure 3.32 *Flooding in Towyn in 1990* ▼

CASE STUDY

Coastal flooding caused by tropical cyclones

Tropical cyclones or hurricanes have been responsible for some of the world's most devastating coastal floods. In recent years, low-lying areas of countries such as Bangladesh, the USA, and Myanmar (Burma) have all suffered tremendous loss of life and damage to property and farmland as a result of these extreme weather events.

- Much of Bangladesh is a low-lying river delta, and millions of its people live on land that is close to or even slightly below sea level. This makes most of the country very vulnerable to severe river and coastal flooding. For example, in 1970, Cyclone Bhola killed up to 500 000 people. And, in 1991, 138 000 people were killed and 10 million were left homeless.
- In 2005, Hurricane Katrina caused widespread devastation and loss of life in the Mississippi delta region of the USA. Much of New Orleans was inundated by floodwater when the city's flood defences were breached (Figure 3.33).
- In 2008, Cyclone Nargis brought widespread destruction to the low-lying Irrawaddy delta region of Myanmar (Figure 3.34).

Figure 3.33 *New Orleans before and after Hurricane Katrina in 2005* ▲

Figure 3.34 *The impacts of Cyclone Nargis, Myanmar (Burma), 2008* ▼

Social	Economic	Environmental
• About 140 000 people were killed. • 2.3 million people were left homeless, because 95% of the buildings in the Irrawaddy delta were destroyed. • Because of supply problems, food prices went up, so many people went hungry. • Stagnant, filthy water created health issues (e.g. cholera, malaria)	• The estimated cost of the damage was $10 billion. • Businesses closed and people lost their jobs and incomes. • Crops were destroyed, reducing government income, so more food had to be imported	• Sewage pipes burst, flooding rice fields with raw sewage. • Saltwater inundated fields of food crops. • The ecosystems of the Irrawaddy delta were significantly affected by saltwater inundation. The shape of the delta was altered radically.

ACTIVITIES

1 Study Figure 3.31.
  a Describe the human factors that contributed to the scale of the Towyn flood disaster.
  b Create a table to consider the short-term and longer-term impacts of the Towyn flood. Divide them up into economic, social and environmental, as in Figure 3.34. Add any additional ideas of your own.
  c How might global warming lead to concerns about the reliability of the new flood defences at Towyn?
2 Study Figure 3.33. Describe the likely economic, social and environmental impacts of the flooding in New Orleans. Use the Internet to help you add some facts and figures to your account.

Internet research

Make an extended study of the causes and impacts of coastal flooding due to tropical cyclones in Bangladesh or Myanmar (Burma). Focus on the physical and human causes of the flooding and on the economic, social and environmental impacts. Include maps and annotated photos.

In this section you will learn about:
- the issues of coastal protection
- the options and strategies
- hard-engineering measures of coastal protection
- soft-engineering measures of coastal protection

Protecting the coast: the issues

Protecting the coast from cliff collapse and flooding is a major challenge that raises many issues. Even though it's often technically possible to protect a stretch of coastline, that may not always be the best option:

- Coastal protection can be extremely expensive. Small individual projects can easily cost several million pounds. With recent economic constraints on public spending, difficult decisions have to be made. What level of priority should be given to paying for coastal protection, compared to education, healthcare, etc.?
- Some stretches of coastline actually have strong ecological or geological reasons for being left unprotected. Natural erosion or flooding creates important habitats that would be destroyed if coastal protection measures were introduced. The Jurassic Coast is a World Heritage Site, due to the valuable fossils found in its crumbling cliffs. Coastal protection would stop the cliffs crumbling and keep these fossils hidden.
- In the past, some coastal defences were installed with little or no consideration given to their knock-on effects further down the coast. For example, a sea wall built to stop cliffs eroding in one place, will reduce the amount of sediment entering the sea. This reduction in sediment then stops beaches further down the coast from being replenished – making the cliffs there more vulnerable to erosion (Figure 3.35). The problem is not solved, just shifted down the coast.
- With sea level expected to rise – and some scientists predicting increased storminess due to global warming – serious issues have arisen about the level of coastal protection that will be needed in the future. Perhaps some low-value stretches of coastline should just be left to erode naturally?

Figure 3.35 *Coastal erosion on the Holderness coast in Humberside* ▼

Protecting the coast: the options

There are four strategies of coastal protection:

A **Hold the line** involves maintaining the current position of the coastline – as shown on a map. In other words, ensuring that the high-tide line marked on a map stays in the same position. Defence measures like sea walls can be used to 'hold the line'.

B **Advance the line** involves extending the high-tide line (the coastline) out to sea. This is most commonly done by encouraging the build-up of a wider beach – by beach nourishment, or the construction of groynes.

C **Managed retreat** involves allowing the coastline to retreat – but in a controlled way. This is an increasingly popular option that often involves the deliberate breaching of dilapidated earth flood banks, built many years ago to protect low-quality farmland from flooding. Today, the cost of repairing these earth flood banks is more than the value of the land behind them, so it makes sense to allow the sea to flood the area in a controlled manner.

D **Do nothing**, which involves – doing nothing! Just let nature take its course and allow the sea to erode cliffs or flood low-lying land at will. This strategy makes a great deal of sense on coastlines with limited development or low-value land.

Protecting the coast: who decides?

In England and Wales, the coastline is divided into 11 major sediment cells (Figure 3.36). These sediment cells are largely self-contained stretches of coastline. Their natural processes of erosion, transportation and deposition have little impact outside each cell. This means that the impacts of any man-made changes to the coast, such as the construction of sea walls, are also contained within each cell. So, sediment cells make ideal planning units.

A **Shoreline Management Plan** (SMP) has been written for each of the 11 sediment cells in England and Wales. Each SMP is an extremely detailed document. It contains information about coastal processes, ecological considerations, and human uses. Many different pressure groups and organisations have been involved in the development of the 11 SMPs.

Coastal protection is a major aspect of any SMP. Local authorities are directly responsible for coastal protection, so they are the organisations that implement the final SMPs. In England and Wales, the Environment Agency has a responsibility for flood protection, so it works closely with the local authorities. In 2010, the Environment Agency first published online maps indicating areas at risk from coastal erosion. It already produced maps showing areas at risk from flooding (page 47).

Coastal management in Scotland is overseen by the devolved government there. Local planning authorities are responsible for implementing coastal protection measures. Local authorities also have responsibility for coastal protection in Northern Ireland. Here, there is pressure for an integrated planning strategy – similar to the SMPs in England and Wales.

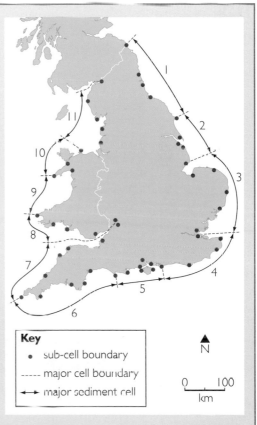

Key

- sub-cell boundary
----- major cell boundary
↔ major sediment cell

N

0   100
km

Figure 3.36 Sediment cells in England and Wales ▲

Protecting the coast: hard or soft engineering?

Hard engineering involves artificial (often concrete), man-made structures – designed to interfere with natural coastal processes.

Figure 3.37 The main forms of hard engineering ▼

	Description	Advantages	Disadvantages
Sea wall	A concrete or rock wall, placed at the foot of cliffs or at the top of a beach. It usually has a curved face to reflect waves back into the sea. Cost: £6000 a metre.	Effective at stopping the sea. Often has a promenade for people to walk along.	Can be intrusive and unnatural to look at. Very expensive to build, and has high maintenance costs.
Groynes	Groynes are timber or rock structures, built at right angles to the coast. They trap sediment being moved along the coast by longshore drift – building up the beach. Cost: £5000-£10000 each (at 200-metre intervals).	The built-up beach increases tourist potential and gives greater protection to the land behind it. Although a form of hard engineering, groynes work with natural processes to build up the beach. Not too expensive.	By interrupting longshore drift, groynes starve beaches along the coast of sediment – often leading to increased erosion elsewhere. The problem is shifted not solved. Groynes are unnatural, and rock groynes in particular can be unattractive.
Rock armour (rip-rap)	Rock armour consists of large, tough rocks dumped at the foot of a cliff, or the top of a beach. It forms a permeable barrier to the sea – breaking up the waves, but allowing some water to pass through. Cost: £100000-£300000 for 100 metres.	Relatively cheap and easy to construct and maintain. Often used for fishing from, or for sunbathing by tourists.	The rocks are usually from other parts of the coast, or even from abroad, so, they do not fit in with the local geology and look out of place. Can be very intrusive. Can be dangerous for people clambering over them.
Revetments	Revetments are wooden or rock barriers placed at the foot of a cliff or the top of a beach. They break up the energy of the waves. Cost: Up to £4500 a metre.	Relatively inexpensive.	Intrusive and very unnatural looking
Gabions	Gabions are wire cages filled with small rocks that can be built up to create walls. Often used to support weak cliffs. Cost: £5000-£50000 for 100 metres (depending on height).	Relatively inexpensive and flexible. Encourages upper beach stability. May eventually blend in with the landscape, as soil collects between the rocks and vegetation starts to grow.	Look very unsightly to begin with. The metal cage can rust and become broken and dangerous.
Barrages	Large-scale engineering projects, involving the construction of a partly submerged wall in a bay (e.g. Cardiff Bay), or estuary. Sluice gates control water flow in and out of the bay. Sometimes used to generate tidal power, e.g. Rance, France. Cost: £200 million (Cardiff Bay).	Can offer huge benefits for the multi-purpose use of a bay or estuary (e.g. electricity generation, harbour and marina, tourist developments, etc.)	Very expensive and potentially damaging to natural habitats, because shoreline processes and environments are disturbed.
Offshore breakwater	A partly submerged rock barrier designed to break up waves before they reach the coast. Cost: Similar to rock armour, depending on the materials used.	Effective permeable barrier.	Visually unappealing. Potential navigation hazard.

Soft engineering involves a 'softer', more environmentally friendly, approach. It often uses natural materials, and tries to work with natural processes.

Figure 3.38 *The main forms of soft engineering* ▼

	Description	Advantages	Disadvantages
Beach nourishment	The addition of sand or pebbles to an existing beach – to make it higher or wider. The sediment is usually dredged from the nearby seabed, so it blends in with the existing beach material. Cost £300 000 for 100 metres.	Relatively cheap and easy to maintain. Looks natural and blends in with the existing beach. Increases tourist potential by creating a bigger beach.	Does need constant maintenance, because the natural processes of erosion and longshore drift might deplete the beach.
Dune regeneration	Marram grass can be planted to stabilise sand dunes and help them become re-established. Areas can be fenced off to keep people off newly planted dunes. Cost: £200-2000 for 100 metres.	Maintains a natural coastal environment. Provides important wildlife habitats. Relatively cheap and sustainable.	Time consuming to plant the marram grass and fence off areas. People do not always respond positively to being banned from certain areas.
Marsh creation	This is a form of managed retreat, by allowing low-lying coastal areas to become flooded by the sea. The land becomes a salt marsh. Cost: varies, depending on the size of the area and the need to compensate landowners.	Relatively cheap and often involves land reverting to the way it was before being managed for agriculture. Provides a buffer to powerful waves – creating a natural defence. Creates an important wildlife habitat.	Agricultural land is lost. Farmers or landowners need to be compensated.
Land-use management	Even if some areas of the coast will eventually be eroded or flooded, land-use management can minimise the impact. For example, caravan parks on cliff tops are appropriate, because the caravans can easily be moved and re-sited.	An appropriate behavioural approach to coastal management that is essentially sustainable.	Some people might not want to have land uses restricted at the coast. Difficult to implement retrospectively.

ACTIVITIES

1 Work in pairs or small groups to consider the issues affecting coastal protection.
   a Which of the issues do you think is most important? Why?
   b Are there any circumstances where wildlife and geology should be more important considerations than people?
   c Can you think of any other issues affecting coastal protection?
2 Study Figure 3.39.
   a Describe the methods of coastal protection shown in the photo. Use Figures 3.37 and 3.38 to help you.
   b Lyme Regis is an important tourist destination. Do you think the new coastal defences have improved the amenity value of the resort?
   c Do the defences fit in with the existing environment, or are they intrusive?
   d Assume that no maintenance is carried out on the defences. If you return in ten years' time, what changes do you think you might see and why?

Figure 3.39 *Coastal defences at Lyme Regis, Dorset. Today it is common for coastal protection schemes to involve a mixture of both hard and soft engineering.* ▲

In this section you will learn about:
- hard and soft engineering at Walton-on-the-Naze
- managed retreat at Abbotts Hall Farm

Skills

In this section you will:
▶ use an OS map and an aerial photo

CASE STUDY

Walton-on-the-Naze

Setting the scene – the need for coastal protection

Walton-on-the-Naze is a small seaside town on the Essex coast (Figure 3.40). It was originally a farming settlement – and was located several miles inland from the coast. However, rapid coastal erosion brought the coastline closer and closer to the town. By about 1800 its function was starting to change, and it eventually became a popular seaside resort.

The Naze itself is a peninsula of land stretching for about 3 km north of the town. Originally farmland, and then a golf course, the Naze became an important lookout point during the Second World War. It is now open land used by the public for walking and watching wildlife.

The Naze is also the location of an important historic monument, called the Naze Tower (Figure 3.41). It was built in 1720 as a lighthouse to guide ships sailing to and from the nearby port of Harwich. However, this tower is now under threat from rapid cliff erosion, which could see it fall onto the beach in less than 50 years.

The cliffs of the Naze are over 20 metres high and rich in fossils. Because of their geological importance, they have been designated as a Site of Special Scientific Interest. This affects what steps can be taken to protect them.

The rapid rate of erosion along this stretch of coast is a result of two factors combining:

- The rocks that make up the cliffs (clays, overlain by sands and gravels) are extremely weak and prone to slumping and collapse.
- High-energy waves from the North Sea pound this coast, particularly during winter storms.

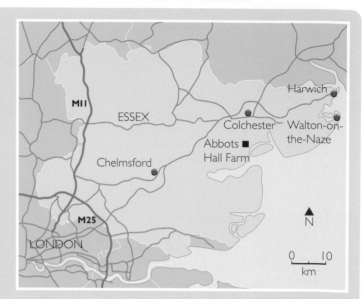

Figure 3.40 *The location of Walton-on-the-Naze and Abbotts Hall Farm* ▲

Figure 3.41 *The crumbling cliffs of the Naze and the Naze Tower* ▼

Protecting the Naze coast

Look closely at Figure 3.42. The current coastal protection zone stretches for about 4 km – from the resort of Walton-on-the-Naze to about halfway along the Naze coast (near the Naze Tower). The northern half of the Naze is currently unprotected.

Since the 1930s, major engineering works have taken place to protect the seafront at Walton-on-the-Naze (Figure 3.43).

- The cliffs have been regraded to make them shallower and less vulnerable to collapse.
- A major sea wall has been built at the foot of the cliffs.
- Wooden groynes have been constructed along the coast to interrupt the northerly drift of sediment and help build up the beach.
- The cliff has been planted with vegetation to anchor the soil and reduce the likelihood of slumping.

Since their construction, the defences have protected the resort from erosion and flooding. However, they are now quite old, and are beginning to need significant repair and reinforcement. For example, the sea wall has begun to be undermined by the sea. This is called **scouring**, and is a major problem with sea walls in general.

- A 130-metre length of the lower sea wall has now been protected by a three-metre-wide wedge of material called Elastocoast. Elastocoast consists of coarse pebbles bound together with polyurethane to produce a strong porous material. It is both cost effective and aesthetically pleasing to look at. The aim of the Elastocoast is to provide a seal at the base of the sea wall to prevent further erosion. The cost of this maintenance work was £500 per metre.
- In 1998, a granite rock revetment was built to help reduce the rate of cliff erosion.
- But, despite these efforts, cliff erosion and collapse is still occurring. For example, in 2000, a major landslip resulted in a hectare of land collapsing onto the beach. The cliff continues to erode at a rate of 1.4 metres a year.
- So, it was decided in 2010 to spend £1.2 million constructing additional defences to help protect the Naze.

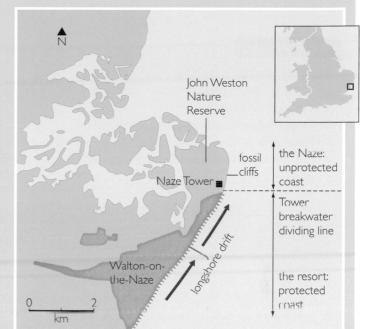

Figure 3.42 The extent of the coastal defences on the Naze coast ▼

Figure 3.43 Protecting the coast at Walton-on-the-Naze ▼

Interpreting an OS map and an aerial photo

Figure 3.44 is an OS map extract of the Naze coast.
Figure 3.45 is an aerial photo of the same area.

ACTIVITIES

1 What direction is the photo looking in?
2 Describe the coastal protection measures to the south of the pier.
3 How effective are the groynes proving to be? Justify your answer.
4 What is the likely purpose of the extended groyne at X?
5 Use the map and the photo to describe the coastal defences to the north of the pier at Walton-on-the-Naze.
6 Use the map to describe the human uses of the Naze.
7 Do you think that these uses justify protecting the whole of the Naze from erosion?
8 With sea level likely to rise in the future, what are the arguments for and against managed retreat in the northern part of the Naze?

Figure 3.44 *An OS map extract of the Naze coast* ▼

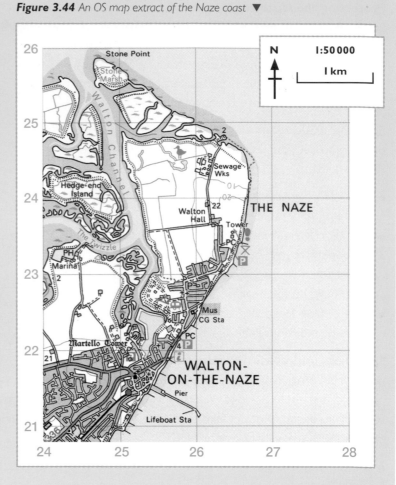

Figure 3.45 *An aerial photo of the Naze coast* ▼

CASE STUDY

Abbotts Hall Farm

In the late 1990s, the sea wall at Abbotts Hall Farm needed repairing. The Essex Wildlife Trust (which owned the land) – together with the Environment Agency – decided that the land currently protected by the sea wall should be allowed to flood. Because it was relatively low-value agricultural land – and with sea level likely to rise in the future – the creation of a new area of salt marsh seemed to be the best alternative. And the Essex Wildlife Trust was keen to establish more salt marshes anyway.

Counter walls were created on either side of Abbotts Hall Farm, to protect the neighbouring farmland. Then, in 2002, the old sea wall was breached and the sea was allowed to flood the area (Figure 3.46). In addition to creating important bird habitats, the newly created marshes form important fish nurseries for bass, herring and 14 other types of fish. The project has been so successful that it is now used as a demonstration project for managed retreat.

Figure 3.46 *Managed retreat at Abbotts Hall Farm, Essex* ▲

ACTIVITIES

1 Study Figure 3.41.
   a What evidence is there on the photo that the cliffs are eroding rapidly?
   b Outline the reasons for the rapid retreat of the cliffs on the Naze.
   c What are the arguments for and against protecting the Naze from further erosion?
2 With reference to Figure 3.43 describe the coastal defence measures that have been adopted to protect Walton-on-the-Naze from coastal erosion and flooding.
3 Study Figure 3.46. Do you think the managed retreat scheme at Abbotts Hall Farm has been successful? Explain your answer.

Internet research

Conduct your own extended study of managed retreat. You could choose to extend your study of Abbotts Hall Farm by accessing the website at:
http://www.wwf.org.uk/filelibrary/pdf/mu52.pdf

Alternatively, consider the managed retreat scheme at Tollesbury:
http://randd.defra.gov.uk/Document.aspx?Document=FD1922_7433_TSM.pdf and
http://www.saltmarshmanagementmanual.co.uk/Management/ManCaseStudiesManReal3.htm

A further option is Orplands at:
http://www.saltmarshmanagementmanual.co.uk/Management/ManCaseStudiesManReal2.htm

arches and stacks Resulting from marine erosion, an arch is formed when two caves on either side of a headland become connected. When the roof eventually collapses an isolated rock pillar (stack) is formed

barrages Dams built across a river estuary, often with movable gates that can be raised to provide protection from flooding caused by high tides or storm surges

bars Deposit of sand and/or pebbles usually formed parallel to the coastline, e.g. across a bay or estuary

beach nourishment Involves sand and/or pebbles being added to a beach to increase its size thereby affording greater protection from flooding and erosion

blowholes Narrow funnel-like feature, often an enlarged joint, in a sea cliff. Water from waves breaking at the foot of the cliff can be 'squirted' along the crack to create a small geyser-like fountain at the cliff top

cliff Steep, often vertical rock face where the land meets the sea, often undercut at high tide level to form a notch

coastal system Inter-relationships between components in a coastal environment, often sub-divided into inputs, processes, and outputs

constructive waves Low, spilling waves characterised by a powerful swash and weak backwash and responsible for building up a beach

destructive waves High, plunging waves characterised by a weak swash and powerful backwash and responsible for eroding a beach

drift-aligned beach Beach deposits (sand and pebbles) that have been transferred along a coastline by longshore drift, often accumulating to form a wide beach at a headland where the lateral drift is interrupted

dune regeneration Encouraging sand dune stabilisation, often by planting marram grass or building fences

emergent coast Coastline exhibiting features associated with falling sea levels, e.g. raised beach

eustatic change Variations in relative sea level resulting from changes in the amount of liquid water entering the oceans, e.g. glacial meltwater pouring into the oceans at the end of an ice age will cause sea levels to rise

gabions Wire cages filled with rocks and built-up to protect cliffs or vulnerable structures from erosion or collapse

hard engineering Commonly, built structures such as concrete sea walls designed to resist natural processes

headlands and bays Often alternating features seen in plan-view at the coast, usually representing hard (headlands) and soft (bays) rock outcrops

isostatic change Variations in relative sea level associated with changes in the buoyancy of the land, e.g. at the end of

an ice age, as the weight of the ice is removed, the land will start to rise causing relative sea levels to fall

longshore drift Lateral transfer of sediment along a stretch of coastline in a series of 'zig-zags' representing repeated cycles of swash and backwash

marine erosion Erosion carried out by the sea, including hydraulic action (sheer power of the waves), corrasion (rock fragments being flung at a cliff face), abrasion (sandpaper effect as rock fragments/pebbles are scraped over bedrock), and solution (dissolving of soluble rocks)

marine transportation Transportation of sediment by the sea involving traction (rolling on seabed), saltation (bouncing), suspension, and solution

marsh creation Artificial creation of salt marshes associated with managed retreat, where the sea is allowed to flood a low-lying coastal area previously protected

revetments Semi-permeable often wooden structures positioned on a beach to break up the waves as they approach the shore

rip-rap Piles of large resistant boulders (also known as rock armour) placed at the foot of a cliff or alongside vulnerable structures to break up and absorb wave energy

salt marshes Coastal ecosystem formed on mudflats (e.g. in a river estuary) largely comprising salt-tolerant plants

sand dunes Deposits of sand transported and shaped by the wind inland from the high tide line

sea walls Solid barriers to the sea usually constructed of concrete and positioned on the coast to protect vulnerable land or human developments, such as housing

sediment cells Largely self-contained (closed) sediment systems involving the movement of sediment in a cyclic manner

soft engineering Management approaches that have minimal impacts on the environment and aim to work with natural processes

spits Narrow 'finger' of deposited material (sand and/or pebbles) protruding out to sea from the land. Where a spit joins an island to the mainland, the feature is called a tombolo

sub-aerial weathering Group of weathering processes operating at the coast but not directly related to marine action, e.g. frost shattering in cliffs, action of tree roots

submergent coast Coastline exhibiting features associated with rising sea levels, e.g. flooded river estuary (ria)

swash-aligned beach Beach deposits (sand and pebbles) accumulated in a bay largely by the action of swash

wave-cut platforms Bare and gently sloping rocky surface at the foot of a cliff, sometimes covered by pebbles

Exam-style questions

1 (a) Describe the processes of physical weathering that may be active at the coast. *(4 marks)*

(a) Focus on physical weathering only.

(b) Study Figure 3.1 (page 98). Suggest reasons why this stretch of coastline needs to be defended from the sea. *(4 marks)*

(b) Use evidence from the photo. Focus on reasons why the coast needs defending.

(c) What factors affect the rate of retreat of a cliff? *(7 marks)*

(c) Consider a number of different factors, such as rock type and geological structure. Use diagrams to support your answer.

(d) Discuss the importance of sea level change in affecting the development of coastal landforms. *(15 marks)*

(d) Identify the landforms associated with rising and falling sea levels. Discuss the importance of sea level change in the development of these landforms

2 (a) Describe the characteristics of constructive waves. *(4 marks)*

(a) Stick to description only. Consider using a diagram.

(b) Study Figure 3.8 (page 102). Suggest possible reasons for the cliff collapse shown in the photograph. *(4 marks)*

(b) Consider factors such as slope, rock type, and trigger mechanisms such as rainfall.

(c) Describe the features of coastal deposition caused by longshore drift. *(7 marks)*

(c) Define longshore drift. Use simple diagrams to describe the main features, but link them clearly to the process.

(d) Discuss the advantages and disadvantages of adopting hard engineering solutions to reduce the risk of flooding at the coast. *(15 marks)*

(d) Define hard engineering. Consider both advantages and disadvantages. Focus on flooding rather than cliff collapse. Make use of case studies.

3 (a) Describe the typical characteristics of a sediment cell. *(4 marks)*

(b) Study Figure 3.27 (page 112). What is the evidence that the sea level has fallen? *(4 marks)*

(c) Describe and explain the formation of a salt marsh. *(7 marks)*

(d) Discuss the issues associated with managed retreat. *(15 marks)*

4 Hot desert environments and their margins

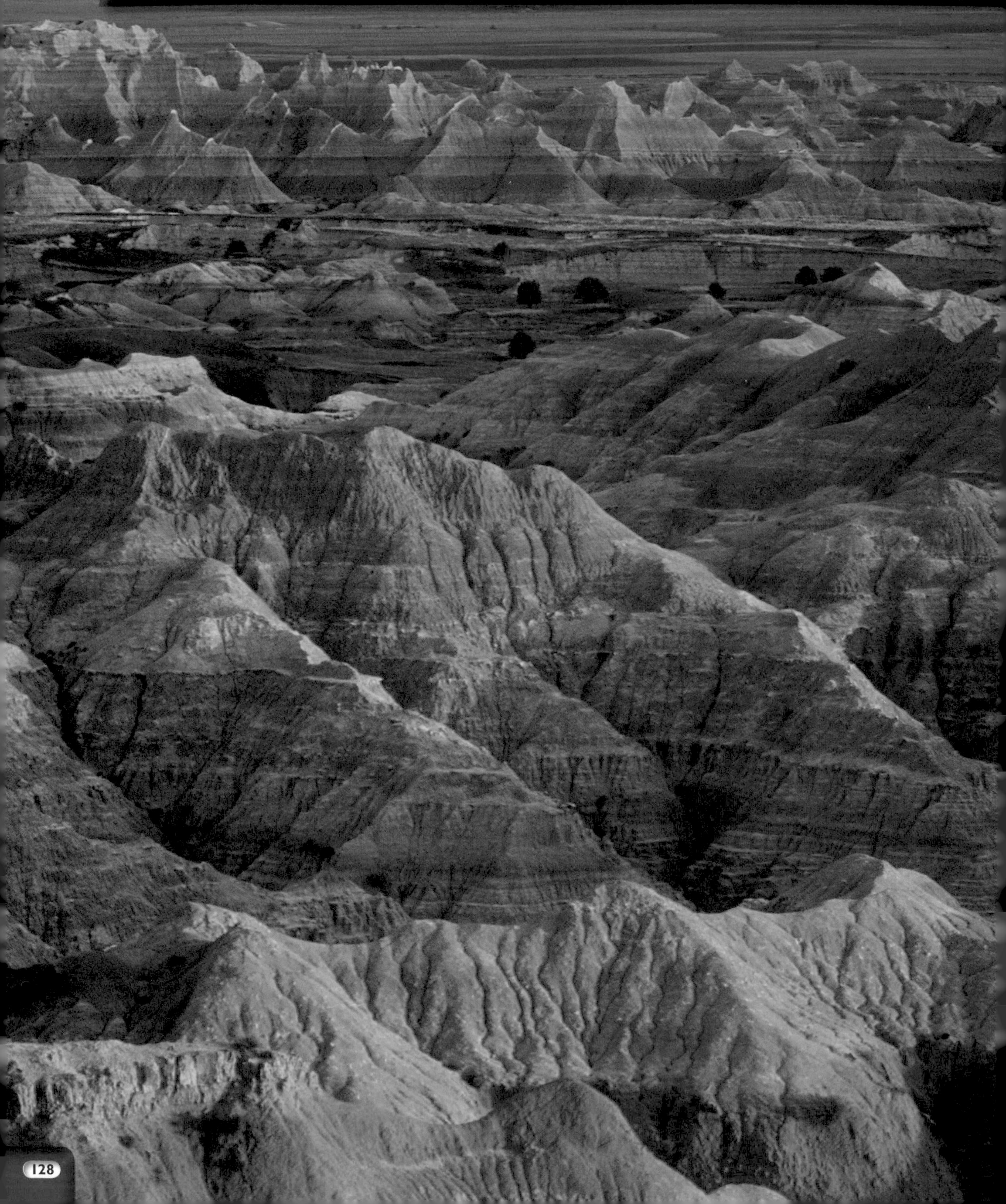

Introduction

Deserts are often considered to be barren, desolate places with no water, no life, and nothing much of interest. They are, in fact, fascinating and often thriving natural environments providing opportunities for agriculture, industry, renewable energy production, and tourism.

Actively weathered and eroded, deserts display extraordinary and unique landforms, often set in spectacular landscapes. On account of their exceptional beauty, large tracts of desert across the world have been designated as national parks.

In this chapter you will learn about the location and causes of deserts, and you will study the processes and landforms associated with wind and water action in them. You will consider some of the issues affecting human activity in deserts and will learn about the importance of sustainable management.

Books, music, and films

Books to read

The Grapes of Wrath by John Steinbeck

Music to listen to

'In God's Country' by U2

Films to see

Back to the Future Part III
Holes
Indiana Jones

◀ *The Badlands at sunset, from Sage Creek Basin Overlook, Badlands National Park, South Dakota*

Why are there no plants or trees on the hills in the photo?
How high are the hills?
Why have the hills been so severely eroded?
Why are the flat areas covered by grass? (Is it grass?)
What will this landscape look like in 100 years' time?

About the specification

'Hot desert environments and their margins' is one of the three Physical Geography option topics in Unit 1 – you have to study at least one.

This is what you have to study:

- The location of hot deserts and their margins (arid and semi-arid). The characteristics of hot deserts and their margins – which means their climate, soils, and vegetation.
- The causes of aridity – so, atmospheric processes relating to pressure, winds, continentality, relief, and cold ocean currents.
- Arid geomorphological processes – specifically, mechanical weathering.
- The effect of wind – its role in erosion (through deflation and abrasion), transportation (through suspension, saltation, and surface creep), and deposition.
- The effect of water, including the sources of water (exogenous, endoreic, and ephemeral). The role of flooding.
- Landforms. Those resulting from wind action: yardangs, zeugen, and sand dunes. And those resulting from water action: pediments, inselbergs, mesas and buttes, salt lakes, alluvial fans, wadis, and badlands.
- Desertification. The distribution of areas at risk of desertification. The physical and human causes of desertification. The impact of desertification on land, ecosystems, and populations.
- A case study of desertification in the Sahel. This should include:
  - the struggle for human survival, with reference to the energy or fuel-wood crisis, water supply issues, and the impact on food supply and farming and livelihoods; and
  - the coping or management strategies adopted, including the role and impact of external aid.
- The management of hot desert environments and their margins. This means considering and evaluating land use and agriculture strategies in areas such as the Sahel, which should then be contrasted with the development of areas such as south-western USA or southern Spain. The implications and potential for sustainability should also be considered.

In this section you will learn about:
- the locations of hot deserts
- the characteristics of hot deserts (landscapes, climate, soils and vegetation)

Skills

In this section you will:
▶ interpret and draw climate graphs
▶ conduct Internet research

For most of us, the image of a desert is either a romantic one – involving nomads leading camel trains over huge golden sand dunes – or the heart-wrenching images of starving children in Africa. While both of these scenes do exist, the vast majority of deserts are dry, barren and stony. They have isolated shrubby plants, and little in the way of human activity. Perhaps surprisingly, semi-deserts – where conditions are slightly less extreme – are actually havens for wildlife, particularly birds.

Despite the obvious harsh conditions, some desert environments are breathtakingly beautiful. Many people are attracted by their peace and tranquillity (Figure 4.1). They are one of the few places on Earth that can still be described as unspoilt wilderness, and their beauty has inspired writers and poets throughout the ages (Figure 4.2).

He who would describe a night in the desert should be, by the grace of God, a poet. For how can its beauty be described, even by one who has watched, revelled and dreamed through it all?

And what a night it is, which here in the desert, after all the burden and discomfort of the day, soothes every sense and feeling! In undreamt-of purity and brightness, the stars shine forth from the dark dome of heaven. The light of the nearest is strong enough to cast slight shadows on the pale ground.

With full chest, one breathes the pure, fresh, cooling and invigorating air ... not a sound, not a rustle, not even the chirping of a grasshopper interrupts the current of thought and feeling. The majesty, the sublimity of the desert is now for the first time appreciated. Its unutterable peace steels into the traveller's heart.

Figure 4.2 *An extract from A.E. Brehm's* From North Pole to Equator *(1895). Alfred Brehm was a naturalist who travelled and spent time with local people in many different parts of the world. On one of his journeys, he joined a desert caravan (camel herders and traders) in North Africa. It was on this trip that he became inspired by the harshness and beauty of the desert, as this short extract illustrates.* ▲

Figure 4.1 *A Jordanian desert landscape* ▼

What is a hot desert?

Hot deserts cover 25-30% of the Earth's land surface. The key characteristic of a hot desert is its dryness (aridity). By definition, a true **arid desert** will have less than 250 mm of rainfall a year. But, in many cases, deserts have far less rainfall than that. In fact, some of the world's deserts go for many years without any rainfall at all! **Semi-arid deserts** – usually found on the margins of arid deserts – have an annual rainfall of 250-500 mm. As a comparison, London's average rainfall is about 625 mm a year.

Another main characteristic of hot deserts is that they have a very high rate of potential evaporation, which often exceeds precipitation. This means that there is little water available for plants and animals. As a result, the organisms living in deserts and semi-deserts have adapted to cope with life in such an extreme environment.

Figure 4.3 *A Saharan sand sea* ▲

The Sahara is the world's largest desert, and covers an area of 9 million km². It contains a huge variety of landscapes – from dramatic mountain ranges to vast stony plains. Famously, it also includes huge expanses of continuously moving sand dunes – known as sand seas (Figure 4.3).

The three main types of transitional landscape in the Sahara, working roughly from north to south, are:

- **hammada** – largely barren and rocky ground, including the northern mountains
- **reg** – a gravel/pebble plain, forming a transitional zone between the hammada and the erg
- **erg** – predominantly sand dunes, and forming just 28% of the Sahara.

Sand only comprises about 25% of the world's total desert area, and just 1% in the deserts of the USA.

Where are the world's hot deserts?

Figure 4.4 locates and names the world's main arid and semi-arid desert areas. As it shows, many of the world's deserts are found close to the Tropics of Cancer and Capricorn. Other desert areas are found in continental interiors (e.g. the Gobi in Asia), or adjacent to mountain ranges (e.g. the Atacama near the Andes in South America). Deserts are transitional features – the aridity gradually becomes less extreme as the distance from the centre of the desert increases. The causes of aridity will be discussed in Section 4.2.

Figure 4.4 *Hot deserts around the world* ▼

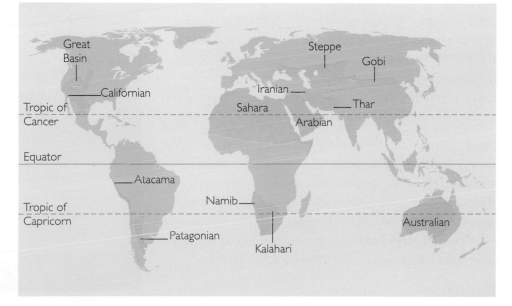

Desert climate, soils and vegetation

Desert climate

Figure 4.5 is a climate graph for In Salah, Algeria. It shows a typical climate for a hot, arid desert. There is hardly any rainfall, and average temperatures are extremely high in the summer. You might be surprised that temperatures in the winter are not particularly high, but remember that a desert is basically a reflection of a lack of rainfall, rather than very high temperatures.

Deserts have a number of other climatic characteristics that will not appear on a climate graph:

- They have high sunshine values, because of the absence of cloud. This is largely because deserts are located in zones with high atmospheric pressure, where the air is sinking and becoming warm and dry.
- The lack of cloud cover also leads to large diurnal (daily) temperature variations – of up to 30°C. In the winter, frosts occur at night – and it can even snow!
- Strong desert winds and sandstorms can occur, especially when there are significant variations in temperature from place to place.
- Thunderstorms can be triggered by intense convective activity.
- Some coastal deserts, e.g. the Atacama in South America, can be plagued by fog rolling in off the sea. In parts of Peru, an almost constant drizzle affects the Andean hillsides during the winter. But this produces less than 50 mm of precipitation on the ground (Figure 4.6).

Desert soils

Soil is a mixture of organic and inorganic material. Deserts tend to have very poorly developed and thin soils, because of the lack of organic matter. It's usually too dry in deserts to support extensive vegetation. Even when any of the scarce vegetation dies, the lack of water stops it rotting and producing humus to enrich a developing soil.

Desert soils tend to be very dry and sandy. They are highly porous and permeable, which leads to poor moisture retention. However, despite their apparent poor quality, the lack of rainfall means that leaching (the dissolving and removal of plant nutrients) is minimal. So, given appropriate irrigation and careful soil management, crops can be grown quite successfully in the desert (Figure 4.7).

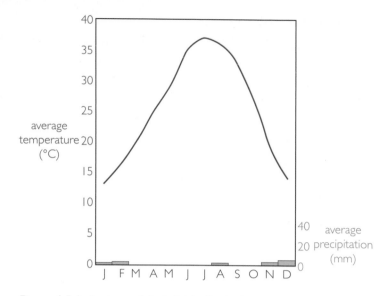

Figure 4.5 *A climate graph for In Salah, Algeria* ▲

Did you know?

The highest temperature ever recorded was 57.8°C at El Azizia in the Libyan Desert in 1922. The appropriately named Death Valley in California comes a close second, with a temperature of 56.6°C in 1913. There are virtually no plants in Death Valley, and it's so hot that occasionally dead birds are reported to fall out of the sky!

Figure 4.6 *A fog-bound Atacama Desert* ▼

Desert vegetation

Deserts are hostile environments for plants. Apart from the low and unreliable rainfall, the very high rates of evaporation, searing temperatures and strong sunshine mean that plants have had to adapt to survive. Plants adapted to the very dry desert conditions are called **xerophytes**.

- Desert plants tend to be sparsely distributed. They typically take the form of isolated thorny shrubs.
- Many desert plants have long tap roots, up to 15 metres long, to enable them to 'tap' water sources deep below the ground surface.
- Leaves tend to be small and waxy, or thorny, to reduce water loss (transpiration).
- Some plants called **succulents** (e.g. cacti) can expand to store water in their leaves or stems after a rainfall event (Figure 4.8).
- In isolated places – called oases – where water is close to the surface, larger trees like date palms can be found, along with grasses.

Figure 4.7 Pivot irrigation in the Saudi Arabian desert. A single water source, pumped from an ancient underground aquifer, rotates on a pivot to water crops in a circle. ▲

ACTIVITIES

1 Design an information poster to describe the locations and characteristics of hot deserts. Use information from this Section, plus extra Internet research, and include the following aspects:
  - A location map of the world's desert areas. This could go in the centre of your poster. Name the main deserts.
  - Brief descriptions of the characteristics of deserts (landscapes, climate, soils, and vegetation).
  - Thumbnail photos to illustrate some of the points (e.g. landscapes).

2 Study Figure 4.9 below. This climate data is for a semi-arid desert environment.
  a Use the data to draw a climate graph for Hall's Creek. Look at Figure 4.5 if you need help.
  b Calculate the total annual rainfall. How does it compare with the accepted definition of a semi-arid region?
  c Which months experience the highest temperatures?
  d During which four months does most rain fall in Hall's Creek?
  e Despite high rainfall totals in these four months, the area remains very dry. Can you suggest why?
  f The highest temperature recorded at Hall's Creek was 44°C in January and the lowest was –1°C in July. Why do deserts experience such extremes of temperature?

Figure 4.8 Prickly pear cacti in the New Mexico desert, USA ▼

Internet research

Use the Internet to investigate plant adaptations to desert environments. Consider two or three different plants from around the world with different adaptations. Download a photo of each one and then add your own labels to describe how that particular plant copes with the hostile desert conditions.

Figure 4.9 Climate data for Hall's Creek, northwest Australia ▶

	J	F	M	A	M	J	J	A	S	O	N	D
Temperature (°C)	30	29	28	26	21	19	18	21	24	28	31	31
Precipitation (mm)	137	107	71	13	5	5	5	2	2	13	35	79

In this section you will learn about:
- the causes of aridity (the global atmospheric circulation system, cold ocean currents, and continental interiors)

What causes aridity and the formation of deserts?

Look back at Figure 4.4 (page 131) to remind yourself about the locations of the world's hot desert areas. In this section you are going to find out why hot deserts are located in those places.

Most of the world's hot deserts are found in a broad – but discontinuous – belt between latitudes 20-25 degrees. But they often extend some degrees north and south of that. They include the Sahara in North Africa and the Arabian Desert in the Middle East, and – in the Southern Hemisphere – the Kalahari Desert in southern Africa and the Great Victoria Desert in Australia.

The global atmospheric circulation system

The main cause of these hot deserts is **the global atmospheric circulation system**. This results in a broad zone of sinking air in these latitudes (Figure 4.10). As the air sinks, it becomes dry and warm – leading to high pressure on the ground (an anticyclone). Under these conditions, cloudless skies dominate the weather (Figure 4.11). They account for the aridity (lack of rainfall), the high sunshine totals, and the great extremes of temperature experienced in these deserts. A high evaporation rate is also common in these conditions, which increases the level of aridity.

Cold ocean currents

Some deserts occur on coastlines with cold ocean currents, e.g. the Atacama Desert in South America (Figure 4.13). Cold air – being denser than warm air – has a tendency to sink and stay close to the ground. The resulting conditions of high atmospheric pressure increase the aridity – any rain-formation is suppressed by the air's tendency to sink. Despite the fog and drizzle often associated with the cold and moist sea air (Figure 4.6), the amount of overall precipitation is very small.

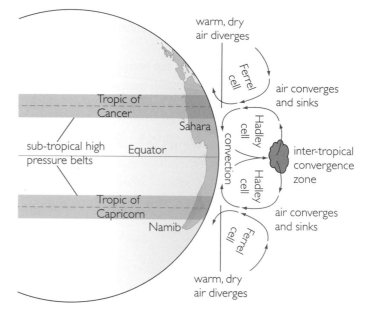

Figure 4.10 *The impact of the global atmospheric circulation system on the locations of hot deserts* ▲

Figure 4.11 *Typical weather conditions in the desert* ▼

Continental interiors

Aridity is also higher in continental interiors, where the influence of moist airstreams from the oceans is reduced. Go back to Figure 4.4 and look at the locations of the Gobi Desert and the southern part of the Australian desert. There is less influence in those locations from the sub-tropic high-pressure belts (Figure 4.10). Occasional storms there also bring welcome rain (Figure 4.12). Deserts like this tend to be semi-arid, rather than arid.

The rainshadow effect

Some deserts are influenced by **the rainshadow effect**. This is caused when moist oceanic air dumps huge amounts of rainfall on the windward side of a mountain range. It does this when it's forced up and over the mountains by the prevailing winds (Figure 4.13). Then, when the air sinks on the leeward side of the mountains, it warms and dries up – creating arid conditions.

Often, several of the above factors combine together to explain the aridity experienced in particular deserts. For example, the Atacama Desert in South America (Figure 4.13) is caused by a combination of:

- the global atmospheric circulation system (lying close to the Tropic of Capricorn)
- the effect of a cold ocean current (the Peruvian current)
- being on the leeward side of the prevailing south-east trade winds.

Figure 4.12 *Even the Gobi Desert has rainstorms – sometimes!* ▲

Figure 4.13 *The causes of aridity in the Atacama Desert* ▼

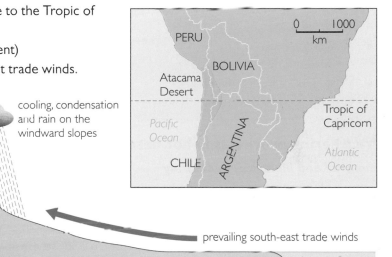

high atmospheric pressure - sinking air warms and 'dries' up

cooling, condensation and rain on the windward slopes

Andes Mountains

rainshadow

on-shore winds are chilled so they are unable to pick up moisture

Atacama Desert

prevailing south-east trade winds

Pacific Ocean

Atlantic Ocean

cold ocean current

A cross-section through South America at 20°S

ACTIVITIES

1 Study Figure 4.10.
   a With the aid of a simple diagram, explain why air sinks at the Tropics to form sub-tropical high pressure belts, or zones.
   b Give a reasoned description of the typical weather conditions experienced under anticyclonic conditions in these belts.
   c Explain why the global atmospheric circulation system leads to aridity in the Tropics.
   d Suggest reasons why some parts of the Tropics are not deserts.
2 Study Figure 4.13. Describe, with the aid of a simple diagram, how the rainshadow effect causes aridity in the Atacama Desert.

Internet research

Prepare a case study about one of the following deserts: Namib, Kalahari, Gobi, or the Great Basin in California. Explain what factors have caused your chosen desert to form in that location. Use diagrams to help you.

Wikipedia would be a good starting place, but conduct a general Internet search as well, so that you can double-check the information and add to it.

In this section you will learn about:
- weathering processes in hot deserts
- wind processes in hot deserts
- the role of water in hot deserts

Look at Figure 4.14. At first glance, there is very little evidence of any landscape-forming processes there. For instance, there are no rivers in the landscape. In fact, there appears to be no water at all. Also, there is no obvious movement of any kind – nothing seems to be happening. Yet this landscape was formed in the past by natural processes – and they are still changing it today.

As a start, consider these questions:

- Why are there deep clefts in the cliffs?
- Where did the sand come from?
- How have the cracks in the rock in the foreground been enlarged?
- Why is there a concentration of vegetation in the centre of the photo?

Figure 4.14 *A hot desert landscape in Jordan* ▲

Weathering processes in hot deserts

Chemical and biological weathering

Because of the absence of water, and a lack of rotting organic matter, chemical and biological weathering are not very significant in hot deserts. Nevertheless, the presence of some moisture and solutes can contribute towards the break-up of a rocky surface prior to erosion.

Mechanical weathering

Mechanical (or physical) weathering involves the gradual disintegration of rocks without any chemical change taking place. In a desert environment, **insolation weathering** is probably the most important type of mechanical weathering. It involves the expansion and contraction of a rock's surface layer, due to intense temperature fluctuations. Eventually, the outer 'skin' peels off in a process called **exfoliation** (Figure 4.15).

The rock expands when it heats up during the day, and contracts when it cools down at night. Deserts experience huge temperature fluctuations – both daily (a range of up to 40°C) and seasonally. Scientists believe that the presence of some moisture helps to promote this weathering process. The moisture is most likely to involve dew, which forms at night when the temperature plummets and the air cools enough to become saturated (its **dew-point temperature**).

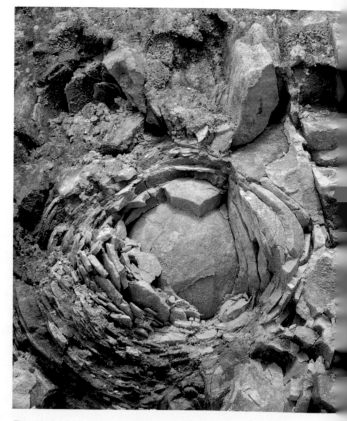

Figure 4.15 *Insolation weathering and exfoliation in action* ▲

Freeze-thaw weathering is not very common in deserts. However, it can sometimes occur during the winter months, when the temperature may fall below 0°C. In the past – when places like the Arabian Desert had wetter climates – freeze-thaw could have been quite significant.

Salt crystallisation is an important form of mechanical weathering in the desert. When it rains, the water dissolves salts held in the soil. But the high evaporation rate in the desert then draws the rainwater back to the surface by capillary action, and the salt crystals are deposited on the ground surface (Figure 4.16). Over time, the salt crystals can expand – causing stresses in rocks (just like ice in freeze-thaw weathering). Over a long period of time, a salt crust called **duricrust** can develop on the ground surface. Salt weathering is a major problem in some developed desert environments, where it damages concrete and tarmac runways and roads.

Figure 4.16 *Salt crystallisation in Death Valley, California, USA* ▲

Wind processes in hot deserts

Wind erosion

Wind is common in the desert, and is capable of carrying out two types of erosion:

Abrasion. If the wind is strong enough, and blows in a dominant direction, it is capable of picking up small particles of sand and 'sand-blasting' the lowest part of exposed rock outcrops (up to about a metre high). This type of erosion can result in peculiar mushroom-like structures (Figure 4.17).

Figure 4.17 *A 'mushroom' landform in the Negev Desert in southern Israel* ▶

Deflation is erosion of the desert floor by the wind, which often involves the removal of light sand – leaving behind a pebbly surface (Figure 4.18). Over time, the desert surface is lowered and often starts to look like a cobbled pavement, because only the heavier stones are left behind. Strong eddies (localised winds) can hollow out the desert surface to produce a **deflation hollow**. Moisture that collects in a particular place can increase the rate at which a deflation hollow forms, by helping to break up the surface. One of the world's largest deflation hollows is the Qattara Depression in Egypt – over 50000 km² in size and eroded to 134 metres below sea level!

Figure 4.18 *The effects of deflation* ▶

Wind transportation and deposition

Wind transports sediment in a similar way to water in a river. Larger particles are rolled along the desert floor by the process of traction, or **surface creep** (Figure 4.19). Slightly smaller particles are 'bounced' along by the process of **saltation**. This is very common in desert environments – it usually happens within 2-3 cm of the ground surface, where particles constantly knock into one another. Fine dust is picked up and carried in **suspension**. It can then be deposited many kilometres away (where it's referred to as **loess**). Occasionally, red dust from the Sahara is deposited as far away as southern Britain.

However, there are two important differences between transportation by wind and by water:

- The particles carried by wind become very rounded – almost spherical. This is because the attritional impacts are stronger in the air. The cushioning effect is less than with particle transport by water.
- Wind sorts particles much more effectively than water. This means that the sand particles in a particular place in the desert are all likely to be the same size.

Figure 4.19 The processes of wind transportation ▼

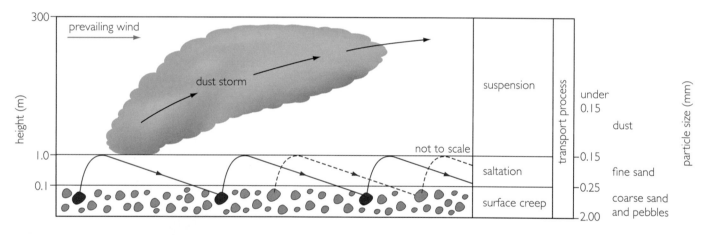

Water processes in hot deserts

Water is the most important force of erosion in a desert. There are four main sources of water:

- **Exogenous rivers** have their sources in mountains outside desert regions. They contain enough water to flow continuously, despite the high evaporation rate when they reach the desert. Examples of exogenous rivers include the Colorado, the Indus and the Nile.
- **Endoreic rivers** flow into deserts and then terminate – usually in a lake or inland sea, e.g. the River Jordan drains into the Dead Sea.
- **Ephemeral river**s are rivers or streams that flow intermittently in desert regions. They might flow after storm events, or they might be fed by snowmelt from adjacent mountains in the spring. Their flow rates can vary dramatically, and in times of flood they can be

powerful forces of erosion.

- Rainfall events in the desert – despite being infrequent – tend to involve torrential convectional storms. They unleash large amounts of water in a very short period of time, which can cause **flash floods**. The sun-baked soil leads to large amounts of overland flow, which is capable of carrying out significant erosion – particularly in mountains where steep gradients increase the flow rate. During high-magnitude events like this, huge amounts of sediment can be washed out of the mountains and deposited as vast alluvial plains on the lowlands below.

(See Chapter 1 pages 18-21 for more about the processes of erosion, transportation and deposition by rivers.)

CASE STUDY

Imlil flash flood, Morocco (1995)

In August 1995, a torrential downpour dumped 70mm of rain in just a few hours near the village of Imlil, in the semi-arid foothills of the Atlas Mountains in Morocco (Figure 4.20). The resultant flash flood swept boulders the size of lorries down the valley of the River Reraya and into the village. For much of the year, there is little or no water in the river (Figure 4.21). But – for just a few hours after the rainstorm – it became a raging torrent.

About 150 people were killed by the flood. Crops of maize, alfalfa and grass were destroyed, and irrigation channels were blocked with silt. Walnut trees – an important cash crop for the local Berber people – were swept away in the flood. It takes about 15 years for a walnut tree to mature, so their loss was a huge blow.

The Imlil flash flood demonstrates how low-frequency – but high-magnitude – events can cause massive changes to the physical and human geography of a desert environment.

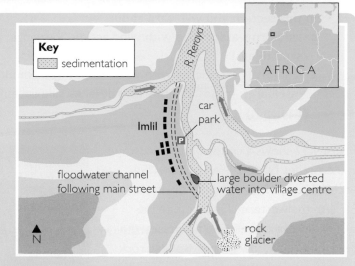

Figure 4.20 *A flash flood hit Imlil on 17 August 1995* ▲

Figure 4.21 *The valley of the River Reraya at Imlil* ▶

ACTIVITIES

1 Study Figure 4.15.
 a With the aid of one or more simple diagrams, describe the processes responsible for the weathering shown in the photo.
 b Under what conditions might the rate of weathering of the rock in the photo be increased.

2 Study Figures 4.17 and 4.18.
 a Draw a simple outline sketch of the mushroom-shaped feature in Figure 4.17. Include the desert floor in your sketch. Allow plenty of space around it for labels.
 b Use Figure 4.19 to add the processes of wind transportation to your sketch.
 c Now add detailed labels to describe the processes of wind transportation and erosion. In particular, make sure that you explain clearly the process responsible for the erosion of the mushroom-shaped feature.
 d What is the evidence of deflation in Figure 4.18?

3 Study Figure 4.21.
 a What evidence is there that the River Reraya is an ephemeral river?
 b What evidence is there that the river occasionally experiences high flow?
 c Describe and explain the vegetation in the bottom of the river valley. Contrast it with the vegetation higher up the valley sides.
 d What other factors shown on the photo increase the likelihood of flash flood events in the area?
 e Comment on the site of the settlement. Suggest advantages and disadvantages.
 f What were the immediate and longer-term impacts of the Imlil flood? Consider social, economic and environmental impacts.

4 Answer the following essay question: 'What is the role of water in hot desert environments?' Use the information in this section, plus any other information that you can find in books, articles or on the Internet.

In this section you will learn about:
- landforms associated with wind action (yardangs, zeugen, sand dunes)
- landforms associated with water action (pediments, inselbergs, mesas and buttes, salt lakes, alluvial fans, wadis, badlands)

Skills

In this section you will:
- use the Internet for research

Figure 4.22 shows a classic desert landscape in the Arches National Park, Utah, USA. Arches is one of several desert National Parks in the western USA. Despite the lack of any obvious erosive processes, the deep valleys, narrow gorges, isolated rock outcrops, and sandy/gravelly deposits have all been formed by the action of wind and water. In this section, you are going to study the common landforms found in hot desert environments.

Landforms associated with wind action

Yardangs

Figure 4.23 shows a large area of elongated ridges, separated by deep grooves cut into the desert surface. These parallel ridges (like the hulls of upturned ships) are called yardangs. They are formed by the process of wind abrasion. Their elongated and parallel orientation is due to the prevailing wind blowing in a single direction. Yardangs will not develop if a desert is affected by multi-directional winds.

Incredibly, yardangs can vary in size from just a few centimetres in height and length, to several kilometres in length and hundreds of metres in height. The Iranian yardangs in the photograph can reach heights of up to 75 metres. Although yardangs are common in most deserts, large-scale 'mega-yardangs' are concentrated in the Tibesti Mountains of the central Sahara.

Figure 4.22 *The 'Wild West' landscape of the Arches National Park in Utah, USA* ▲

Figure 4.23 *A yardang field in the Dasht-e Lut Desert in Iran* ▶

Zeugen

Zeugen are weirdly shaped, isolated rock outcrops – up to 30 metres high (Figure 4.24). They are selectively eroded by wind abrasion (see page 137). Zeugen are typically made up of sedimentary rocks – usually sandstone – laid down in horizontal layers. This layering generally makes the zeugen flat-topped and stepped in profile.

Because most abrasion is concentrated within a metre or so of the desert floor, zeugen often have a slightly narrower – more eroded – lower portion. This concave section can make the feature wave-like in appearance, or – in extreme cases – look like a mushroom (their slang name is 'mushroom rocks').

Even though wind abrasion is the dominant erosive process, it seems likely that other processes might be active too, such as water erosion and weathering (the concentration of moisture, in the form of dew).

Sand dunes

In sandy deserts, the wind can form vast and beautifully symmetrical sand dunes (Figure 4.26). And, just like coastal sand dunes, desert sand dunes usually start with deposition on the leeward side of an obstacle – like a rock. As more

Figure 4.24 *Zeugen (or mushroom rocks) in the desert* ▲

and more sand is deposited, it becomes shaped by the wind.

Sand dunes are rare features in American deserts, but they form extensively in the Sahara and Australia. Vast expanses of sand are called sand seas (Figure 4.3). The Great Eastern Erg in Algeria is estimated to cover an area the size of France!

There are two common types of sand dune – barchans and seifs.

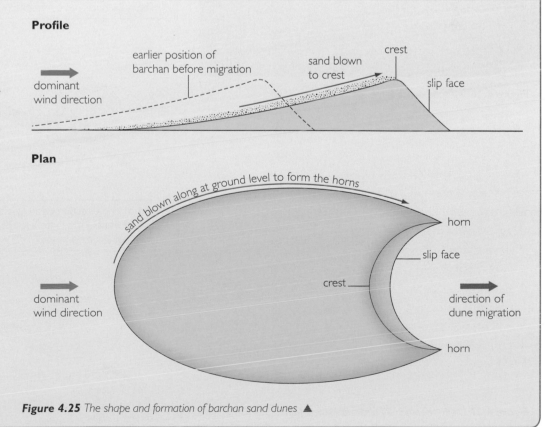

Barchans are crescent-shaped sand dunes. They are often found individually in deserts where there is a relatively limited supply of sand – but a strongly dominant wind direction. For instance, they are quite common in the Atacama Desert.

As Figure 4.25 shows, the sand is blown up the gentle windward side of the barchan, before sliding down the steeper sheltered side (or slip face). Barchan dunes advance, or migrate, several metres a year.

Profile

dominant wind direction

earlier position of barchan before migration

sand blown to crest

crest

slip face

Plan

sand blown along at ground level to form the horns

dominant wind direction

crest

horn

slip face

direction of dune migration

horn

Figure 4.25 *The shape and formation of barchan sand dunes* ▲

Seifs are longitudinal, elongated sand dunes – common in sand seas (Figure 4.26). In places, they can stretch for several hundred metres. They form parallel to the prevailing wind direction.

As Figure 4.27 shows, sometimes barchans can develop into seifs.

Figure 4.26 *Seif sand dunes in the Sahara Desert* ▲

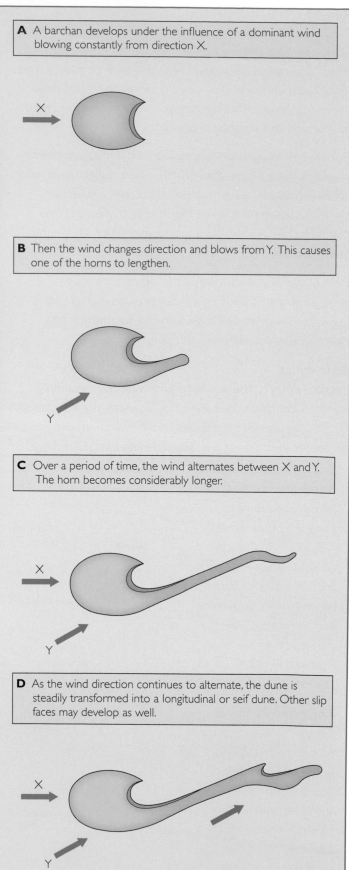

A A barchan develops under the influence of a dominant wind blowing constantly from direction X.

B Then the wind changes direction and blows from Y. This causes one of the horns to lengthen.

C Over a period of time, the wind alternates between X and Y. The horn becomes considerably longer.

D As the wind direction continues to alternate, the dune is steadily transformed into a longitudinal or seif dune. Other slip faces may develop as well.

Figure 4.27 *The development of a seif sand dune from a barchan* ▶

Landforms associated with water

Running water, in the form of rivers and sheetflow, is the most powerful erosional agent in the desert – even though it might be completely absent most of the time. You have already seen on page 139 how flash flooding at Imlil resulted in a huge amount of sediment being carried several kilometres down a Moroccan river valley.

Figure 4.28 shows a typical desert landscape that contains many features associated with erosion, transportation and deposition by water.

Wadis and alluvial fans

Wadis are steep-sided, dry river valleys – usually displaying evidence of previous high-discharge events (e.g. braided channels, a large amount of coarse sediment, and steep river banks). In the USA, the term **canyon** is used to describe larger river valleys that are much deeper and wider than wadis. The Grand Canyon is an obvious example.

Wadis are formed by river erosion, when occasional high-discharge events – like flash floods – wash out sediment from the surrounding mountains. The vast amount of sediment leads to rapid river erosion through processes like corrosion. People who spend time in deserts are advised never to set up camp in a wadi if it looks like rain!

The sediment washed out through a wadi is deposited at the base of its mountain range or plateau – forming a delta-like **alluvial fan** (Figure 4.28). As the water spreads out at the mouth of the wadi, energy is lost and sediment is rapidly deposited. Over time, as water re-works this vast store of sediment, the alluvial fan displays clear sorting patterns. The coarser sediment becomes concentrated closest to the plateau or mountain range, while the finer sediment is washed out onto the desert plain. Alluvial fans can extend for several kilometres away from the mountain edge – and can reach thicknesses of up to 300 metres!

Figure 4.28 Landforms associated with water action in deserts ▼

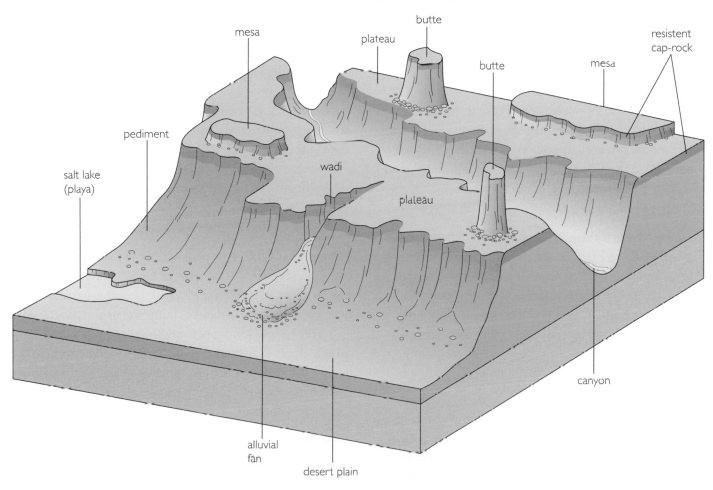

Figure 4.29 *Mesas and buttes in Monument Valley, Arizona, USA* ▲

Mesas, buttes and inselbergs

Mesas, buttes and inselbergs are all relic features – at different scales – that become isolated from a plateau or desert mountain range by river erosion over a long period of time. Mesas and buttes (Figure 4.29) commonly form in horizontally bedded sedimentary rocks, where a tougher cap rock prevents erosion of the underlying strata.

Mesas are large plateau-like features, often bordered by steep wadis or canyons. **Buttes** are smaller pinnacles of rock, and represent an advanced stage of landscape development. Both features are usually surrounded by flat desert plains. Mesas and buttes have extensive scree slopes, formed by mass movement (rockfalls) and mechanical weathering.

Inselbergs are more-rounded relic landforms. They develop in rocks like granite, where there is an absence of layering and variable rock strengths. Some scientists believe that inselbergs may have been formed during past climates when higher levels of humidity would have led to more chemical weathering. This is because chemical weathering tends to produce smooth edges to exposed rocks.

Pediments and playas (salt lakes)

A **pediment** is a gently sloping erosional rock surface at the foot of a desert mountain range. It's often blanketed by sediment (e.g. alluvial fans) washed down by rivers from the nearby mountains. Pediments cause a great deal of debate about their formation. It's possible that they form in much the same way as wave-cut platforms on the coast, when the sea cliffs gradually retreat. Alternatively, intense scouring by rivers concentrated at the foot of the desert mountain range might be responsible.

When water flows over pediments and their surface deposits, it carries loose material further and further out onto the desert floor. Much of the water percolates through the coarse sediment, but some of it might collect in a large hollow (or depression) to form a **salt lake** or **playa**. Rapid evaporation then leads to the formation of a salty crust. In some deserts, the accumulation of salt is so great that it can be exploited commercially, e.g. the Chott el Djerid in southern Tunisia.

Badlands

The term **badlands** is used to describe semi-arid landscapes that have been intensively carved by heavy rainstorms to create very dramatically eroded landscapes (Figure 4.30). In the USA, the Badlands National Park was established to protect this dramatic type of landscape.

Badlands (so-named because the land is of little agricultural value) develop where the rocks are relatively weak and impermeable. They are characterised by having incredibly high drainage densities – although, for much of the time, the rivers are dry. The erosion rate following a period of heavy rain is very rapid. A lot of deposition also takes place.

Badlands can also result from poor land management, where marginal land has been overgrazed – leading to soil degradation and the loss of vegetation. Under such circumstances, the heavy rainstorms quickly turn productive land into a badland!

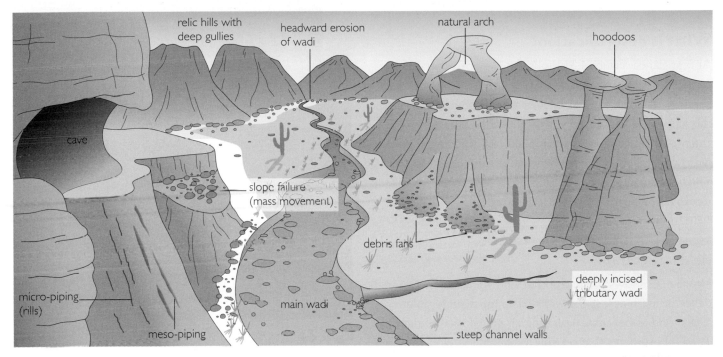

Figure 4.30 *The typical features found in a badlands landscape* ▲

ACTIVITIES

1   Study Figure 4.22 and work in pairs to answer the following questions.
   a   Which of the desert landforms described in this section can you see in the photo?
   b   What evidence is there that the rock is sedimentary sandstone?
   c   How does this rock contribute towards the formation of the features identified in (a) above?
   d   Imagine that there was a heavy rainstorm lasting for a few hours. Consider how the landscape would be affected both during the storm and in the following days.
2   Consider the various landforms resulting from wind action. To what extent is wind strength and wind direction important in the formation of these landforms?
3   Study Figure 4.28.
   a   What is the difference between a wadi and a canyon?
   b   What is the difference between a mesa and a butte?
   c   Is it true to say that erosional landforms are found in the mountains, while depositional landforms are found on the desert plain. Justify your answer.

Internet research

There are a large number of arid landforms in this section. Use the Internet to find some photos to illustrate the characteristics of the various landforms. Add detailed labels – electronically or by hand. Keep them for use as flashcards to help with your revision.

   d   To what extent is the arid landscape in the diagram controlled by geology (rock type, arrangement, orientation, etc.)?
   e   How and why do you think the landscape will change over the next 1000 years?
4   Study Figure 4.30.
   a   What is the evidence that this landscape is being actively eroded at the present time?
   b   Which group of processes do you think are dominant in this environment – wind or water? Explain your answer.
   c   How might overgrazing have made this landscape more vulnerable to the processes of erosion?
   d   Could anything be done to make this landscape commercially valuable?

In this section you will learn about:
- the locations of areas at risk of desertification across the world
- the causes of desertification
- the impacts of desertification on people and natural environments
- management strategies to combat desertification

What is desertification?

At first glance, Figure 4.31 looks just like any other desert photo – lots of sand. But if you look more closely, you can see signs of change in this environment. In the recent past, this landscape looked very different to the way it looks today. As you can see, there is plenty of dead vegetation, plus the remains of the animals that once grazed the area. The land in the photo is turning into a desert. This is called **desertification**.

How widespread is the risk of desertification?

Figure 4.32 shows the location of land vulnerable to desertification across the world. The most vulnerable areas tend to be located on the margins of the hot deserts. The UN estimates that roughly a third of the world's land surface is currently affected by desertification.

The process of desertification involves the destruction of ecosystems and habitats. Land that was once marginal is turned into an unproductive wasteland – vegetation dies and soil becomes exposed and eroded.

Figure 4.31 *Desertification in Sudan* ▲

Figure 4.32 *Land vulnerable to desertification worldwide* ▼

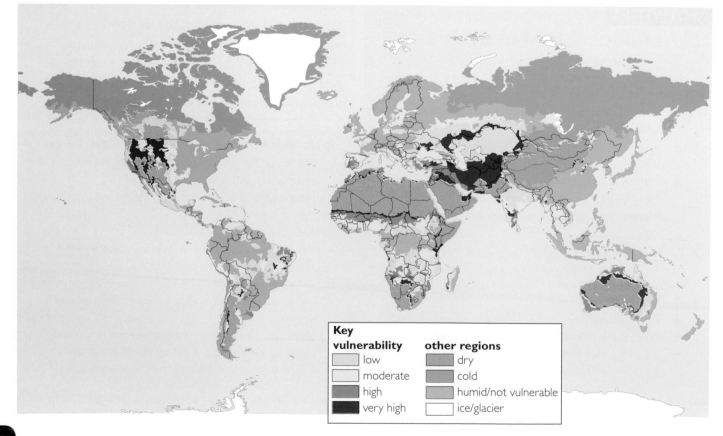

Key

vulnerability	other regions
low	dry
moderate	cold
high	humid/not vulnerable
very high	ice/glacier

What causes desertification?

Natural causes

The main natural cause is connected to climate, which has changed throughout geological time – altering global temperature and rainfall patterns. For instance, there is evidence that – as recently as 8000 years ago (around the end of the last Ice Age) – the climate in North Africa and the Middle East was much wetter than it is today. This evidence includes large aquifers (groundwater reserves) lying beneath desert countries like Egypt and Jordan, as well as fossil plant remains and archaeological evidence (such as ancient rock art). Natural climate change turned this region into desert thousands of years ago – long before humans had any major impact.

Climate worldwide is still changing today, but now there is serious international concern that human actions are worsening natural global warming and climate change. For example, serious droughts have become more common in many parts of Africa over the last few decades (Figure 4.33).

Temperature and rainfall patterns worldwide have certainly been changing. But the changes have not been gradual or consistent – they have been erratic, and have involved extremes of drought and flood. The climate is becoming more unpredictable and more variable. Only time will tell how much of this is a direct cause of current and future desertification.

Figure 4.33 *The effects on livestock of a recent drought in Kenya* ▲

Human causes

People are not likely to deliberately damage the land on which they depend for their survival. However, circumstances can lead to people's actions tipping the delicate balance and inadvertently contributing towards the process of desertification. Most commonly, this involves:

- **over-cultivation**. Intensive farming on marginal land can reduce soil fertility and damage its structure. The lack of organic matter makes it crumbly and more likely to be washed or blown away. It also reduces its capacity to retain moisture.
- **over-grazing**. Marginal grassland has a sustainable carrying capacity – the number of animals that can be supported without causing any long-term damage. If this number is exceeded, the system becomes unsustainable and the vegetation and soil deteriorate. If it continues, desertification can result.
- **over-irrigation**. If plants are appropriately irrigated, little water should be wasted. However, if land is over-irrigated, salinisation can occur. This creates an impermeable and infertile salty crust on the surface, which (according to UNESCO) is a key feature of desertification.

Other human activities that can damage the soil and vegetation (leading to soil erosion and ultimately desertification) are: road building, deforestation and inappropriate tourism (Figure 4.34).

Figure 4.34 *Safari vehicles in Kenya damaging the soil* ▲

Desertification in the Badia, Jordan

The Badia is a vast, stony desert region in eastern Jordan (Figure 4.35). It is sparsely populated by Bedouin – who herd camels, sheep and goats. In the past, the Bedouin lived nomadic lives. But today they generally live in permanent settlements. Annual rainfall in the Badia is less than 150 mm – much of it from torrential storms. The temperature can soar to over 40°C in the summer, but in winter it can fall below freezing.

The Tal Rimah rangelands are an area of gently rolling hills, close to As Safawi. They are sparsely vegetated with thorn shrubs and grasses. Despite being marginal semi-arid land, the rangelands have been grazed sustainably by local farmers for hundreds of years. However, following the first Gulf War in 1991, local farmers bought hundreds of cheap sheep from Iraq and began grazing them at Tal Rimah. This over-grazing tipped the ecological balance and the land quickly became desertified and unproductive. In the end, it had to be abandoned.

Then, in 2002, the Tal Rimah Rangeland Rehabilitation Project was set up (with financial support from the charity USAID and the US Forestry Service). After lengthy discussions with local herders, a plan was agreed to turn the desertified wasteland back into sustainable, productive grazing land. Stone walls were built to control and retain the limited water – called water harvesting. The water was diverted along the contours created by the walls into shallow ditches, where drought-tolerant shrubs (e.g. atriplex) were planted (Figure 4.36). The plants:

- provided appropriate grazing for animals
- encouraged biodiversity by creating new habitats
- helped to hold the soil together and prevent erosion.

By 2008, some sheep had been re-introduced at Tal Rimah – to help assess the Project's progress. Careful stock management will be essential in the future if the rangelands are to be kept sustainable. But the initial signs are encouraging. Between 2004 and 2008, the number of plant and animal species at Tal Rimah increased from 21 to 54. Flowering plants attracted butterflies and other insects. Birds are now nesting in the bases of the shrubs. mammals and reptiles are also returning to the area.

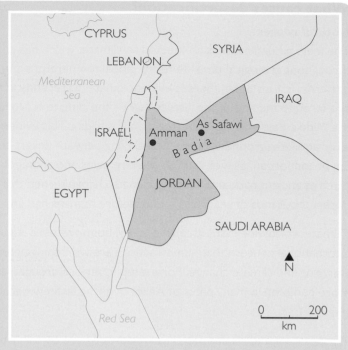

Figure 4.35 *The Badia region of eastern Jordan* ▲

Figure 4.36 *Atriplex planted in a shallow ditch at Tal Rimah* ▼

The Sahel

The Sahel is the name given to the vast semi-desert on the southern edge of the Sahara. It stretches right across the African continent, and includes parts of Mauritania, Mali, Niger, Chad and Sudan (Figure 4.37).

The people who live in the Sahel are some of the poorest people on Earth. They also have to cope with some of the most hostile natural environments. A combination of low and unreliable rainfall – and searing summer temperatures – means that much of the Sahel is barely able to support life. Many Sahelian people are nomads – herding their animals across the parched grassland (Figure 4.38). In such a fragile environment, they are living on the very edge of existence. So desertification is a huge issue. Long droughts and serious famines are common. In 2010, the world's attention turned to Niger, where many thousands of people faced starvation when their harvests failed.

Causes of desertification in the Sahel

- **Population increase**. Currently, about 260 million people are thought to live in the Sahel region. But the population there is estimated to be doubling every 20 years. The rate of population increase (3% a year) is greater than the increase in food production (2%). So, the rapidly increasing population is putting additional pressure on the land to be more productive. This is leading to over-cultivation and over-irrigation.
- **Firewood**. Energy sources are limited in the Sahel – with most people having to rely on wood for cooking. With limited availability and increasing demand from the extra people, the land has been stripped of its few trees. Residents of the town of Zinder (in northern Niger) travel up to 200 km to collect firewood! Not only is this travelling distance unsustainable, but the clearing of the trees exposes the soil to wind and rain erosion.
- **Overgrazing** has also led to a reduction in the remaining vegetation. When the desperate animals resort to eating roots as well as leaves, the plants cannot recover and they die. The soil then becomes increasingly exposed – leading to more soil erosion.

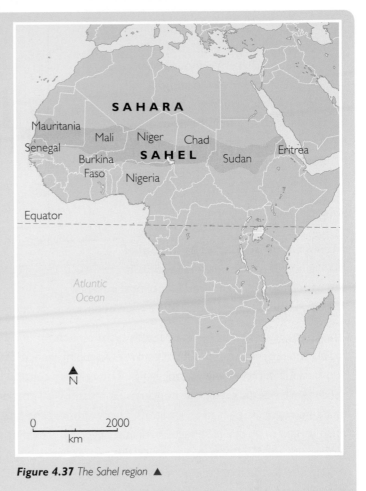

Figure 4.37 The Sahel region ▲

Figure 4.38 A nomadic herder with his zebu cattle in Niger ▼

- **Drought**. The Sahel region suffers from unreliable rainfall, so drought is common there. Without water, vegetation starts to die and the soil turns to dust – making it easier for the wind to erode it. Records show a significant decrease in rainfall since the 1970s (Figure 4.39). Computer models suggest that droughts may well become more severe and frequent as global warming continues.

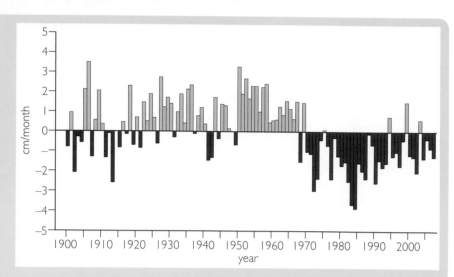

Figure 4.39 Rainfall trends in the Sahel, 1900-2007 (the figures are for June to October each year, and show trends above or below the average) ▲

Management strategies in the Sahel

Managing the issue of desertification in the Sahel has been hindered by political instability and the poverty of the people. Nevertheless, there have been some developments.

The United Nations has encouraged a system of community self-regulation. Local communities take over the maintenance of water sources and surrounding grazing lands. This enables cultural habits, subsistence needs, economic concerns, and ecological conservation, to be managed as part of an integrated programme. The UN has encouraged projects in Senegal, Mali and Niger by providing free veterinary and health care. However, their success has been limited, mainly due to the extreme poverty of the people and the relative powerlessness of local community organisations.

Figure 4.40 A solar cooker (in the foreground) being used by Sudanese refugees in Chad ▼

Other strategies employed in parts of the Sahel include:

- the use of fences or tree lines to retain soil and reduce the effects of wind erosion
- water harvesting, using lines of rocks to control water flow and retain soil (Figure 4.41)
- careful management of irrigation, e.g. the use of drip-irrigation, which reduces waste and prevents excessive salinisation
- promoting the use of solar ovens – by a number of charity organisations – to reduce the need for firewood (Figure 4.40)
- the adoption of a 'green wall' project in Senegal in 2005, which involves planting a 15-km-wide tree belt to combat the spread of the desert
- a massive tree planting programme in Sudan and Ethiopia, supported by the charity 'The Eden Reforestation Project'. Over 1 million trees are being planted in Ethiopia's rift valley, which has become a dust bowl following years of deforestation

◀ **Figure 4.41** *Building stone lines (called diguettes) in Burkina Faso. The lines slow down the water runoff when it rains, giving the water time to soak into the hard soil. They also help to prevent soil erosion. The use of diguettes has increased crop yields in Burkina Faso by 50%, on average – more in drier years.*

ACTIVITIES

1 Study Figure 4.32.
   a On a blank world outline map plot the dry regions (deserts) and the main areas of 'very high' vulnerability to desertification.
   b Use an atlas to label the deserts and name some of the regions/countries at high risk of desertification.
   c Why do most of the high-risk areas appear to be located at the margins of deserts?
   d Are there any exceptions? Suggest possible reasons for them.

2 Study Figure 4.34. Suggest how pressure from tourism can contribute towards desertification.

3 Study Figure 4.36 and the case study of desertification in Jordan.
   a What caused desertification in the Tal Rimah rangelands?
   b Suggest some possible impacts of the desertification on the environment and on local communities.
   c How is the desertification process now being reversed?
   d Why is it important to involve local communities at all stages when trying to solve the problem of desertification?

4 Study Figure 4.38.
   a Why is it in the interest of the nomadic herder in the photo to manage the land in a sustainable way?
   b What might cause the nomads to overgraze the land?
   c What will happen to them if the land becomes desertified?

5 Study Figure 4.39. The long-term average rainfall is shown by the horizontal line at 0 cm/month. The anomalies show the variations either side of this average.
   a Describe the rainfall trends in the following three periods:
      ● 1900 – 1950
      ● 1950 – 1970
      ● 1970 – 2007
   b How does this data help to explain the recent increase in desertification?
   c Assume that the current rainfall trend continues into the future. How could the affected communities respond to prevent desertification occurring?

Internet research

1 Consider the impact of desertification on another part of the world. An interesting example is the area around the Aral Sea in central Asia. You will find plenty of information about this and other regions on the Internet. Try to find out why your chosen area is suffering from desertification.

2 Investigate the contribution of aid projects to improving life in the Sahel. Consider the work of Oxfam in Mali by accessing the website at: http://www.oxfam.org.uk/resources/countries/mali.html

Managing hot deserts and their margins

In this section you will learn about:

- the management needs of desert environments and their margins
- contrasts between management in the Sahel and the USA
- managing the Mojave Desert in the USA

Skills

In this section you will:

- conduct Internet research

Why do hot deserts and their margins need to be managed?

What image comes to mind when you think about hot deserts? For most people, the term 'hot desert' conjures up images of barren, desolate landscapes – with sand dunes, oases and camels. It's hard to believe that a landscape like this needs any form of management at all! However, deserts and their margins offer tremendous potential for development, and managers there have to cope with a number of pressures that often conflict (Figure 4.42). Sustainable management for the future is essential if the fragile balance in hot desert environments and their margins is to be maintained.

Pressure	Details
Ecological	Despite their harsh conditions, hot deserts provide viable habitats for a wide range of birds and animals. Desert margins are rich in wildlife, but these ecosystems are very fragile and are easily harmed.
Indigenous peoples	The different peoples who have lived in hot desert environments, and their margins, for many generations are part of those fragile ecosystems. Just like the wildlife, they have basic needs like water, food and shelter. So, local communities need to be part of the management planning mechanism.
Economic	Deserts are rich in natural resources, e.g. oil in the Middle East. Abundant deposits of 'evaporates', such as gypsum and salt, also provide important raw materials for industry. All of these resources need to be exploited appropriately.
Water	Water is the most precious resource in hot desert environments. It's in short supply and needs very careful management. For example, deep aquifers need to be exploited sustainably to prevent the stored groundwater running out in the future.
Agriculture	Over-grazing, over-cultivation and over-irrigation can all lead to desertification. Both subsistence and commercial farming place demands on the natural environment that need careful management if they are to be sustainable.
Retirement homes	Increasingly, particularly in richer countries, older people are keen to move to the margins of hot desert environments when they retire – to enjoy the hot, sunny weather. This puts additional pressures on these fragile environments, e.g. increasing the demand for water.
Tourism	Adventure tourism is one of the fastest-growing sectors in the tourist industry. Many people choose to visit hot deserts and their margins for trekking, climbing, nature watching, fossil hunting, etc. They all need accommodation, water, food and roads.

Figure 4.42 *Pressures on hot deserts and their margins* ▲

Management contrasts – the Sahel and the USA

There are significant contrasts in the management of hot deserts and their margins between poor areas (e.g. the Sahel) and rich areas (e.g. the south-western USA). On pages 149-151 you looked in detail at some of the issues facing the people who live in the Sahel region of Africa. You also examined a number of management strategies. However, when compared with rich countries like the USA, a number of contrasts can be identified (Figure 4.43).

The Sahel is a vast international region. Reaching agreements between different countries is hugely challenging. This often results in piecemeal and disjointed strategies. Some countries are more proactive than others in their attempts to manage the environment. For example, several countries have agreed a conservation programme for endangered desert antelopes. Some countries in West Africa also cooperate in monitoring and controlling locust swarms that can devastate crops.

By contrast, the USA is a single country, so there can be more 'joined-up thinking' between managers and stakeholders.

Political instability and the lack of a decision-making infrastructure is another problem faced by several countries in the Sahel. Some governments and government departments have limited power or control, so there might be a lack of cooperation. The links between central government, local government and local communities can be tenuous or non-existent.

By contrast, there is a strong system of government in the USA, and American organisations and agencies are used to working together. A strong law-and-order system also helps to enforce management decisions. The USA also has well-respected environmental pressure groups.

Poverty is a massive issue in the Sahel – both for entire countries and for individual communities. National governments in this region might prefer to spend their limited money on development projects elsewhere, e.g. industrialisation, energy provision, or improvements to cities. The sparsely populated Sahel could seem a less-attractive investment – a bit of a lost cause. Individual communities and people living in the Sahel have little spare money or time to spend on improvement projects themselves.

This is a dramatic contrast with the USA – where deserts are seen as resources to be exploited. To a large extent, money is no problem if a management project is deemed necessary or desirable.

The Sahel is a region where conditions are becoming increasingly harsh. Rainfall is becoming less reliable (Figure 4.39), and desertification is a major problem.

By contrast, in the USA these trends are less of a threat, and desertification is kept largely under control.

Pivot irrigation in the Mojave Desert in the USA – a technological solution to managing and exploiting the desert

Figure 4.43 *Management contrasts between rich and poor areas* ▲

CASE STUDY

Mojave Desert, the USA

The Mojave Desert is named after the Mojave tribe of Native Americans. It's mostly located in south-east California, but also extends into the neighbouring states of Nevada, Utah and Arizona (Figure 4.44). The Mojave is a stunningly beautiful environment, with dramatic canyons, sand dunes and colourful rock formations. It also incorporates the Joshua Tree and Death Valley National Parks (Figure 4.45).

Figure 4.44 *The Mojave Desert in the south-western USA* ▶

Managing the Mojave Desert

The Mojave Desert is managed by several public organisations, including the Bureau of Land Management (BLM), the National Park Service (NPS), and the Department of Defence (DoD) – as well as a number of private landowners. They all work together to manage the desert successfully and sustainably.

The following extract from a Department of Defence report sums up the challenges that lie ahead for the management of the Mojave.

'Land managers in the Mojave Desert today are faced with multiple challenges. Expanding economic development is causing increasing pressure on natural resources, while the public demands objective and effective management strategies. Diverse groups seek to achieve conflicting goals that make multiple demands on fragile, exhaustible resources. These goals include establishing and expanding National Parks, creating wilderness areas, protecting threatened and endangered plants and animals, developing recreational areas, and expanding economic development.'

Figure 4.45 *The Death Valley National Park* ▲

Recent management issues

- **Urban expansion**. Since the 1990s, the population of this region has grown dramatically. It is expected to triple in the next 20 years. The largest city is Las Vegas (population 2 million), with a further 1 million people living on the edge of the desert in Greater Los Angeles. There is considerable demand in the area for retirement homes and commercial developments – all of which require land, a water supply and service provision. A new airport is planned to serve Las Vegas. But, with a shortage of suitable private land for construction, the airport will probably have to be built on public land. Money from the sale of this land will then be used to purchase ecologically sensitive private land elsewhere.

- **Water supply**. The Colorado River flows through the eastern part of the region, and is a major source of water. Water also comes from sub-surface aquifers. A major water-transfer scheme, called the California Aqueduct, also transports water from the Sierra Nevada Mountains in central California to the Mojave – via a network of canals, tunnels and pipelines. The Mojave is a major producer of the crop alfalfa, and depends very heavily on irrigation (Figure 4.43).

- **Military training**. The Department of Defence owns a large amount of land in the Mojave, and has several military training bases there. It has been heavily involved in the development of the West Mojave Coordinated Management Plan – a comprehensive integrated plan, developed with the BLM, to conserve the biological resources of the area. The DoD has also established a scientific database (the Mojave Desert Ecosystem Program) to assist with the sustainable management of the environment.

- **Conservation**. The Mojave Desert contains a number of species on the US Endangered Species List, including the Mojave desert tortoise (Figure 4.46). Between 1996 and 2006, about $93 million was spent on various projects to successfully rescue this species from the brink of extinction.

- **Waste management**. In a recent initiative, a clean-up of public lands near Barstow (Figure 4.44) attracted more than 500 local volunteers. They collected all sorts of rubbish from the desert, including abandoned cars and washing machines! In 2009, a plan to create a massive landfill site in an abandoned mine close to the Joshua Tree National Park was halted by the Court of Appeal. The landfill would have received 20 000 tons of waste from Los Angeles **every day** for an estimated 117 years!

- **Tourism**. Every year, millions of visitors are attracted to honeypot locations, like the Death Valley and Joshua Tree National Parks. This creates massive management challenges for the authorities. They need to balance the demands of the visitors with conservation of the natural environment. In response to a call for more off-road driving, the Bureau of Land Management has now designated large areas of the Mojave for the use of off-road vehicles (Figure 4.47). This action is intended to prevent damage to sensitive environments elsewhere in the Mojave. However, in 2009, an alliance of environmental organisations successfully won a court order to prevent further expansion of off-road routes in the desert. The Judge found that the BLM had failed to conduct adequate impact studies when it extended its routes by 5000 miles in 2006. The Judge ruled that there had been an inadequate consideration of alternatives, to limit damage to sensitive lands.

Figure 4.46 *The Mojave desert tortoise* ▲

Figure 4.47 *Off-road biking in the Mojave Desert* ▲

Internet research

There are many other deserts throughout the world, in both rich and poor areas. Use the Internet to find out more about management issues and strategies for one or more deserts of your choice. Contrast rich and poor areas. Present your research in the form of an information poster.

ACTIVITIES

1 Study Figure 4.42 and work in pairs for this activity.
   a Use the information about the Mojave Desert in this section, together with extra Internet research, to complete a table similar to Figure 4.42 describing the main pressures on the Mojave.
   b Then use the information about the Sahel in Section 4.5, again with extra Internet research, to compile a similar table for the Sahel. You will probably want to alter some of the pressure headings. For example,

   there is not much demand for retirement homes in the Sahel!
   c Evaluate the results of your research to compare the pressures on these two desert environments.

2 What problems have to be overcome in the Sahel if management policies are to be successfully formulated and implemented there?

3 To what extent do you think the Mojave Desert is being sustainably managed? Explain your answer.

abrasion Process of wind erosion concentrated close to the ground surface involving particles of sand being picked up by the wind and blasted against rock outcrops

alluvial fans Delta-like, fan-shaped feature of deposition formed when a mountain river (often flowing through a wadi) deposits alluvium as it reaches the desert plain

aridity Dryness (lack of precipitation) is the main defining characteristic of deserts, with true deserts having less than 250 mm precipitation a year. Semi-arid deserts have 250-500 mm a year

badlands Barren semi-arid landscape with thin soils and severely weathered and eroded rock outcrops, often reflecting unsustainable management practices, such as over-grazing

barchans Crescent-shaped sand dunes with a gentle windward slope and steep leeward (sheltered) slope

buttes Flat-topped isolated rock outcrops surrounded by the desert surface

canyon Largely American term given to a very deep and steep-sided river valley, e.g. the Grand Canyon

deflation Process of wind erosion of the desert floor responsible for removing finer particles and leaving a predominantly rocky surface. In sand deserts, surface hollows or depressions (deflation hollows) may be formed by intensive wind erosion

desertification Formerly productive land that has become a desert, often due to a combination of natural factors, such as drought, or human factors, such as over-grazing

duricrust Crusty salt-rich surface deposit formed by extensive salt crystallisation

endoreic River that flows into a desert from outside, but is terminated (usually in an inland lake or sea) in the desert

ephemeral River that only flows intermittently, usually after a storm event; for much of the year the river channel is dry

exfoliation (see insolation weathering)

exogenous River that has its source outside a desert area but has sufficient discharge to maintain its flow despite high rates of evaporation

flash flood Extreme discharge event, usually confined to a river channel or wadi following a torrential storm

inselbergs Isolated severely chemically weathered rock outcrops (typically granite) characterised by having a rounded top

insolation weathering Type of mechanical weathering involving alternating extremes of heating and cooling of exposed rock surfaces causing outer layers to 'peel' away (exfoliation or 'onion-skin' weathering)

loess Fine-grained (suspended) wind deposits often carried great distances away from desert source regions

mesas Flat-topped (horizontally-bedded sedimentary rock) plateau-like features bordered by wadis or canyons, typically found in USA desert regions

pediments Gently sloping erosional rock surface at the foot of a mountain range, often blanketed by deposits of alluvial fans

rainshadow effect Some deserts are located on leeward (sheltered) sides of mountain ranges (e.g. the Atacama Desert is sheltered by the Andes). Rainfall is concentrated on the windward slopes (as moist air is forced to rise up and over a mountain range) leaving the drier sheltered slopes in a 'rainshadow'

salt crystallisation Process of mechanical weathering resulting from the growth and expansion of salt crystals precipitated by the evaporation of salt-rich water

salt lake (playa) Inland salt-rich lake often fed by endoreic rivers that may periodically dry up completely

saltation 'Hopping' or bouncing motion of wind transportation

sand dunes Wind deposition features in sandy deserts (see barchans and seif dunes)

seif dunes Elongated sand dunes commonly found in areas of extensive sand (sand seas)

surface creep Wind transportation process involving particles being pushed or rolled across the ground surface

sustainable development Developments (often involving improvements to quality of life) that do not do any lasting damage to the environment

wadis Steep-sided, often flat-bottomed valley infrequently occupied by a river (following a storm), but most commonly dry

yardangs Elongated ridges separated by deep grooves cut into the desert surface largely resulting from wind action, with the ridges running parallel to the prevailing wind direction

zeugen Mushroom-shaped rock outcrops with evidence of severe wind erosion close to the ground surface

Exam-style questions

1 (a) Study Figure 4.4 (page 131). Describe the distribution of hot deserts. *(4 marks)*

(a) Look for general patterns. Use the map to identify examples.

(b) Study Figure 4.41 (page 151). Explain why this is a good example of a sustainable management strategy. *(4 marks)*

(b) Clarify the meaning of sustainable. Use evidence from the photo to support your answer.

(c) Describe the action of weathering in hot desert environments. *(7 marks)*

(c) Describe the different processes of weathering, but consider their importance too (e.g. preparing a surface for erosion).

(d) To what extent is the action of water more important than wind in the formation of desert landforms? *(15 marks)*

(d) Consider the role of both water and wind. Make sure you address the question by comparing the importance of the two.

2 (a) Describe the characteristics of vegetation in hot desert environments. *(4 marks)*

(a) Focus on the adaptations that enable plants to survive.

(b) Study Figure 4.24 (page 141). Explain the formation of the feature shown in the photograph. *(4 marks)*

(b) Ensure you relate the characteristics of the feature to precise processes. Use correct geographical terminology.

(c) Describe the typical features associated with a badlands landscape. *(7 marks)*

(c) Use specific terminology for the features. Describe the overall landscape. No need to explain.

(d) Discuss the advantages and disadvantages of human activity in hot desert environments. *(15 marks)*

(d) Identify a range of different human activities. Use case studies where appropriate. Make sure you discuss both advantages and disadvantages.

3 (a) Describe the characteristics of a wadi. *(4 marks)*

(b) Study Figure 4.13 (page 135). Explain how the rainshadow effect contributes to the formation of the Atacama Desert. *(4 marks)*

(c) Describe the role of wind in the formation of sand dunes in hot desert environments. *(7 marks)*

(d) To what extent is desertification the result of human practises rather than natural processes? *(15 marks)*

Population change

There are five born every second

Where do you think this baby was born?

What are the chances of this baby surviving until it is one year old?

How long do you think it will live?

What do you think its life will be like?

How many children do you think this woman will have?

Introduction

Since the 18th century, the world's population has been growing rapidly. Many have speculated about the number of people the earth can support. Followers of Thomas Malthus believe the earth is doomed without population control and that rapidly growing population is a major cause of poverty, whilst those of Danish economist Ester Böserup believe the earth can support ever-increasing numbers and that a large population is actually a necessity for economic growth.

In this chapter you will learn how different countries – Uganda, UK, China, and Australia – face different population issues. You will also learn how population change can affect local areas, with examples from east and west London, the rural-urban fringe in Essex, and the relatively remote and rural Cornwall.

Books, music, and films

Books to read

Tribes of the Orange Sun by Gene Shiles
How Many People Can the Earth Support by Joel Cohen

Music to listen to

'Earth Song' by Michael Jackson
'Bengali in Platforms' by Morrissey

Films to see

I Am Legend
Survivors (BBC tv series)

About the specification

'Population change' is the compulsory Human Geography topic in Unit 1 – you have to study this topic.

This is what you have to study:
- Population indicators – the vital rates or indicators (birth rate, death rate, fertility rate, infant mortality rate, life expectancy, migration rate, and population density) for countries at different stages of development.
- Population change: the demographic transition model (five stages), and its validity and applicability in countries at different stages of development.
- Population structures at different stages of the demographic transition. The impact of migration on national population structure. The implications of different structures for the balance between population and resources.
- The social, economic, and political implications of population change. Attempts to manage population change to achieve sustainable development, with reference to two case studies of countries at different stages of development.
- The ways natural population change and migration affect the character of rural and urban areas.
- Settlement case studies comparing two (or more) of the following areas: an inner city area, a suburban area, an area of rural-urban fringe, and an area of rural settlement. These case studies should include reference to characteristics such as housing, ethnicity, age structure, wealth, employment, and the provision of services.
- The implications of the above for social welfare.

In this section you will learn about:
- how global population growth is changing

Skills
In this section you will:
▶ learn how to identify arithmetic and exponential/geometric growth

Seven billion and counting

This chapter is about population change. In the middle of 2010, there were just over 6.8 billion people on Earth. The six billionth person was born in 1999 – and the seven billionth will probably be born sometime in 2012. In thirteen years, one billion people will have been added to the global population – all requiring food, water and shelter. Like you, they will also want access to technology, education, and opportunities to improve their lives.

Demographers (who study population) have been fascinated by this rapid population growth for ages. In 1968, Paul Ehrlich published a book called *The Population Bomb*. He estimated that, up to 8000 BC, it had taken the human race a million years to double in size. Yet, by 1968, the global population was 3.5 billion and its doubling time had fallen to 44 years. This population growth trend is not exact, and is constantly changing, but in broad terms it resembles what mathematicians call **exponential growth**. Ehrlich referred to it as a population 'explosion'.

Year	Number of people
1500	0.5 billion
1804	1 billion
1927	2 billion
1960	3 billion
1974	4 billion
1987	5 billion
1999	6 billion
2012 (est.)	7 billion

Figure 5.1 *The surge in global population since 1500, according to the United Nations* ▲

SKILLS BOX

Exponential growth

Exponential growth is also known as geometric growth. In it, the thing in question (e.g. population) doubles at each stage (1:2:4:8:16, etc.). This is in contrast to arithmetic growth, where something grows by the same amount each time (1:2:3:4:5, etc.).

Exponential growth is also known as the Malthusian growth model, after Thomas Malthus. In 1798, Malthus wrote an influential essay. In it, he argued that population grows exponentially, while food production grows arithmetically (see page 216). As a logical result of this difference in growth patterns, Malthus argued that eventually the population will always outstrip the food supply – causing starvation. He called this a 'natural check' on population growth, because the population would then fall back to a manageable level, before the whole growth process started again. Other 'natural checks' include war and disease.

Many people worry that the global population is growing too fast. For decades, people have talked about the world being 'overpopulated'. The three scenarios in Figure 5.2 show different 'guestimates' for the global population level by 2050. Scenario 1 shows the population continuing to increase rapidly. Scenario 2 shows a gradual slowing in the growth rate. Scenario 3 assumes that drastic action will have been taken to reduce population growth.

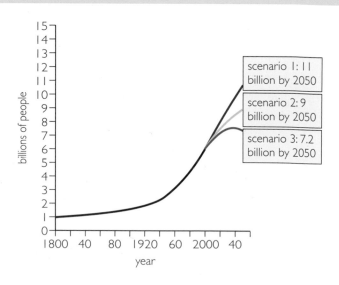

scenario 1: 11 billion by 2050

scenario 2: 9 billion by 2050

scenario 3: 7.2 billion by 2050

Figure 5.2 *Different 'guestimates' for the global population level by 2050, according to the United Nations* ▶

In 1999, the United Nations (UN) estimated that the global population would grow from 6.1 billion in 2000 to 8.9 billion by 2050 – an increase of 46%. During this period, the UN also thought that the average annual population growth rate would slow to 0.77% – less than half the 1.76% average between 1950 and 2000. By 2045, population growth was expected to have slowed to just 0.33%.

Another Italy each year

But even with a slowing growth rate, an average of 57 million people would be added to the global population every year up to 2050. That's roughly the population of Italy. Project that over 50 years, and the total increase of over 2.8 billion is more than double China's current population! Much of this increase will take place in the world's Low Income Countries (LICs) – the populations of which are estimated to grow by 58% by 2050, compared with just 2% for the world's High Income Countries (HICs).

Did you know?

Not a single country in the European Union (EU) is now producing enough babies to stop their population declining.

ACTIVITIES

1 a Make a copy of Figure 5.1 and add two extra columns. Head them: 'Years taken to add 1 billion' and 'Doubling time'. Then calculate and complete the number of years in each case.
   b How far is the trend shown by the data in Figure 5.1 'exponential'? Are doubling times changing?
2 Study Figure 5.2.
   a Using your answers to Activity 1, which 'guestimate' of global population level by 2050 seems the most likely?
   b Suggest reasons why it's difficult to estimate future population change.
3 Study Figure 5.4 and explain why – for centuries now – many people have assumed that the world is 'over-populated'.

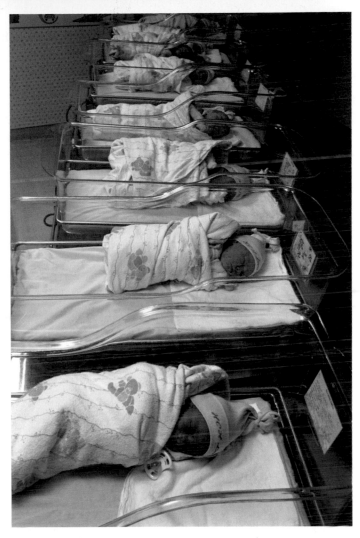

Figure 5.3 What kind of world will these extra mouths – and 57 million others born in 2010 – face? ▲

"Come on, you can always fit one more."

Figure 5.4 How many more can the world fit in? ▶

In this section you will learn about:
- the Demographic Transition Model and how it helps in understanding population change

Skills

In this section you will:
▸ learn how to calculate population vital rates

How far is population growth a good or a bad thing? Malthus (Figure 5.5) argued that population growth will always outstrip food supply – with negative consequences like famine and war. Although he wrote his essay on population in 1798, it still has many supporters. Influential global groups, such as the Club of Rome and the charity Population Concern, support the basic belief that parts of the world are over-populated and are not able to feed their people properly.

In 1929, Warren Thompson (a US demographer) published an influential article in the *American Journal of Sociology*. In it, he outlined his theory of **demographic transition**. This theory has since become known as the **Demographic Transition Model (DTM)**. He argued that population, food supply and economic development are all linked, and that changes in one will bring about changes in the others.

Thompson's theory of demographic transition

Like Malthus, Thompson believed that population growth causes problems. He believed that economic development would increase if population growth was reduced, and vice-versa. Basically, with fewer people demanding resources, there would be more available for individuals and countries to use for economic development. It is an attractive, simple theory.

Thompson classified all countries into one of three groups, according to population and wealth:

- **Group C countries** had high **birth** and **death rates**. They were the world's poorest countries (see Figure 5.6). High death rates, high **infant mortality**, and low **life expectancy** were caused by hunger and disease, resulting from poverty.
- **Group B countries** had high birth rates but falling death rates. As industrialisation took place, the increasing wealth would lead to falling death rates. Rapid population growth could be expected in these countries, but industrialisation would create enough new wealth to support it.
- **Group A countries** had low birth and death rates. They were the world's wealthiest countries.

Thompson assumed that countries would eventually progress from Group C to Group A, if they industrialised. He identified four stages in this process. Since then, a fifth stage has been added to the model (see Figure 5.7, pages 164-165).

Figure 5.5 *Thomas Malthus. His 1798 essay on population has had far greater and longer influence than he could possibly have imagined.* ▲

Figure 5.6 *Famine relief in Somalia in 2008. Thompson believed, like Malthus, that famines like this were due to over-population. In fact, in 2006, the United Nations' FAO reported that recent food shortages were more likely to have been caused by civil war and climate change than by population pressure.* ▼

Understanding vital rates

A number of different factors affect population growth – collectively known as 'vital rates'. This refers to the vital statistics needed to calculate population change (often measured per 1000 people). There are two categories within vital rates – **crude rates** and **refined rates**. Throughout this section, all data used are crude rates.

- **Birth rate** is the number of births in a year – expressed as a rate per 1000 population.

  Calculated by dividing the total number of births in a year by the population, and multiplying by 1000.

- **Death rate** is the number of deaths in a year – expressed as a rate per 1000 population.

  Calculated by dividing the total number of deaths in a year by the population, and multiplying by 1000.

- **Rate of natural increase** is the number of people in a year by which a population increases or decreases – expressed as a rate per 1000 population.

  Calculated by subtracting the death rate from the birth rate. The result is a rate per 1000, which can then be converted into percentage growth or decline by shifting the decimal point.

- **Doubling time** is the time it takes for a population to double.

  Calculated by dividing 70 by the percentage change. For example, 2% annual increase has a doubling time of 35 years.

- **Infant mortality rate** is the number of deaths of infants (before their first birthday) in a year – expressed as a rate per 1000 live births.

  Calculated by dividing the annual number of deaths of infants aged up to 1 by the number of live births, and multiplying by 1000.

- **Life expectancy** is the number of years that a person in a given population can expect to live – usually expressed 'at birth'. Often, this figure is also broken down by gender, because women tend to live longer than men.

  Calculated by estimating the statistical likelihood that an individual will reach a particular age. By averaging 'the average age of death' now and in the past, estimates can project how these might change in future. Countries with a low life expectancy tend to have a high infant mortality rate, which reduces the average age of death significantly.

- **Dependency ratio** is the proportion of the population not in work (i.e. children aged under 16 and retired people aged 65 and older) who are dependent on those in work (people aged 16-64). It is normally expressed as a percentage.

  a Calculate the percentage of the population aged under 16 and over 64 (the dependents).

  b Calculate the total percentage of those in work, i.e. aged 16-64.

  c Divide (**b**) into (**a**), and multiply by 100 to get a percentage.

- **Fertility rate** is the average number of children expected to be born to a woman over her lifetime – assuming that she survives from birth through to the end of her reproductive life. In the UK the average is 1.66, but in Afghanistan it is 5.5.

- **Replacement level** is the number of children needed per couple to maintain a population size. It is normally judged to be 2.1 – to allow for deaths early in life.

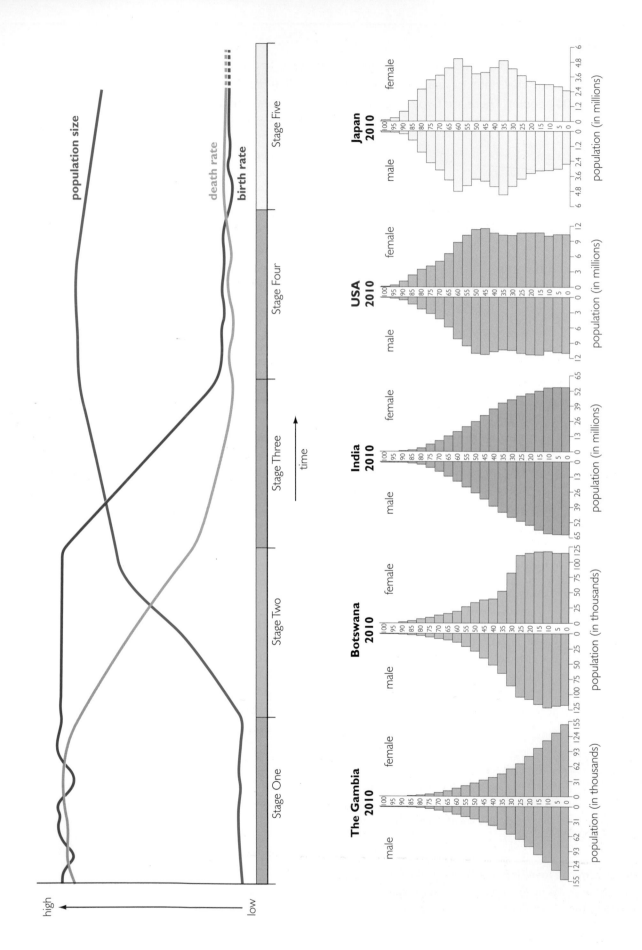

Figure 5.7 *The five stages of the Demographic Transition Model. Each stage also includes the age-sex structure (or population pyramid) for an example country at that stage.* ▲ ▼

Stage One

Thompson described this as a rural, pre-industrial society – existing by **subsistence farming**, with little surplus food. Regular famines, and the lack of safe water and proper sanitation, lead to disease and hunger and a high death rate. The birth rate stays high to compensate for this. Population growth is slow.

Look at the population pyramid for Stage One. The sharp reduction in age groups reflects two things – high infant mortality and low life expectancy, because each age group is smaller than the one below it.

Stage Two

Economic development begins to occur at this stage. Increasing wealth leads to an improved food supply and better sanitation – so the death rate falls and life expectancy increases. The death rate starts to fall before the birth rate does. This produces a rapid increase in population. The birth rate remains high, because – in industrialising countries – the use of child labour gives children an economic value to a family.

Look at the population pyramid for Stage Two. The infant mortality rate is still fairly high at this stage, but the age-sex structure is more balanced in the early years, as life expectancy begins to increase.

Stage Three

Birth and fertility rates both fall at this stage, because parents increasingly decide to provide a higher quality of life for their children – including a good education and more material possessions. Because each child costs more as a result, parents make the decision to have fewer children. Increasing urbanization also leads to smaller families (the lifestyle and housing in urban areas is more expensive than in rural areas, so families have fewer children). Tighter child labour laws also help prevent children from working, so child labour is often replaced by education. In the short term, children become less of an economic asset for their families. Increasing education for girls also leads to later marriage and a lower fertility rate.

Look at the population pyramid for Stage Three. Population growth begins to slow down, and the age-sex structure develops a more even profile, as life expectancy increases.

Stages Four and Five

In **Stage Four**, the birth rate falls to a similar level as the death rate – and the population increase stabilises. However, theorists have now proposed a **Stage Five**, where the birth rate (and therefore the fertility rate) continues to fall below the replacement level. This leads to a declining – and ageing – population, and an increasing dependency ratio as the working population gets smaller. The death rate initially stays low – but it might start to climb as lifestyle diseases increase, e.g. because of low exercise levels and fatty and salt-rich foods.

Many countries in this situation have encouraged the immigration of younger working people – to try to combat the increasing dependency ratio and raise the birth and fertility rates (see Sections 5.5 and 5.6).

Look at the population pyramids for Stages Four and Five. The age-sex structure increasingly bulges upwards and narrows at the bottom, as life expectancy increases and the birth rate falls.

What's wrong with the DTM?

The DTM is not perfect. All theories reflect the people who devise them, and the conditions in which they work. Thompson spent his life studying changes in European and North American populations, and then applied them to all countries. His conclusion was that American population trends were related to the food supply. He was a Malthusian at heart. He also assumed that there was a pre-determined path to prosperity for all countries.

The following criticisms have been made about the DTM as a catch-all explanation for every country's population trends:

Links between birth and death rates

Not all diseases are linked to the food supply, e.g. HIV/AIDS (Figure 5.8). The DTM assumes that an increasing death rate (e.g. because of the spread of HIV/AIDS) would lead to a higher birth rate to compensate for it. However, despite the dramatic increase in death rate caused by HIV/AIDS in some African countries, there is no evidence of the birth rate increasing as a result.

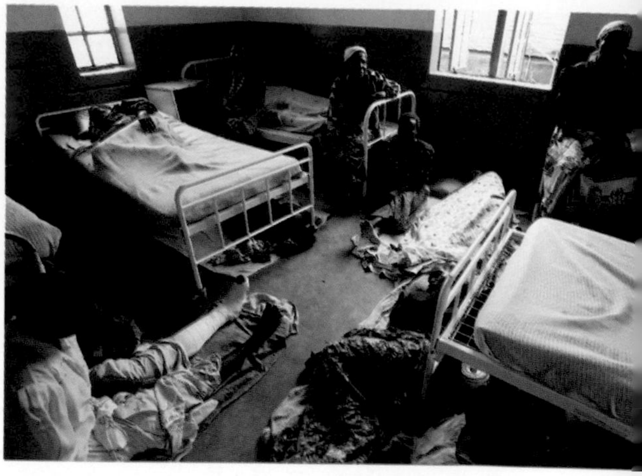

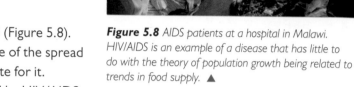

Figure 5.8 *AIDS patients at a hospital in Malawi. HIV/AIDS is an example of a disease that has little to do with the theory of population growth being related to trends in food supply.* ▲

Links between population and economic development

The DTM assumes that population trends follow food supply. In fact, population trends are more likely to mirror industrialisation. European industrialisation after 1800 resulted in falling death rates (Stage Two), followed by declining birth rates (Stage Three) – all within a century. More recently, Brazil industrialised rapidly and passed through DTM Stages Two and Three quickly. Meanwhile, many African countries have had rapidly falling birth rates (as in Stage Three), but without industrialisation first. In none of these cases has food supply determined the birth rate.

Differences within countries

Countries never have uniform populations. In most countries, there are large variations in birth, death and fertility rates:

- between urban and rural populations. Usually, the birth rate in urban families is smaller than in rural families.
- within cities. In British cities, there is a wide variation in the death rate. For example, there is a 14-year difference in life expectancy between women living in wealthy areas of western Sheffield and men living in poorer areas to the east.
- between high- and low-income groups (Figure 5.9). Although overall life expectancy increases as countries develop, high-income families still tend to live longer than low-income ones.

	Examples of occupation	1972–76		2002–05	
		Men	Women	Men	Women
Non-manual					
Professional	Doctor, chartered accountant, qualified engineer	71.9	79.0	80.0	85.1
Managerial and technical	Manager, journalist, teacher	71.9	77.1	79.4	83.2
Skilled non-manual	Clerk, cashier, retail staff	69.5	78.3	78.4	82.4
Manual	Supervisor of manual workers, plumber, electrician, goods vehicle driver				
Skilled manual		70.0	75.2	76.5	80.5
Partly skilled	Security guard, machine tool operator, care assistant	68.3	75.3	75.7	79.9
Unskilled	Labourer, cleaner, messenger	66.5	74.2	72.7	78.1

Figure 5.9 *Contrasts in life expectancy between different income and occupation groups* ▲

Political intervention

- Political and social changes can be critical in reducing the birth rate. For example, providing education for girls (Figure 5.10) leads to a lower fertility rate, because the girls marry later and are more informed.

- China's one-child policy (see pages 180-182) was a deliberate attempt to reduce the country's high birth rate. This policy helped to cut short artificially DTM Stages Two and Three in China.

- By contrast, some governments (e.g. those of Singapore and Australia) have tried to increase their countries' birth rates. The aim of these deliberate government interventions is to increase the size of the future working population and reduce the country's dependency ratio.

- Government migration policies also disturb the natural population trends of the DTM. Economic growth in HICs has been fuelled by immigration policies designed to reduce the dependency ratio and increase the birth rate.

- Specific, targeted global aid projects have also helped to reduce the infant mortality rate in many of the world's poorest countries, e.g. UNICEF's child vaccination campaigns in the 1990s and 2000s.

Figure 5.10 *Equatorial College School, in Uganda, is a secondary school financed by University College School in north London. Educating girls' and improving their status is vital in bringing down fertility rates.* ▲

Boserup's theory

Malthus's theory was challenged by the Danish economist Ester Boserup (1910-1999). Boserup believed that the Earth has fewer limits than Malthus allowed for, and that the DTM may be wrong. She believed that:

- rather than being perceived as a problem, population growth is a pre-requisite for economic growth – countries actually need population growth to drive their economic development

- people are intelligent – as demand for food or resources increases, we will always find new ways to supply it

- new or better technology is invented whenever it's needed. If old resources run out, new ones will replace them (e.g. oil and gas have already replaced coal in many countries, and now fossil fuels are being increasingly replaced by renewables, like wind and solar power).

Internet research

Go to the US Census Bureau website, which provides demographic indicators and population pyramids (age-sex diagrams) for all countries for the years 1950-2050: http://www.census.gov/ipc/www/idb/informationGateway.php

1 Research two countries that you think could represent each Stage of the DTM. Feed back your findings to the class.

2 Can you find any countries with population trends that seem to fit none of the Stages of the DTM? What possible reasons can you find for this?

ACTIVITIES

1 Draw a large copy of the DTM in Figure 5.7 and annotate it with what happens at each stage to birth rate, death rate, and life expectancy.

2 In pairs, draw up a table to show (**a**) the strengths of the DTM as a theory about population change, and (**b**) its drawbacks.

3 How far is the DTM (**a**) valid, (**b**) invalid as a theory? Use evidence to support your views.

4 In pairs, discuss whether (**a**) there should be a sixth stage in the DTM, showing countries that increase immigration, or (**b**) if this 'missing stage' is evidence that the DTM is a weak theory.

In this section you will learn about:
- population change in Uganda
- the challenges facing a country with a youthful population

Unlike many stereotypical images of Africa, Uganda is green, fertile, and has plenty of water (Figure 5.11). Its population (33.4 million in 2010) is just over half the size of the UK's – in the same land area. So, its **population density** is about half that of the UK. Its people are also fairly evenly spread across the whole country (Figure 5.12), because it's a rural society.

In theory, Uganda should be a wealthy country, capable of supporting a large population. It has hydroelectric power, important reserves of metal ores, and fertile soils that produce its main exports of coffee, tobacco, sugar cane and tea. But, in 2009, Uganda had a **Gross Domestic Product (GDP)** per capita of $1300 (PPP$). That is just 3.7% of the UK's $35 200.

This big difference in wealth is because Uganda's economy is based almost entirely on the export sale of **primary products**, such as tea (Figure 5.11). The value of these products varies, because global supplies and prices vary. Therefore, the government's tax income from exports also varies. There are also very few wealthy companies and individuals in Uganda for the government to tax. This affects what it can afford to spend on things like education and healthcare. And this, in turn, has an impact on population – birth, death and fertility rates, as well as infant mortality.

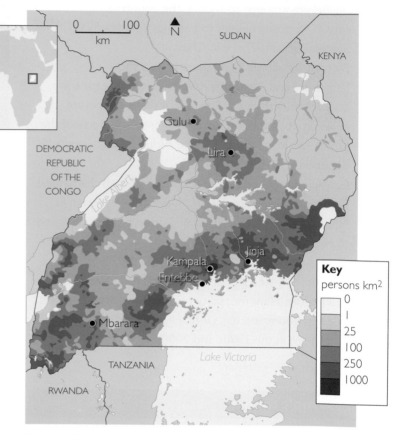

Figure 5.12 *Uganda's population density and key settlements* ▲

Figure 5.11 *A tea plantation in Uganda. Uganda's regular rainfall pattern and fertile volcanic soils make this rich productive farmland.* ▶

SKILLS BOX

Line graphs

A line graph joins a series of data points. It's used to show trends over time, e.g. increases and decreases in population.

- The x-axis is usually a time variable – generally years. These should be plotted proportionately. For example, a graph showing data for 1980, 1990, 2000 and 2005, should have half the space between 2000 and 2005 than it has between 1990 and 2000.

- The y-axis is used to plot data, starting at 0 and increasing to the maximum value plotted.

- The points are joined by straight lines, which help to identify a sequence.

Line graphs can also be multiple – showing more than one set of data (e.g. comparing changes in birth and death rates with different lines on the same graph).

Uganda's population

Uganda has the second fastest growing population in the world (Figure 5.13). It has a population doubling time of just 19 years.

In 2010, Uganda's:
- population was growing by 3.56% a year.
- birth rate was 47.5 per 1000 (the world's second highest). The birth rate has been high for decades, because 87% of the population is rural (Figure 5.14), and family sizes in rural areas are always much larger than in cities.
- death rate was 11.9 per 1000 (similar to the UK's 10). Like many Low Income Countries (LICs), the death rate has fallen sharply in the last 20 years. This is because of UNICEF's global programme to provide child vaccinations against common killer infections, treatments for the biggest childhood killer (diarrhoea), and improved treatments for malaria.

Uganda's population structure

Uganda has one of the world's most **youthful populations**. In 2010, 50% of Ugandans were aged under 15 – 16.7 million out of a population of 33.4 million. The **median age** was 15 – making it the second youngest country in the world. Young age groups dominate its age-sex structure (Figure 5.15).

◄ **Figure 5.13** Uganda's population 1960–2010 (in millions)

Date	Population
1960	6.8
1970	9.8
1980	12.6
1990	17.6
2000	23.3
2010	33.4

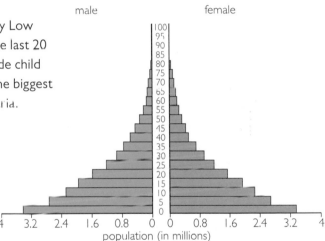

Figure 5.15 Uganda's age-sex structure in 2010 ▲

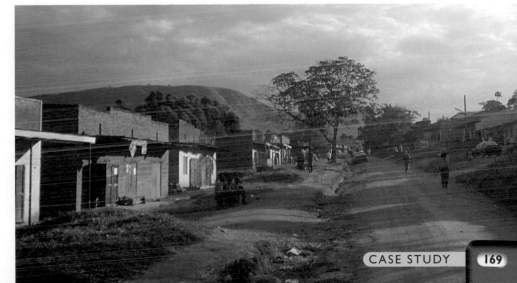

Figure 5.14 The village of Kyarutunga in south-western Uganda, near Mbarara. Ugandan villages like this have among the highest birth rates and population growth in the world. ▶

Bar charts

Bar charts are used to present data for discrete amounts e.g. rainfall, numbers of people from a survey, etc. They take one of the following forms:

- Changes in one variable over time, e.g. monthly rainfall totals for one year.
- Differences between two or more variables at one point in time, e.g. age and gender of HIV patients.
- Variations in, or frequency of, data, e.g. infant mortality rates for particular countries.

Note: Time periods on bar charts are discrete, whereas on line graphs time is continuous. So, any variable that needs to show a continuous pattern (e.g. population) is always drawn as a line graph.

There are two components to bar charts:

- A quantitative axis (vertical for column graphs and horizontal for bar graphs) to show the value or frequency of the data.
- A qualitative axis to show categories of data, e.g. region names, occupational groups or years.

The two main types of bar charts are bar graphs and column graphs (Figure 5.16).

Age-sex graphs (population pyramids)

An age-sex graph is a particular kind of bar graph designed to show the structure of a population – the numbers (or percentages) of males and females, broken down into age groups. These normally cover five years, starting 0-4, 5-9, and working up to 100+. A central axis divides male data from female. This vertical axis also shows the age scale. The horizontal axis either has numbers or percentages related to population.

Age-sex graphs are useful for interpreting a lot of information about a population (Figure 5.17):

- The number, or percentage, of people in each age group.
- Which age groups are largest and smallest.
- Whether birth rates are rising or falling.
- Whether the infant mortality rate is high or low, by comparing the 0-4 age group with the others. If infant mortality is high, the 5-9 age group above will be much smaller than the 0-4 age group.
- Whether life expectancy is high or low, by comparing all of the age groups. If life expectancy is low, the next age group up will always be smaller than the one below it.
- It is also possible to infer whether life expectancy is high or low by comparing the proportion of people in the older age groups.

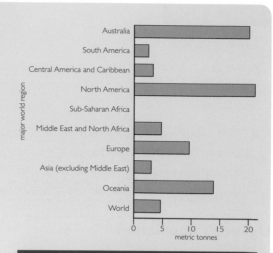

Bar graph
- Drawn horizontally.
- Used to compare the size of one variable (e.g. infant mortality) in different places, **or** different variables at a single point in time.

Figure 5.16 Bar and column graphs ▲ ▼

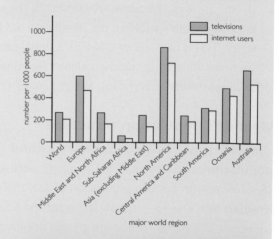

Column graph
- Drawn vertically.
- Used to show one or more variables. In Geography, these would normally be either over periods of time, or by variation between places.

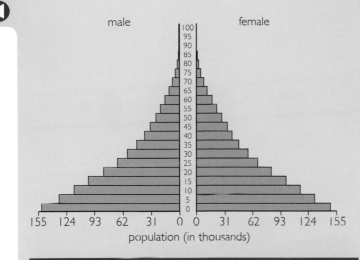

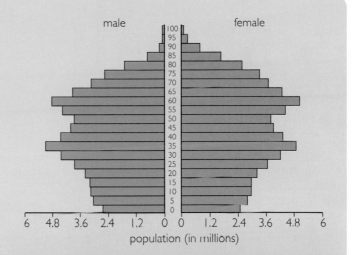

Stage One of the DTM

- Broad 0-4 age group – indicates a high birth rate.
- Smaller 5-9 age group – may indicate high infant mortality.
- All age groups are smaller than the group below – indicates a low life expectancy/high death rate.

Stage Five of the DTM

- Narrow 0-4 age group – may indicate a low/declining birth rate, and probably an ageing population.
- 5-9 age group is larger – may indicate low infant mortality and/or birth rate is declining.
- Age groups vary in size, with bulges – indicating periods of 'baby boom' and variable birth rates.
- Older age groups are quite large – may indicate long life expectancy, and ageing population.

Figure 5.17 *Comparing two age-sex diagrams* ▲ ▶

Describing graphs

Use four steps when describing graphs (including age-sex diagrams), to identify the main features of the data:

a General patterns – what is the general distribution like?

b Which are the highest and lowest values?

c Examples – give examples to illustrate the general distribution.

d Anomalies – identify examples that stand out as different from the general distribution, or are exceptions. Identify them and describe how they differ.

Life expectancy and health

- By 2010, Uganda's infant mortality rate had nearly halved to 63.7 in 20 years. But it was still in the world's worst 20% of countries for this statistic.
- Ugandan life expectancy at birth in 2010 was 53 years (compared to the UK's 79). Again, Uganda had one of the lowest statistics in the world.
- The spread of HIV/AIDS reduced the life expectancy in Uganda – just as it was starting to improve. But, despite this, Uganda's government was the first in Africa to attract international aid to develop HIV/AIDS education programmes. In the early 1990s, 20% of Uganda's population was HIV-positive. By 2010, it was only 5.4%.

But even without the HIV/AIDS factor, Ugandans only have a 62% chance of reaching the age of 40 (in the UK it's 98%), and a 40% chance of reaching 60 (in the UK it's 91%). Amongst the reasons for this are that:

- 80% of babies are born with no skilled health workers in attendance. Most are born at home, or in small clinics (Figure 5.18).
- 24% of Ugandan families are undernourished.
- Several diseases still cause premature death, e.g. malaria, typhoid and cholera.
- Living conditions are poor for many. Only 64% of Ugandans have access to safe water, and 43% to an improved sanitation system.

Figure 5.18 *A maternity ward in Katine, Uganda. Few Ugandan children are born in hospitals like this.* ▲

Education

Primary education in Uganda is free and universal (i.e. for boys and girls equally). But there are very few government secondary schools, so most families have to pay fees if they want their children to be educated beyond primary school. For rural families with an annual income of just £200 from growing crops like coffee, there is little left over for children's education. School fees of £20 a term are too expensive for most rural families to afford. Even for those children who do go to secondary school, few of them go on to university. The Ugandan government funds just four universities – a quarter of the total. To share resources, each district is allocated a number of university places, but that means one place for every 30 000 school students!

Girls' education

Few girls receive secondary education, because families will spend their limited money on educating their sons. In primary school, the ratio of girls to boys is 1:1. By age 11 it is 1:2, and by 16 it may be 1:6 or even 1:10. Only 17% of girls attend secondary school beyond the age of 13. Girls marry as young as 13 or 14 in rural communities, and have their first child soon after – hence Uganda's high fertility rate of 6.7. Girls are economic assets for their families, because they attract a dowry when they marry.

Women are the poorest Ugandans. By tradition, they rarely own land and usually work as landless labourers, when work is available. They have little control over their earnings, or whether they will have a career. They return to work soon after giving birth. Maternal mortality rates are high, and unhealthy mothers have babies that are more likely to die in their first five years.

Yet education guarantees change. By staying on at school – even university – and delaying marriage until they are older, educated girls are likely to select a career, work before and during marriage, select their own marriage partner, and have children later. Fertility rates are lower among professional women in Uganda, and infant mortality rates among educated women are almost as low as those in HICs.

Figure 5.19 *Primary schooling in Uganda is free. The next target is secondary schooling.* ▲

Cancelling Uganda's debts

Debt cancellation for Uganda came in two phases – in 2000 and 2005 – when the G8 (the world's richest nations, such as the USA, the UK, and Germany) cancelled most of Uganda's debts. However, they imposed conditions that the money saved had to be spent on poverty reduction, education and healthcare.

Debt cancellation has had major impacts in Uganda:

- Spending on healthcare has risen by 70%, including the abolition of fees for basic healthcare.

- Spending on education has also risen by 40%. Free primary schooling is now universal, with girls benefiting the most (Figure 5.19). Before debt relief, there were 20% fewer girls than boys in primary school. Five million extra children now attend school. Enrolment rates for primary schooling increased from 62.3% in 2000 to 92% of girls and 94% of boys by 2009.

- 2.2 million extra people now have access to clean water. Fetching water is traditionally the responsibility of girls, and has often been one of the reasons why they could not go to school.

ACTIVITIES

1 Draw up a table of population indicators for Uganda – birth rate, death rate, etc.

2 Study the Skills box about line graphs. Now draw an accurate line graph to show Uganda's population growth since 1960, using Figure 5.13 for the data.

3 Study the Skills box about bar charts.
   a Copy an accurate version of Figure 5.15, showing Uganda's age-sex structure.
   b Annotate your age-sex graph to identify and explain its key features.
   c In pairs, decide – with reasons – which Stage of the DTM you think Uganda is in.

4 Explain why:
   a Ugandans have a low chance of reaching age 40 or 60
   b Ugandan girls are likely to marry early

5 Explain how debt cancellation in Uganda might affect:
   a life expectancy
   b girls' education
   c girls' chances of marrying later in life
   d Uganda's future age-sex profile

6 In pairs, identify population priorities for the Ugandan government for the next 20 years. Feed back your ideas in class.

In this section you will learn about:
- population change in the UK
- the challenges facing a country with an ageing population

In 2005, the UK's population reached the milestone figure of 60 million. By 2010, an extra 1.3 million people had already been added to that total. 61.3 million people – crowded into just under 250 000 square kilometres – means a population density of 243 people per square kilometre (Figure 5.20).

In 2009, the UK's GDP per capita (PPP$) was $35 200 – making it one of the world's largest and most influential economies. The UK has fertile soils and a long industrial background, but it's the booming services sector and **'knowledge' economy** that now dominates the country's wealth creation. The service sector earns 75% of the UK's GDP (compared with just 1.2% from agriculture and 23.8% from industry/ manufacturing).

However, despite its current size and influence, the British economy is facing an uncertain future. The UK has an **ageing population**, with increasing numbers of people moving from work into retirement. In 2010, the average life expectancy in the UK for anyone reaching their sixtieth birthday was another 23 years! The country has the serious problem of working out how to support millions of people who might be retired for 20 years or more. This problem is not helped by the fact that the UK's birth rate is falling, and has been for decades. Where are the workers going to come from to support the pensioners of the future? What will happen to the UK's dependency ratio? Food for thought!

Skills

In this section you will:
- develop skills in drawing flow lines from data

Did you know?

Macau has a population density of 18 500 people per square kilometre. That's crowded!

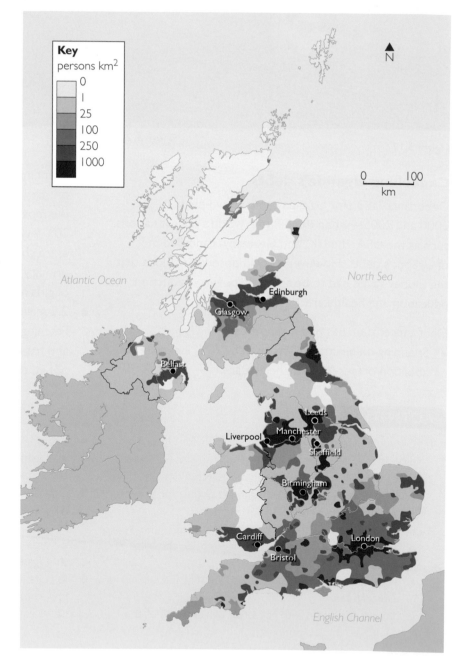

Key
persons km^2
- 0
- 1
- 25
- 100
- 250
- 1000

Figure 5.20 *The population density of the UK* ▶

The UK's population structure

For almost every British family, there has been a consistent fall in family size since 1900. The stereotypical suburban '2.4 children' per family was coined in the 1950s – having been about 5 in 1900. Now it's about 1.7. At the same time, people are living longer. The effect of this is to create an ageing population in the UK, as shown by Figures 5.21 and 5.22.

In 2010:
- only 16.5% of the UK's population was aged 0-14 (compared with Uganda's 50%).
- an almost equal proportion of the population was aged over 65 (compared with Uganda's 2.1%).
- the UK's median age was 40.5 – one of the highest in the world.

Increasingly, however, older age groups are beginning to dominate the UK's age-sex structure (Figure 5.22). This structure currently looks like Stage Four of the DTM, but it might soon look more like Stage Five if current population trends continue (see page 164).

In 2010:
- the UK's natural increase in population was just 0.7% a year – one of the world's slowest.
- the birth rate was 10.7 per 1000. In many High Income Countries (HICs), including the UK, birth and fertility rates have been falling for decades.
- the death rate was 10 per 1000. It has fallen steadily over the last century. Life expectancy at birth in the UK is now almost 80 (Figure 5.23). This is because of free universal healthcare via the NHS, improved treatment for the biggest causes of death (heart disease and cancer), and a better lifestyle due to increasing wealth.

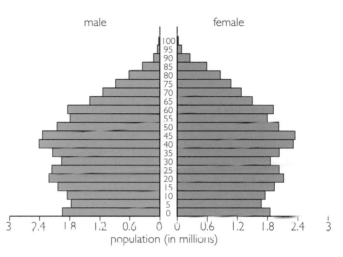

Figure 5.21 *The UK's age-sex structure in 2010* ▲

	Mid 1960s	2010
Fertility rate	2.40	1.66
Number of births each year	1 million	655 000
Population aged under 16	14.7 million	11.5 million
Population aged 65 or over	7.2 million	10 million

Figure 5.22 *The UK's rapidly ageing population* ▲

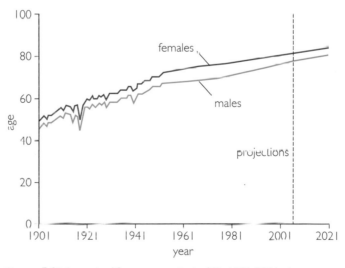

Figure 5.23 *Improving life expectancy in the UK, 1901-2021* ▲

The falling fertility rate

The UK's fertility rate first fell below the replacement level of 2.1 in the 1970s. It has remained below that crucial level ever since (Figure 5.24).

The drop in fertility rate has been partly caused by the improved status of women in British society. More women in the UK now go to university than men. As a result of this higher education, many women decide to pursue professional careers before marrying and having children in their thirties, rather than their twenties. Many of these women then continue their careers alongside raising a family.

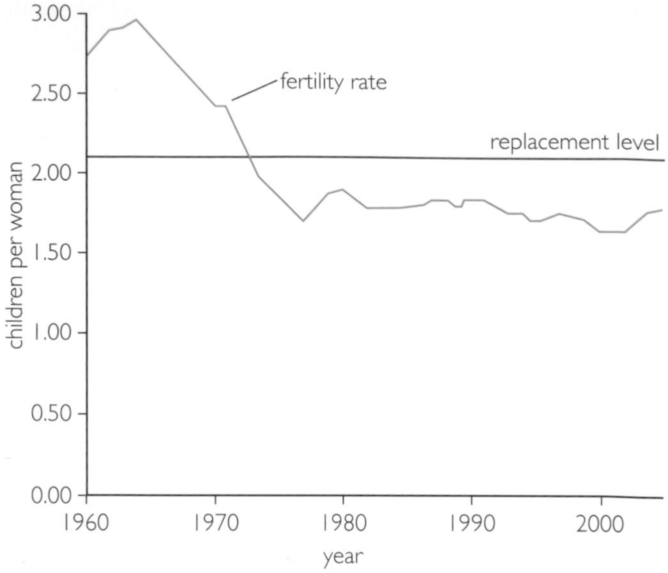

However, because many women now start their families later in life, they have to face the consequence that they will be less fertile at that point. The chances of conception decrease for a woman in her thirties, making it more likely that family size for older mothers will be smaller – especially if they want to return to work for periods between having children.

Figure 5.24 *The changing fertility rate in the UK, 1960-2005* ▲

As well as many British women starting their families later in life, increasing numbers are also deciding never to have children at all. In 2009, research by the London School of Economics showed that 10% of women aged 45 were intentionally child-free, and that 25% of younger women were also child-free at that point. The researchers suggested that both figures will increase in the future.

One conclusion is inescapable. At present, the UK's population will not replace itself by natural change. Many politicians quietly believe – because the issue is sensitive with public opinion – that immigration must be encouraged to maintain a strong workforce and economy (see page 178).

Did you know?

Two terms are used to describe women who choose not to have children – child-free and childless. We chose child-free for this book; were we right?

Internet research

Is childbearing a personal, political, or religious matter? In many EU countries, including the UK, many women are choosing to have fewer or no children. But this view is not universal – attitudes towards childbearing vary across the world.

In small groups, research fertility rates and attitudes towards childbearing across a range of countries. Then design a presentation called: 'Factors explaining differences in fertility rate between countries'.

Include in your class sample a few:
- HICs, e.g. USA, Canada, France, Japan
- MICs, e.g. Brazil, Chile, Argentina, St Lucia, South Africa, Thailand
- LICs, e.g. Haiti, Angola, D.R. Congo

- countries where Islam dominates, e.g. Iran, Afghanistan, Saudi Arabia
- countries where Catholicism dominates, e.g. Mexico

For each country, use the CIA World Factbook to research fertility rates and trends (Google 'CIA World Factbook'). Then add the following phrases, plus the name of your country (shown as X), in a Google search:

- changing fertility rates in X
- women's rights in X
- attitudes towards marriage in X
- attitudes towards contraception in X
- your own choice of search phrases

The ageing trend

'Ageing' is also known as the '**greying**' of the population (Figure 5.25). By 2014, for the first time, there will be more people aged over 65 in the UK than under 16. Not only are more people surviving into retirement, but many are living to a ripe old age (Figure 5.26). There are already 1.2 million people aged over 85 in the UK.

	Males		Females	
	1981	2002	1981	2002
Life expectancy	70.9	76.0	76.8	80.5
Healthy life expectancy	64.4	67.2	66.7	69.9
Years spent in poor health	6.4	8.8	10.1	10.6
Disability free life expectancy	58.1	60.9	60.8	63.0
Years spent with disability	12.8	15.0	16.0	17.5

Figure 5.26 *Recent changes in life expectancy and the implications for health* ▲

Figure 5.25 *The Who recorded 'My Generation' in 1965. The Zimmers re-recorded the song in 2007. The band got together as part of a TV documentary to highlight the isolation felt by many elderly people.* ▲

The challenges of an ageing population

An ageing population creates both benefits and problems:

Benefits	Problems
Many 'retired' people still do part- or full-time work – to keep themselves active and to supplement their pensions. Over 25% of the employees at DIY chain B&Q are aged over 55. Retired people are useful for companies, because they can work flexible hours and do not carry costs such as National Insurance. They also have a lot of skills, experience and knowledge, which can be passed on.	Many elderly people only have the basic State pension to live off. If the price of food or fuel rises, it's harder for those on fixed pensions to compensate for it. Poverty amongst the elderly is a major challenge for the government – and for non-governmental organizations, like Age UK, which campaign on their behalf.
Many businesses have developed to service the '**grey pound**' (wealthy retired people, such as those with generous private pensions). For example, DIY outlets, gardening centres, and specialist holiday companies.	The World Health Organization (WHO) warns that ageing populations worldwide will lead to increasing heart disease, cancers, and diabetes. Treating these is expensive and provides a real challenge for the NHS – and the voluntary sector (e.g. Macmillan Cancer), which relies on donations of money and time from volunteers.
The growing private health sector also regards the ageing population as an opportunity. Firms providing private healthcare insurance, e.g. BUPA, or companies such as Boots (who offer healthcare services), regard the elderly as good business.	An ageing population also creates a need for new housing, e.g. smaller properties with no stairs, wider doorways and lower kitchen units for those with limited mobility. And **sheltered accommodation** for those who need carers on site.
The investment of pension funds by fund managers provides a valuable source of money for companies and organizations in both the private and public sectors. For example, many British pension funds invest in big companies like BP.	A growing pensions crisis is emerging. State pensions in the UK are paid for by National Insurance (NI) contributions from those in work. As life expectancy increases, more people are claiming pensions for longer. The ratio of people in work compared to over 65s is falling. In 2000, there were 3.7 people in work for every person aged over 65. By 2040, this ratio will have fallen to 2.1. The government will be receiving less tax revenue from the smaller number of workers, just as demand for pensions and healthcare is rising.

Is immigration the answer?

Migration into the UK – together with most HICs – is at an all-time high. But, until 2000, more emigrants left the UK (for places like Australia and Spain), than immigrants arrived.

Immigration into the UK has come from two main groups:

- Since the late 1940s, there has been a steady flow of immigrants from Commonwealth countries such as India, Pakistan and Caribbean nations like Jamaica, together with smaller flows from Africa.
- In 2004, the EU was enlarged with ten extra countries. Because of EU rules about the free movement of labour in member states, there was a surge in immigration to the UK from Central and Eastern Europe. Polish migrants constitute the UK's largest single migration ever.

Alongside social changes in the UK's population – such as falling birth, death and fertility rates – immigration has had a real impact, especially in the last 30 years:

- Immigration from India, Pakistan and the Caribbean during the 1950s/60s, coincided with a baby boom in the UK. As a result, natural increase was responsible for 98% of the UK's population growth – and net immigration just 2%.
- But, since the 1980s, net immigration has contributed more and more of the UK's population growth. As the fertility rate has fallen, so has natural increase.
- From 2001 to 2008, net immigration added 180 000 people to the population each year, compared with 90 000 from natural increase – a ratio of 2:1.

Recent immigrants from the EU tend to be young and well qualified. In 2008, about 70% of EU immigrants who registered for work in the UK were aged 18-35, and most were graduates. Family related migration is also becoming more common. Those arriving with dependent children increased from 4% in 2004 to 15% by 2009. Birth statistics show that the number of children born in the UK to mothers from other EU countries has increased substantially since 2004.

Immigrant populations tend to have a younger age structure than the white British population (Figure 5.27). For example, large numbers of Black Africans migrated to the UK in the mid-1980s as children. Now, many of them have their own young families, which explains the large percentage of under 16s.

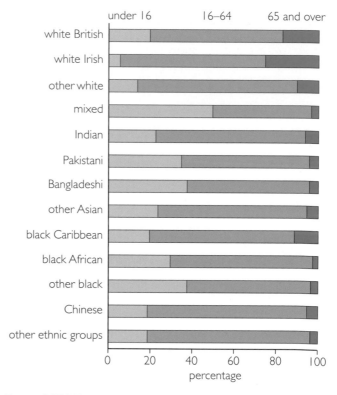

Figure 5.27 *UK population structure by ethnic group (2001)* ▲

The contribution of recent migrants

It's easy to understand policies that encourage immigration, because of the benefits that migrants bring – particularly to an ageing population. Quite simply, they cost less than the host British population – and contribute more. Recent migrants registering for work are 60% less likely than UK-born residents to receive benefits or tax credits, and 58% less likely to live in social housing. They also contribute tax revenues in proportion to their representation in the population, despite the fact that they often earn lower wages than UK-born workers.

SKILLS BOX

Drawing flow lines

Flow lines are a way of turning statistics into mappable data. For example, they help to identify **source nations** (where migrants originate from) and **host nations** (where they move to) in the migration process. Figure 5.28 shows the number of migrants entering and leaving the UK on a typical day in 2006.

The arrows are used to show both the direction of movement and the volume of people moving. The width of the arrow represents the volume of people, so a scale is used where a certain number represents 1 mm thickness. In Figure 5.28, the scale is roughly 1 mm = 50 people. So 100 people would be shown by a 2mm thick arrow, and so on. Two colours have been used – one for those entering the UK and one for those leaving. Where numbers are too small to present realistic flow lines, dotted lines can be used below a certain number.

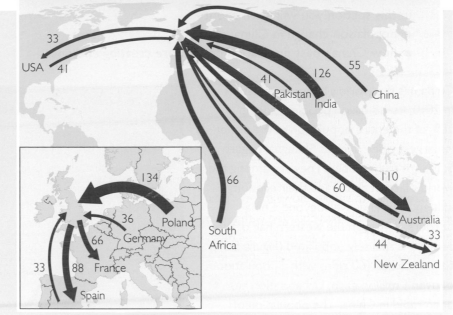

Figure 5.28 *The average number of migrants entering and leaving the UK on a typical day in 2006, and the main countries involved* ▲

Figure 5.29 *The changing pattern of migration into the UK, 2002-2006 (the number of overseas nationals entering the UK and being allocated a National Insurance number, in thousands)* ▼

2002/2003		2003/2004		2004/2005		2005/2006	
India	25.0	India	31.3	Poland	62.6	Poland	171.4
Australia	18.9	South Africa	18.4	India	32.7	India	46.0
South Africa	18.6	Australia	17.1	Pakistan	20.3	Lithuania	30.5
Pakistan	16.8	Pakistan	16.8	South Africa	19.3	Slovakia	26.5
France	13.8	Portugal	14.0	Australia	16.6	South Africa	24.0
Philippines	11.8	China	13.3	Lithuania	15.6	Australia	23.8
Spain	11.7	France	13.1	France	13.3	Pakistan	22.3
Zimbabwe	10.3	Spain	11.9	China	12.6	France	17.2
Iraq	10.1	Poland	11.2	Portugal	12.2	Latvia	14.2
Portugal	9.8	Philippines	10.7	Slovakia	10.5	Germany	13.3

ACTIVITIES

1 Why is sustained population growth essential for the UK's economy?

2 Explain why:
   a the UK's fertility rate is now so low
   b increasing numbers of women are choosing not to have children

3 Explain why:
   a immigration might be the only way to sustain the UK's economic growth
   b politicians regard this as a sensitive issue

4 Study the data in Figure 5.29.
   a Add up the total number of migrants to the UK (between 2002 and 2006) from each country listed.
   b Using a world map, construct flow lines to show these migrants.
   c Describe the pattern shown on the map, and give reasons for it.

5 In pairs, think about the advantages and disadvantages to the UK of continuing to admit migrants in large numbers. Discuss your ideas with the rest of the class.

In this section you will learn about:
- population change in China
- how China's past population policies are creating challenges

China has the world's largest population. In 2010, its 1.3 billion people made up 20% of the entire population of the planet! Its largest cities, such as Shanghai (Figure 5.30) are truly 'global'. They are among the world's largest cities, and their economic influence extends well beyond China's borders. Although the western half of the country is sparsely populated (Figure 5.31), China still has 160 cities with a population of over 1 million each. Urbanisation has been rapid in China. The population of Guangdong Province – where urban economic growth has been fastest – grew by 37.5% between 2000 and 2010.

China already has the world's fastest-growing economy. In August 2010, it overtook Japan to officially become the world's second largest economy (behind the USA). Even during the banking crisis of 2007-8, China's annual economic growth rate only slowed to 7%. By 2010, the growth rate was back in double figures. It is estimated that, by 2035, China will overtake the USA as the world's largest economy. But there have been warning signs from the Chinese government that its economic growth could depend on what happens to China's population over the next 20 years or so.

China's population structure

Many geography students study China because of the government's famous one-child policy, which was introduced in 1982. The Chinese government feared that rapid population growth would lead to poverty, so it ordered that every Chinese family should only have one child. Parents who have more than one child can have their wages reduced and be denied some social services.

Figure 5.30 *One of China's booming cities – Shanghai* ▲

Figure 5.31 *Population density in China* ▼

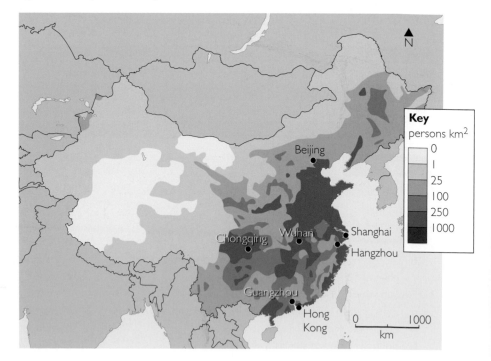

Key

persons km²

| 0 |
| 1 |
| 25 |
| 100 |
| 250 |
| 1000 |

Despite this, there have been abuses of the system – allegedly caused by corruption (e.g. bribery of officials) – which have allowed some families to have more than one child. Nevertheless, since 1982, relatively few Chinese families have had a second child. As a result, by 2010, China's fertility rate had dropped to 1.54. Many people now question whether China's increasing prosperity and urbanisation might have reduced its population growth rate anyway.

China's annual population growth rate in 2010 was 0.49%. This was because its:

- birth rate had dropped to 12.17 per 1000
- death rate was 6.9 per 1000
- political isolation from the rest of the world meant that migration had little effect.

The effects of the one-child policy

Population policies, such as the one-child policy, can never act quickly. When introduced, there are already large numbers of existing children and young people in the system, who have yet to marry. So, although the original aim of the one-child policy was zero population growth, that will not happen until 2043 at the earliest. Criticisms have also been made that some people have been 'allowed' to have more children if they support the Chinese government. Meanwhile, the one-child policy is presenting China with two major population issues – an ageing population and a gender imbalance.

An ageing population

Like many major economies, China now has an ageing population (Figures 5.32 and 5.33). The older age groups are starting to dominate its age-sex structure, and they will continue to do so into the future.

- China's median age in 2010 was 35.2 – not far away from the UK's 40.5.
- Rising prosperity has caused life expectancy at birth to increase to 74.5 years.
- In 2010, only 17.9% of Chinese were aged 0-14 – again, not far off the UK's figure of 16.4%.
- Although, in 2010, only 8.6% of Chinese were aged over 65, this figure is expected to rise to 12% by 2020, and to 25% by 2050 – by which time it will be the same percentage as in the UK.
- The prediction for the number of Chinese aged over 60 is scary – 162.4 million by 2020, and 320 million by the 2040s (Figure 5.33).
- There is a dependency 'time bomb' ticking. In 2010, 20% of China's working population was aged 50-64, and will retire by 2025 at the latest. But, traditionally, many Chinese retire as early as 51, which is much earlier than the legal retirement age. That is a lot of non-working Chinese to support.

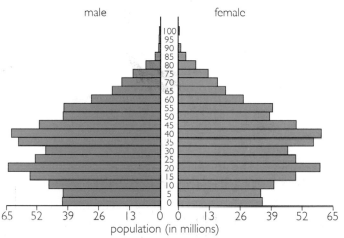

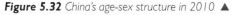
Figure 5.32 *China's age-sex structure in 2010* ▲

Figure 5.33 *China's projected age-sex structure by 2040* ▼

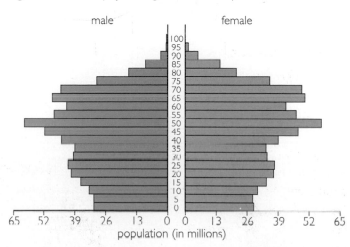

Gender imbalance

Unlike almost all other countries, males outnumber females in China – by 50 million in 2010. In most populations, the usual ratio of male to female births is 105.5:100. China's last national census, in 2000, showed a ratio of 117:100. According to the 2000 census, there were 12.77 million more boys under the age of 9 than there were girls. Potentially, this means that by 2020 there will be 30-40 million young Chinese men who will be unable to marry because there will be such a shortage of women.

The reason for this gender imbalance is the high rate of female infant mortality. In China, infant mortality among females exceeds that of males by 1.5 per 1000. China is the only country of the four studied in this chapter where this occurs (Figure 5.34).

Country	Male rate per 1000	Female rate per 1000
Uganda	67.3	60.0
UK	5.3	4.2
China	15.8	17.3
Australia	5.0	4.3

Figure 5.34 *The infant mortality rate, by gender, for the four countries studied in Sections 5.3-5.6 of this chapter* ▲

Chinese tradition is that a son will provide for his parents in their old age. So, the pressure from the government to have just one child, has led to strong family pressure to kill infant girls (called female infanticide). Although infanticide is illegal in China, it has existed for a long time – but the one-child policy has made the problem worse. The Chinese government has tried to tackle this problem by forbidding the use of ultra-sound to establish the sex of a foetus – thus discouraging selective abortion of female children.

How do population issues threaten China's economy?

On present trends, China's population will peak at about 1.5 billion, and then decrease. Would a decrease be beneficial?

- The Malthusian view says that China's population is already putting pressure on resources and the quality of the environment (Figure 5.35). The one-child policy was introduced to prevent the rapidly growing population outstripping the available resources. So a reduction in population would be a good thing, according to Malthus.
- However, the Boserupian view (see page 167) is that a reduced population could cause an economic slowdown. The Chinese government is concerned that a declining population could threaten China's rapid economic growth. As early as 2016, China's working population (aged 15-64) is expected to peak at 990 million, after which it will probably decline to 970 million by 2030 and 870 million by 2050.

Figure 5.35 *China – fast economic growth, but appalling levels of air pollution* ▼

China faces two other big problems in the future:

Urbanisation and workers' rights

Almost all of China's economic growth (an average of 8-10% a year for over 20 years) has taken place in its cities. But its rate of urbanization (2.7% a year) is slower than the rate of economic growth. In 2008, 43% of China's population was urban. At current rates, 300 million people will migrate from rural to urban areas in the coming 20 to 30 years. But that may not be enough to meet the increasing demand for labour to maintain China's ambitious plans for economic growth.

China's rural-urban migration began in the late 1970s. Since then, 150 million migrants have moved to cities – working on building sites (Figure 5.36) and in service industries. Traditionally, rural dwellers have rights to their land. But, as migrants to cities, they are denied the benefits that come from permanent urban residency (access to education, housing and health) – known as **hukou** – unless they give up their rural land rights. This deters potential migrants, because they don't want to give up the security of their land.

Lack of investment in education

In the drive for economic growth, investment in healthcare and education has been low. In 2009, China spent just 1.9% of its GDP on education (compared to the UK's 5.6%). Only 2% of Chinese aged over 25 are university educated, compared with most HICs where 25% is usual. This lack of a highly educated workforce limits China's potential in the knowledge economy, where employers face problems appointing qualified staff.

Figure 5.36 *A building site in Shanghai. Many construction workers on sites such as this are temporary migrants, who are paid less and have fewer rights and benefits than permanent urban workers.* ▲

Figure 5.37 *Peking University in Beijing. China needs a huge expansion in university education if it is to take part in the knowledge economy.* ▼

ACTIVITIES

1 In pairs, decide which stage of the DTM China had reached by 2010.
2 Draw two spider diagrams. On one, draft out reasons why China could justify its one-child policy. On the second, annotate the problems that have arisen from the policy.
3 Using China as an example, are Malthusian or Boserupian population policies better?
4 In pairs, draft out two action plans for China:
  a One to persuade temporary urban migrants to give up their land rights

  b One outlining how China could increase the percentage of people gaining a university education
For each action plan, outline the potential benefits and problems if it is not followed through.
5 Should China:
  a increase its retirement age, as the UK is doing?
  b abolish its one-child policy?
Write 300 words on each.

In this section you will learn about:
- population change in Australia
- whether future population increase is sustainable

Did you know?

Australia's most populated state – New South Wales – is three times larger than the UK, but with less than 10% of its people!

Australia is an interesting country in so many ways. With an area of 7.6 million square kilometres, it's the world's sixth largest country – over 30 times bigger than the UK. Yet its population is only about a third of the size of the UK's – just 21.5 million in 2010. This means that Australia's population density is one of the lowest of any country (Figure 5.38), and particularly of any High Income Country (HIC). Yet, despite this, demographers have already started asking whether there are too many people living there, and whether they are causing environmental damage.

Most of Australia's interior is baking hot semi-desert, and hostile to human settlement (Figure 5.39). Three-quarters of Australians lives in a narrow coastal belt in the south-east corner – stretching from Brisbane to Adelaide (Figure 5.38). This coastal region also supports most of Australia's productive farmland.

Although a lot of Australia's export wealth comes from its farms and vineyards, a far larger amount comes from its enormous reserves of minerals, e.g. iron ore. Australia's GDP per capita (PPP$) in 2009 was US$38 800 – making it one of the world's richest countries – and its prosperity is rising fast. It actually managed to avoid going into recession in the global financial crisis of 2007-2009 – largely due to high demand from China for iron ore and food to expand its own economy. One argument for increasing the population of Australia is to help maintain its economic growth.

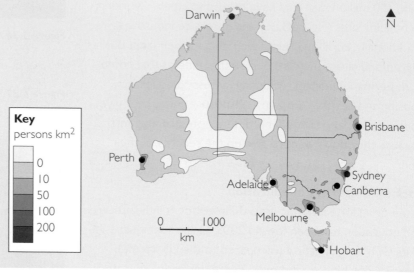

Key
persons km²

	0
	10
	50
	100
	200

0   1000
km

Figure 5.38 *Australia's population density and key settlements* ▲

Figure 5.39 *Australia's semi-desert interior. The reddish-brown colouring gives a clue to the reserves of iron ore lying under large areas of Australia's deserts.* ▼

Australia's population structure

Australia has one of the highest standards of living of any country in the world. But, although its age-sex structure (Figure 5.40) resembles Stage Four of the DTM, it does not fit the increasingly common population patterns of most HICs. Its population data for 2010 showed:

- a high life expectancy of 81.7 (but for aboriginal Australians, it was just 62).
- a low infant mortality rate of 4.67 per 1000 live births (but, again, the rate for aboriginal Australians was three times higher).
- a higher birth rate (12.39 per 1000) than either the UK (10.67) or China (12.17).
- a lower death rate (6.74 per 1000) than the UK (10), and similar to China (6.9).
- an increasing fertility rate of 1.78, but still well below the replacement level.

Look at Figure 5.40. Australia has more young people aged 0-14 (18.4%) than either the UK (16.5%) or China (17.9%), although nothing like as many as Uganda (50%). Australia's median age (37.5 years) is less than that of the UK (40.5), but has increased since 1980 (when the median was just 29.4), so the population is ageing.

The Australian population is growing by 1.3% a year. That is one of the fastest growth rates in the HICs (the UK's rate is just 0.28%). At the *current* growth rate (Figure 5.41), Australia's population is expected to double in 40 years.

What makes Australia's population patterns different from most other HICs is the migration factor. About half of Australia's 1.3% population growth rate is due to natural increase. The other half is down to the country's immigration policy, which aims to increase Australia's pool of skilled and professional people.

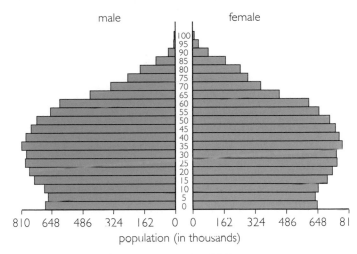

Figure 5.40 Australia's age-sex structure in 2010 ▲

1901	3.8
1921	5.5
1941	7.1
1961	10.5
1981	15.0
2001	19.5
2010	21.5

◄ **Figure 5.41** Australia's population 1901-2010 (in millions)

Arguments for increasing Australia's population

Australia has an ageing population, so it has an increasing dependency ratio. Although only 13.7% of the population is currently aged over 65 (compared to the UK's 16.4%), this proportion is going to keep increasing in future. Basically, Australia's birth and fertility rates are just not high enough to offset the ageing population. As in many HICs:

- Australians are marrying later. The average age of first marriage in 2008 was 32 for men and 29 for women.
- Australian women are also having children later. The average age for a woman having her first child in 1984 was 27, but by 2006 it had risen to 30.

So, in 2005, the Australian government introduced paid parental leave and baby grants (worth AU$5000 per baby in 2009) to encourage parents to have more children and help reverse the low fertility rate.

Promoting immigration

About half of Australia's population growth is due to deliberate large-scale immigration. Australian governments use the Boserup argument (page 167) that immigration drives economic growth, supplies young skilled workers to help offset and support the ageing population, and increases the size of the domestic market.

But the migration is not all one way. Every year, 60 000 well-qualified young Australians emigrate to work overseas, and need to be replaced. Without immigration, Australia would face a skills shortage in its mines, industries and farms.

Australian immigration has had two phases:
- Until the 1970s, **assisted passages** encouraged young families to migrate to Australia. From the 1950s to the 1970s, a million British citizens emigrated to Australia for just £10 a family – called the 'ten-pound passage' (Figure 5.42). Other immigrants included Italians and Greeks.
- Since the 1970s, **skills-based migration** has been used. Prospective immigrants must pass a points-based skills test, based on:
  - employment (to match skills shortages in Australia)
  - educational qualifications (university degrees are encouraged)
  - age (those under 30 are encouraged)
  - the ability to speak some English.

Figure 5.42 Mr and Mrs Henwood – and their eight children – about to leave Southampton for Australia on the 'ten-pound passage'. Over a million 'Poms' migrated to Australia this way. ▲

Arguments against increasing Australia's population

The debate rages about how many people Australia can actually support. Many people regard Australia's environment as fragile, and only able to produce enough food and resources for a limited number of people. They suggest that Australia's economic and population growth has contributed to major environmental problems.

Figure 5.43 A groundwater wind pump. Most farms in the drier parts of Australia obtain water for cattle and sheep using these. ▼

Drought and water supply

Australia is a very dry country (most of it is desert or semi-desert), so water supply is one of its biggest problems. There are three main issues with the water supply:
- The rainfall is unreliable – varying from place to place and year to year.
- The evaporation rate is high. On average, 94% of inland rainfall evaporates due to the high temperatures, 2% drains into the ground, and 4% ends up as runoff. There is little left over for storage.
- So, water inland is mostly obtained by drilling for limited groundwater supplies. It's then pumped up from deep aquifers, using wind pumps (Figure 5.43). Much of this groundwater is ancient and cannot be replaced, so its sources are drying up.

Rural areas in Australia are suffering population decline as a result of several prolonged droughts – the last of which occurred between 2002 and 2008. Animal farming was badly hit. The number of sheep fell from 120 million in 1998 to 100 million by 2008 (cattle from 26.7 million in 1998 to 23.3 million by 2004).

To increase Australia's water storage capacity, the Snowy Mountains Scheme was begun in the 1950s. A number of dams and reservoirs have been constructed in the mountains of the wetter parts of Victoria and New South Wales (Figure 5.44). Today, the Scheme supplies irrigation water for 60% of Australia's food and wine production, together with hydroelectricity.

Declining soil quality

Two soil problems have arisen because of poor land management – soil salinity and soil erosion (see right).

◀ *Figure 5.44*
A dam and reservoir forming part of the Snowy Mountains Scheme

Soil salinity is an increase in the salt content of soil beyond the level at which plants can survive (Figure 5.45). One of its causes is the addition of irrigation water rich in mineral salts, e.g. sodium, to soil. As the water evaporates in the heat, it leaves the minerals behind. These become more concentrated over time, and, as the salinity increases, plants begin to wilt and die.

Soil erosion is the removal of the most fertile soil particles by wind or surface run-off. It takes place where dry areas have been over-grazed – exposing the soil. Cattle and sheep have often been farmed in numbers that are far too great for the fragile land to support. Soil erosion is almost always irreversible.

Figure 5.45 *Land affected by soil salinity. The saturated ground has become too saline for plants to survive – hence the dying trees in this photo.* ▼

SKILLS BOX

Mapping data

Isoline maps

Isolines join places of equal value. In Figure 5.38, population isolines have been plotted to join areas in Australia with equal population densities. The isoline map has then been shaded appropriately within each isoline (as a choropleth map, see opposite). The process is straightforward:

- The values for population density are plotted at known points.
- Isolines are then drawn to link those points with equal values.

Drawing an isoline map

- Study Figure 5.46 on the right. The values are people per square kilometre.
- Two isolines have already been drawn on diagram A – 30 and 35 people per square kilometre. So each isoline joins dots of equal value.
- Where dots are higher or lower than the isoline value, an estimate is made of the path between the points.
- The remaining isolines have been plotted in the same way on diagram B.
- When completed, the final isoline map can then be shaded to produce a choropleth population density map, like Figure 5.38.

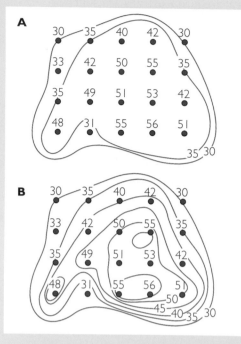

Figure 5.46 *A theoretical population density map, where densities per square km are known at points, which are then used to draw isolines* ▲

Dot maps

Dot maps represent distributions, e.g. people in a country or region. Dots can either be:

- individual
- scaled to represent different numbers e.g. 1 dot = 100 people
- scaled and sized, so that small dots show low values (e.g. 1000 people) and larger dots represent larger values (e.g. 100 000 people).

Atlas maps show population distribution by marking settlements with single dots – towns are shown by larger dots, and cities as larger dots or shaded areas.

State or territory	Population
Australian Capital Territory	351 200
New South Wales	7 099 700
Northern Territory	224 800
Queensland	4 406 800
South Australia	1 622 700
Tasmania	502 600
Victoria	5 427 700
Western Australia	2 236 900

Figure 5.47 *The population of each state or territory in Australia, 2009* ▲

Drawing a dot distribution map

Choose an appropriate dot value for the data you are handling:

- Using Figure 5.47 as an example data set, select a dot value that is suitable for those quantities. For instance, if one dot = 1000 people, you would have to draw 7099 dots in New South Wales!
- Using a value of one dot = 100 000 people would mean 71 dots in New South Wales, but just two in Northern Territory. This would probably be the best compromise dot value.

Drawing a dot map to show geographical distribution gives you three choices:

- If you want to show the population distribution of New South Wales as a whole, draw the dots evenly spaced over the whole state.
- If you want to show that the population of New South Wales is mainly in coastal cities, you could cluster the dots to show this – using an atlas to help you.
- However, Sydney (4.5 million people) would need 45 dots in a small space, so you could use a larger dot value (representing, say, 1 million people) for Sydney, and use smaller dots elsewhere.

Choropleth maps

Choropleth maps use shading to show density. They are used for a range of different maps, such as rainfall totals or population density (Figure 5.38). Monochrome patterned shadings can be used (e.g. dots or hash lines), or different shades of the same colour, or a range of different colours. A key should give the value for each shade or pattern. The lowest values have the lightest shades, and the highest values the darkest. Areas on the map are shaded according to the size of data.

Drawing a choropleth map

- Again using Figure 5.47 as an example data set, divide the data up into the categories that you would use for your shading. For a map with this number of examples, four categories would be enough.
- To decide on the values for your categories, construct a dispersion diagram (see below).
- Rank the data into quartiles, as shown, and decide on shades from dark to light in one chosen colour (using the quartiles).

Dispersion diagrams and quartiles

Use a dispersion diagram to work out the categories for your shading (Figure 5.49). This is a simple procedure:

- Rank the data in size order.
- Divide the data into equal **quartiles** (i.e. quarters), to categorise the values and finalise the key for your choropleth map.

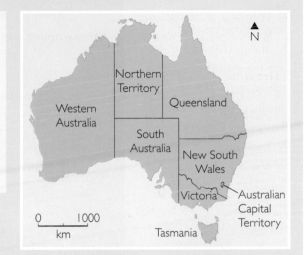

Figure 5.48 *Australia's states and territories* ▲

New South Wales	7 099 700
Victoria	5 427 700
upper quartile	
Queensland	4 406 800
Western Australia	2 236 900
middle quartile	
South Australia	1 622 700
Tasmania	502 600
lower quartile	
Australian Capital Territory	351 200
Northern Territory	224 800

Figure 5.49 *A dispersion diagram for the data in Figure 5.47* ▲

Alternatively, you could divide the data for all eight states into four categories:

6 million or more
4 million – 5 999 999
2 million – 3 999 999
Less than 2 million

In this section you will learn about:
- how population change and migration have affected the character of inner-city areas of east London

For over 30 years, east London has been a place of dynamic change:

- London Docklands, which contain Canary Wharf and other important financial buildings (Figure 5.50), are the world's largest urban regeneration project (Figure 5.51). For years, Canary Wharf has been one of **the** places to do business. Annual salaries in its biggest building average £100 000 per person – even when low-paid shop workers and security guards are included in the average.
- The Westfield Shopping Centre in Stratford is the largest shopping centre in Europe. 8000 jobs were created when it opened.
- The 2012 Olympics and Paralympics in Stratford have firmly placed east London in the global spotlight. The Olympic Park is London's first new park for over a century.
- Stratford International Station – the first stop out of

Figure 5.50 *Canary Wharf, the flagship of the Docklands regeneration project* ▲

London St Pancras on Eurostar's rail link to Paris – is just 2 hours from Paris and even less from Brussels.
- The new £16 billion Crossrail project will link east London with Heathrow Airport and the suburbs of west London.

Canning Town

It seems as if east London is the place to be! Yet it also contains some of the UK's most deprived areas. Just three kilometres east of Canary Wharf is the inner-city suburb of Canning Town (Figure 5.51). In that short distance, wealth and poverty sit almost side by side. Residents in Canning Town suffer **deprivation**, low incomes, poor health, and (despite rapidly falling crime levels) comparatively high crime (see Figure 5.63 on page 196). On a journey along London's Jubilee underground line between Westminster in central London and Canning Town, the life expectancy drops by nine years – more than one year for every station!

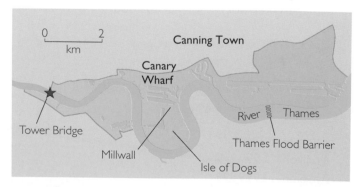

Figure 5.51 *The London Docklands area, in orange, and the locations of Canning Town and Canary Wharf* ▲

However, if you walk around Canning Town, it looks like a perfectly reasonable place to live (Figure 5.52). Most housing there is social housing (for those on low incomes), but all of the decaying, high-rise tower blocks were demolished in the 1990s – and most homes now have gardens. Renovation has also dramatically improved the condition of many of the older buildings.

But appearances can be deceptive. Figure 5.53 shows population data for Canning Town from the 2001 census. Although the data are quite old, poverty rarely shifts quickly – and can soon become a trap. Canning Town has higher than average unemployment (6.7% in 2001) and nearly 1 in 11 works part-time. Few residents, then or now, play any part in the booming 'knowledge economy' in nearby Canary Wharf.

Figure 5.52 *Housing in Canning Town* ▲

Canning Town residents face a number of big problems:

- **Expensive housing**. Although the housing in east London is less expensive than most other areas of the capital, it's still unaffordable for those on low incomes. Also, large family houses are in short supply in Canning Town.
- **Low incomes**. In 2007, the average household income in Canning Town was £23 000, with an average income per person of just £10 000. Only 37.6% are in full-time work, which lowers the average income.
- **Poor health**. The economic prospects of many Canning Town residents are limited by long-term illness or disability (9.1% in 2001), or poor health generally (20.7% in 2001).
- **Low educational achievement**. In 2001, over 43% of working-age adults had no qualifications at all. It's one of the reasons why those in work have low wages – because they're unqualified.

In addition, Canning Town has a high percentage of **minority ethnic groups** who were born outside the UK, and for whom poor English language skills can be a barrier to employment and achievement.

	Canning Town	Newham (including Canning Town)	Millwall (including Canary Wharf)
Ethnicity (%)			
White	61.0	39.4	66.6
Asian or Asian British	6.4	32.5	18.4
Black or Black British	26.9	21.6	5.5
Other	5.7	6.5	9.5
General health (%)			
General health: good	65.0	67.9	75.7
General health: not good, or with a limiting long-term illness	20.7	27.4	11.4
Employment status: aged 16-64 (%)			
Employed full-time	37.6	40.0	52.7
Employed part-time	8.1	7.7	5.0
Permanently sick / disabled	9.1	6.8	4.3
Highest qualification level reached (%)			
No qualifications	43.1	33.6	20.8
Qualified with university degree	16.5	21.3	45.2

Figure 5.53 *Selected 2001 Census data comparing Canning Town with Newham as a whole, and with Millwall (which contains Canary Wharf)* ▲

SKILLS BOX

Using census data

A census is an information-gathering exercise, which many countries use. In the UK, it is held every ten years (Figure 5.53). The data collected include every person's age, sex, ethnicity, employment, educational qualifications, general health, and housing (size, number of residents, owned or rented). They help the government to predict need (e.g. school places) and service provision. The government also uses the data to identify clusters, e.g. areas with high unemployment, poor health and low qualifications. These areas are said to have 'multiple deprivation' – i.e. more than one indicator. Canning Town is one of these.

Quality of life in Canning Town

In 2002, Queen Mary University of London (QMUL) interviewed people in Canning Town about their quality of life (Figure 5.54). The QMUL results matched findings from MORI, which found that:

- 21% of Canning Town residents were dissatisfied with the area (equivalent figures for London and England were 5% and 3%)
- only 10% were very satisfied with it, compared to 51% for England
- education and crime were the two top priorities that people wanted addressed.

A	Is there a strong sense of community here?
Yes	33%
No	64%
B	**How many of your neighbours do you know by name?**
< 4	44%
4-9	33%
10 +	19%
C	**What do you like about living here?**
Nothing	20%
Family / friends / community	17%
It is a quiet area	12%
D	**What do you dislike about living here?**
Crime	32%
Dirty streets	10%
Lack of facilities	9%
Everything	9%
Noise	8%
E	**If you could do one thing to improve the area, what would it be?**
More facilities, including those for children, and community schemes	30%
Reduce crime / more police	18%
Cleaner streets	16%
More green areas / parks	6%
F	**What people said about living in Canning Town:**

◀ *Figure 5.54 The 2002 Canning Town survey from QMUL*

Figure 5.55 The negative multiplier in Canning Town/ Newham ▶

But areas like Canning Town are difficult to improve. Newham Council (which is responsible for Canning Town) gets low income from business rates and council tax, because of the high levels of deprivation and unemployment in the area. As a result, the council has low spending power – but also has to commit a lot of its income to benefits and social care. Without central government help, this creates a **negative multiplier** in which few changes can occur (Figure 5.55).

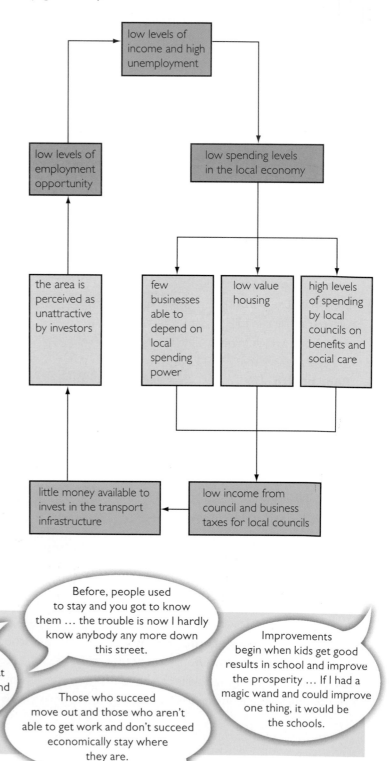

Front gardens are full of rubbish and they don't care where it goes or anything.

In cases where the properties are let and the turnover of residents is quite high, you lack that close community – because people tend not to stay in the same place for very long.

Before, people used to stay and you got to know them ... the trouble is now I hardly know anybody any more down this street.

Those who succeed move out and those who aren't able to get work and don't succeed economically stay where they are.

Improvements begin when kids get good results in school and improve the prosperity ... If I had a magic wand and could improve one thing, it would be the schools.

Improving Canning Town: the CATCH Project

The Canning Town and Custom House (CATCH) Regeneration Project (Figure 5.56) has been funded by central government since the mid-2000s to meet the needs of Canning Town and its population (Figure 5.57). Its aim has been to create neighbourhoods with a mixture of improved employment, services, and owned and rented accommodation. Costing £3.7 billion, the project has focused on four areas:

Housing

- Up to 10 000 new homes have replaced the old high-rise tower blocks – particularly family sized houses that are affordable for either rent or purchase. Newham has many large families and needs appropriate housing.
- Much of the existing housing has either been replaced or renovated, e.g. new windows.

Wealth and employment

- Skills training has been provided for local people.
- Low-cost offices and workspaces have been built for local businesses.
- The new workplaces include shops and a supermarket.
- Improved bus services now offer better transport to workplaces. Few residents have cars.

Education

- There have been major improvements to primary and secondary schools. In 2007, 55.5% of Newham's students achieved five or more GCSEs at A* to C – double the rate of 1996 (27.9%). This local improvement has been greater than the equivalent national improvement in the same period.

Health and welfare

- Improvements to services include a new health centre, library, community centre and upgraded play areas for children.
- Open spaces have been redesigned, and there have been moves to make the streets safer (e.g. using traffic calming).

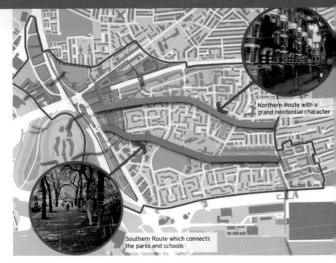

Figure 5.56 *The Canning Town and Custom House Regeneration Project* ▲

Figure 5.57 *The age-sex structure for Newham, the borough that includes Canning Town (from the 2001 census). The specific data for Canning Town match the Newham data almost exactly.* ▼

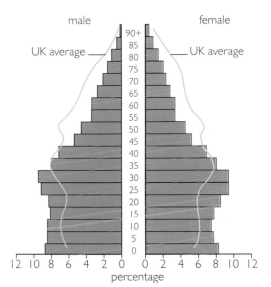

ACTIVITIES

1. Study Figure 5.57 and describe Newham's population structure.
2. In pairs, decide which are the biggest problems facing Canning Town. Use your answers to Activity 1 and Figures 5.53 and 5.54 to help you. Justify your decisions.
3. How are the problems of low qualifications, poor health, low incomes, poor housing and ethnicity linked? Try to design a flow diagram to show the links. Use Figure 5.55 to help you.
4. Copy the table, and complete it using information from this section.

Problem	Reasons for the problem	What the CATCH regeneration project has done to deal with the problem	How well has CATCH tackled the problem?
Housing			
Low incomes			
Low employment			
Local services			
Education and welfare			

5. How might (a) the 2012 Olympics, (b) the new Westfield Shopping Centre in Stratford, and (c) Crossrail help to tackle these problems further?

In this section you will learn about:
- how population change and migration have affected the character of the suburban borough of Richmond upon Thames in south-west London.

The London borough of Richmond upon Thames – with a population in 2008 of 180 000 – is located about 13 km south-west of central London (Figure 5.58). It is an idyllic spot. The River Thames flows gently past Petersham meadows. Riverside pubs attract crowds on sunny days and warm summer evenings (Figure 5.59). People walk, or cycle, along the Thames to Kew Gardens in one direction, or Twickenham and Kingston upon Thames in the other.

Large Georgian and Victorian houses occupy the roads leading away from the river up to Richmond Hill – where some of London's most expensive property can be found. Richmond upon Thames probably represents the best of London living. But it also comes at a price. House prices there in 2010 were some of the highest in London – and therefore the UK. So, who can afford to live there?

Skills

In this section you will:
- sketch and annotate an age-sex diagram
- estimate future changes

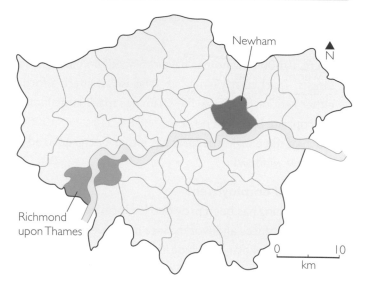

Figure 5.58 *The locations of suburban Richmond upon Thames, one of London's outer boroughs, and Newham in inner London (see Section 5.7)* ▲

Figure 5.59 *The Thames riverside in Richmond upon Thames* ▼

Who lives there?

The population structure of Richmond upon Thames (Figure 5.60) is different from that of inner-city east London (see Figure 5.57 on page 193). Compared with most other London boroughs, in 2001, Richmond upon Thames had:

- quite a high proportion of 0-4 year olds – reflecting a recent 'baby boom' among 30-something women there
- far fewer 10-24 year olds
- a big bulge of people of working age, especially aged 30-44
- fewer 60-84 year olds
- a larger number of over 85s.

The birth rate in Richmond upon Thames (13.9 per 1000) is higher than the UK average (10.67), although the fertility rate is low. This is because a high proportion of the births are to mothers aged over 35 from managerial and professional occupations. The death rate (8.5 per 1000) is below the national average (10), reflecting the high degree of wealth in the borough. Net migration into the borough increased by nearly 10 000 between 2001 and 2010.

The population in 2001 was also over-whelmingly white (Figure 5.61). Richmond upon Thames has one of the least ethnically diverse populations in London, although its diversity is actually similar to the UK average. In 2001, minority ethnic groups comprised just 9% of the borough's population – many of whom were Indian. However, since 2001, the number of new National Insurance registrations for non-UK nationals has risen sharply by about 3000 a year. Although almost all of these new arrivals are white, they have nonetheless increased diversity in the borough. The majority are Polish (17% of new registrations), Australian (10%) and South African (8%), with other European countries, the USA, New Zealand, and India making up the rest.

Who works where?

Most residents of Richmond upon Thames are highly paid managerial and professional workers, who commute into central London to work (Figure 5.62). The borough has one of London's highest concentrations of very skilled workers – only 2.9% of adults have no qualifications, and 25% of the working population has a postgraduate or professional qualification (in addition to a degree). As a result, the average annual income there in 2009 was £46 415 – over 4 times that of Canning Town.

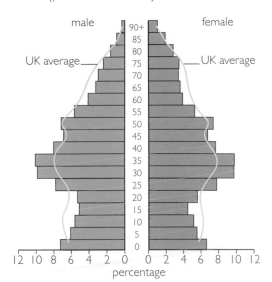

Figure 5.60 *The age-sex structure for Richmond upon Thames (from the 2001 census)* ▼

	Richmond upon Thames	London average	UK average
White	91.0	71.2	90.0
Asian or Asian British	3.9	12.1	4.6
Black or Black British	0.9	10.9	2.3
Other	4.2	5.8	3.1

Figure 5.61 *Ethnicity in Richmond upon Thames (from the 2001 census)* ▲

Figure 5.62 *Selected 2001 census data for Richmond upon Thames* ▼

	Richmond upon Thames	London average
General health (%)		
General health: good	76.3	70.8
General health: not good, or with a limiting long-term illness	18.3	23.8
Employment status: aged 16-64 (%)		
Employed full-time	46.9	42.6
Employed part-time	8.7	8.6
Permanently sick / disabled	2.5	4.6
Occupations of those in work (%)		
Managerial and professional	71.3	56.0
Middle occupations (e.g. skilled, self-employed)	19.6	26.1
Semi-routine and unskilled	9.1	17.9

Modern social profiling

In 2009, the credit consultants, Experian, found that residents were attracted to Richmond upon Thames by its good rail links to central London, high-quality housing, and superb environment. As one of the highest-earning populations in the UK, residents there had high disposable incomes, and invested in high-interest bank accounts and stocks and shares. They also ate out frequently, and were very likely to buy a newspaper like *The Guardian*, *Independent*, *Times*, or *Financial Times*.

On the downside, many spent very long hours at work. They also had high personal and mortgage debt – owing, on average, £48 000 per adult, which is twice the amount owed by the average UK resident.

Experian discovered that Richmond upon Thames has concentrations of three particular social groups:

- **'New Urban Colonists'** (26% of the working population). These are young, mostly single, high-income earners – who spend a high proportion of their income on housing, whether renting or buying. There were 21 times more 'New Urban Colonists' in Richmond upon Thames than in any other UK town or suburb!
- **'Cultural Leaders'** (22%). The people in this group are middle-aged, usually married, very well educated professionals. They work in law, media, medicine or investment banking. Almost all live in expensive family homes. They are 23 times more likely to live in Richmond upon Thames than elsewhere in the UK.
- **'Global Connectors'** (10%). These are mainly middle-aged people, without children, who work in global commercial occupations, e.g. banking. They live in very expensive housing. They are 15 times more likely to live in Richmond upon Thames than elsewhere in the UK.

Measuring quality of life

With Kew Gardens, Richmond Park, and many other parks besides, Richmond upon Thames gives residents more open space per person than any other London borough. It also remains one of London's safest boroughs, with relatively low crime rates (Figure 5.63). The main types of crime there are wealth-related – burglary and motor vehicle crime – and hate crimes of different types.

Services in Richmond upon Thames vary:

- In terms of health, the borough has no NHS hospitals, although the adjacent boroughs of Hounslow and Kingston upon Thames both have large teaching hospitals.
- Its state schools achieve good exam results. In 2008, 75% of the borough's 16 year olds achieved five GCSEs at Grade C or above. Over 25% of students attend fee-paying independent secondary schools.
- Predicted significant increases in the number of over-85s will put a strain on health and social services in the future.

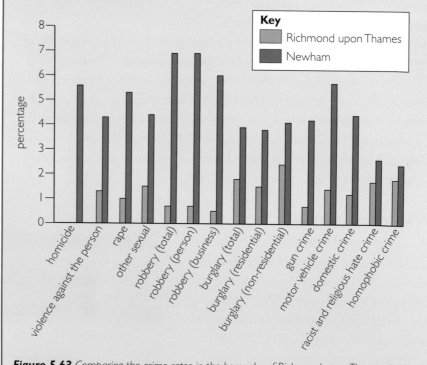

Figure 5.63 *Comparing the crime rates in the boroughs of Richmond upon Thames (blue bars) and Newham (red bars). The bars show each borough's percentage share of all the crimes in London – in each category – during the period May 2009-April 2010. For instance, there were no murders in Richmond upon Thames in this period, but 5.6% of the murders committed in London between May 2009 and April 2010 were committed in Newham.* ▲

Social welfare issues

The image of Richmond upon Thames as a comfortable borough ignores social welfare issues that do exist. Even though many indicators show Richmond upon Thames to be better off than the UK average, 8.3% of its working age population has some degree of disability, and more than 1 in 6 adults has poor health or a long-term limiting illness.

Richmond upon Thames also has significant housing problems, especially with the provision of affordable social housing. The borough has some of London's best-quality housing, but – whether for rental or purchase – it is very expensive (Figure 5.64). It is almost always in the five most expensive areas of London. Land there is so expensive that even small houses or flats are very expensive to build. In 2010, a two-bedroom flat in a house conversion could cost over £450 000 – and over £1 million if it overlooked the Thames!

Richmond upon Thames has the fourth smallest percentage of affordable housing in Greater London – just 12% of the housing there is social. 'Affordable' housing means different things in different areas, but generally it means affordable by the majority of the population. In the 2000s, the provision of affordable housing was intended to make homes available for key workers, e.g. nurses and teachers. But in Richmond upon Thames, most key workers have little option but to live outside the borough – increasing their journey time to work, and its cost.

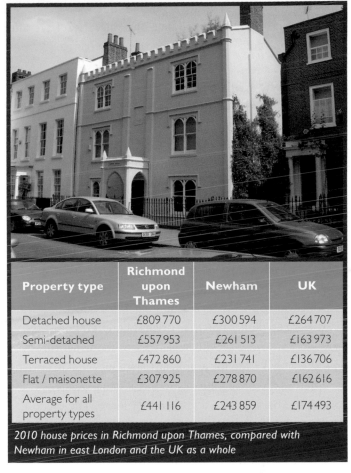

Property type	Richmond upon Thames	Newham	UK
Detached house	£809 770	£300 594	£264 707
Semi-detached	£557 953	£261 513	£163 973
Terraced house	£472 860	£231 741	£136 706
Flat / maisonette	£307 925	£278 870	£162 616
Average for all property types	£441 116	£243 859	£174 493

2010 house prices in Richmond upon Thames, compared with Newham in east London and the UK as a whole

Figure 5.64 *High-quality (and expensive) housing in Richmond upon Thames* ▲

ACTIVITIES

1   In pairs, list the ways in which Richmond upon Thames is better off economically, socially and environmentally than other areas of London.
2   In pairs, discuss the advantages and disadvantages of living in Richmond upon Thames. List your ideas in a table. Which is the longest list?
3   Study Figure 5.55 on page 192. In pairs draw up a 'positive multiplier' diagram to show how Richmond upon Thames is enjoying an 'upward spiral' of economic growth.
4   Study Figure 5.60. Using information from this section, sketch out how you think the age-sex structure of Richmond upon Thames might change by the 2021 census. Annotate your diagram and justify your changes in 200 words.
5   Summarise the three issues that you think Richmond upon Thames borough council should focus on in the next decade. Feed back to the class, justifying your choices.

Population on Essex's rural-urban fringe

In this section you will learn about:
- how population change and migration have affected the character of Essex's rural-urban fringe.

Skills
In this section you will:
▶ identify and annotate features on photographs

Terling is a very traditional-looking English village, near Chelmsford in Essex. The church and village green (Figure 5.65) are a short walk away from the cricket meadow, where teams play in the summer. Nearby is Terling Place, the home of Lord Rayleigh, whose family owns much of the land for miles around – and has done for centuries.

But, beneath the traditional surface, the village's independence is threatened and its services are struggling – the shop is fighting for survival, the doctor's surgery is only open five hours a week, the bus runs just twice a week, and the pub has recently closed. The dairy farms that once surrounded Terling no longer provide jobs – the cows were sold off when milk prices plummeted, and most farm workers were sacked. Grain prices then soared, so now the land is used for arable farming. But outside farming contractors are brought in to do the actual ploughing, sowing and harvesting – and then they leave. The old farmhouses have been sold off and the cow milking sheds and barns now lie derelict.

Yet, despite these obvious signs of problems, Terling's property prices are booming! Traditional cottages there can sell for up to £750 000. This is because Terling is actually a very convenient (as well as attractive) place to live (Figure 5.66). The railway station at Hatfield Peverel is just five minutes' drive away – with half-hourly trains to London. A five-minute drive also takes you to the A12 – the main road through Essex to London and Colchester. Chelmsford is just seven miles away. It is the supermarkets and other convenient shopping and service options of Chelmsford which are helping to put pressure on Terling's own services.

Welcome to the **rural-urban fringe** – a broad area surrounding major cities, which depends on those cities for work and many services (see the box opposite). In the fringe, growing towns and transport networks are expanding into the countryside and its villages. Every day, 650 000 commuters leave places like these in the Home Counties to work in London.

In 1971, Chelmsford had 58 000 residents, but now it has 120 000. This growth has mostly come about through people migrating out of London – in a trend called **counter-urbanisation**. It's a process that works well for many people. A high London salary makes up for the cost of a rail season ticket, especially when housing is much cheaper in Essex than in London. The only other cost is the time spent on crowded trains or congested roads.

Figure 5.65 The green and church at Terling in Essex ▼

Figure 5.66 Terling and Chelmsford on Essex's rural-urban fringe ▼

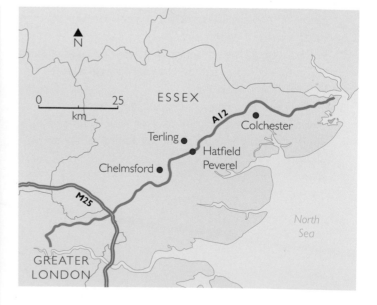

The rural-urban fringe

The rural-urban fringe is the transition between rural and urban land uses. It works at two different scales:

- **Locally**. Urban landscapes, like Chelmsford, transform into rural villages, like Terling.
- **Regionally**. For example, the landscape of urban Greater London changes to rural, but is punctuated by towns like Chelmsford on the way.

This transition is dynamic and changing, rather than fixed. As new housing estates are built on the edges of urban areas, the fringe shifts outwards. What was once rural becomes suburban – hence the term **suburbanisation**.

Both scales exist within Greater London's sphere of influence. Its fringe alters from:

- the **outer suburbs**, to …
- … the **Green Belt**, where little further development is allowed (to prevent urban sprawl), to …
- … **dormitory towns** outside the Green Belt (e.g. Chelmsford), where commuters live and travel to work in the main city, to …
- … **suburbanised villages**, like Terling, lying beyond the dormitory towns, to …
- … the **rural landscape**, where farming still takes place.

The population of the rural-urban fringe

The age-sex structure for Chelmsford (Figure 5.67) is typical of many other dormitory towns around major cities. However, its socio-economic characteristics (Figure 5.68) do differ from other dormitory towns, like Reading or Slough:

- Chelmsford's population is predominantly white (97%), with few minority ethnic groups. The largest age groups are typically adults in their 30s, with pre- and primary-school age children. Most live on recently built estates – in detached or semi-detached houses – with substantial mortgages. Owner-occupation is above average (79%) and social housing below average (12.9%).
- The average income in Chelmsford is high, unemployment is low, and there are few benefit claimants. The majority of residents are in professional and managerial jobs, with company pensions, private healthcare and company cars. Most work in **service industries**, especially banking and finance in central London and London Docklands.

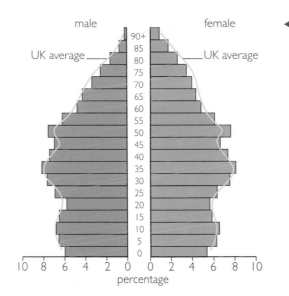

◀ *Figure 5.67* *The age-sex structure for Chelmsford (from the 2001 census)*

Figure 5.68 *Selected 2001 census data for Chelmsford* ▼

	Chelmsford	UK average
General health (%)		
General health: good	73.9	68.8
General health: not good, or with a limiting long-term illness	19.7	27.0
Employment status: aged 16-64 (%)		
Employed full-time	45.5	40.8
Employed part-time	12.7	11.8
Permanently sick / disabled	2.6	5.3
Average annual income in 2009 (£)	30 000	24 292
Highest qualification level reached (%)		
No qualifications	22.6	28.9
Qualified with university degree	20.0	19.9

Changes in the rural-urban fringe

What makes the rural-urban fringe so dynamic is constant change. Pressure for new developments from towns and cities creates problems, as one type of land use takes over from another – even where green belts are supposed to protect the land. There are two main drivers of change:

- **Economic**. Changes to the farming economy, and increasing demands from towns and cities for new buildings and additional workers.
- **Political**. Councils may either allow changes to land use (e.g. new housing estates), or prevent it (e.g. listing buildings or creating conservation areas).

Four different types of landscape can result in the rural-urban fringe from these drivers (Figure 5.69 opposite):

- **Valued landscapes**. Planning constraints are put on new developments, e.g. by establishing green belts, conservation areas, listed buildings.
- **Simplified landscapes**. Changing economic demands result in habitat removal and the establishment of huge 'factory farms' owned by large agri-businesses.
- **Disturbed landscapes**. Planning permission for new developments allows waste disposal, mineral extraction, airports, transport depots, industry, supermarkets.
- **Neglected landscapes**. Changing economies leave some landscapes idle, e.g. industry closures, changes in farming practice, derelict land.

ACTIVITIES

1 Using the following headings, describe the population of the rural-urban fringe in Essex:
   - Age structure
   - Ethnicity
   - Wealth and employment
2 How can a population increase in villages like Terling result in:
   a the threatened closure of the village shops?
   b cuts to bus services?
3 Identify how and why the rural-urban fringe is often a difficult place to live for:
   a low wage earners
   b single-parent families
   c the elderly

Internet research

In pairs, research and list the likely advantages and disadvantages of counter-urbanisation for people moving from London to the rural-urban fringe, e.g. Chelmsford or Terling. Carry this out as follows:

- Select an area in one of the Home Counties surrounding London.
- Use OS Maps and Google Maps and Images to identify the varied appearance of the area.
- Identify and annotate places which qualify as valued, simplified, disturbed and neglected landscapes. What makes them so?
- Look at local council websites (e.g. essex.gov.uk) for population characteristics
- Use the National Rail website (nationalrail.co.uk) to identify frequency of train services, duration of journeys to London, and annual costs of season tickets.
- Identify property prices. For example, use findaproperty.co.uk to compare house prices in your study area with those in London.
- Assess the likely quality of life by researching amenities, leisure and services.

Present your results in class.

Valued landscapes

- Traditional cottages and villages (see below) become more valued (and more expensive!) as new housing estates are built elsewhere in cities.
- Period buildings are listed and conservation areas are established.
- Planning permission for new developments becomes hard to get.
- Country Parks are created, as even more development leads to pressure to conserve the landscape.

Simplified landscapes

- Rural land values rise as the urban area expands.
- With the rising land values, and mounting pressure to maximise production and profit, large agri-businesses take over from small traditional farms.
- Guided by global food prices, they replace labour-intensive dairying with arable farming, e.g. for grain.
- Fields become larger, and hedges, habitats and gateways are removed to allow big machinery to operate (see below).

Disturbed landscapes

- Traffic increases as the urban economy expands and people commute longer distances.
- The landscape quality deteriorates as roads expand and road interchanges take up space in the countryside.
- Planning permission becomes easier to get for commercial activities, like out-of-town shopping (retail space and supermarkets), road widening or by-passes.
- New housing estates 'fill in' the spaces between towns and new road schemes, e.g. Beaulieu Park, Chelmsford.
- Planners try to improve the urban landscape with new developments reproducing traditional styles of architecture.

Neglected landscapes

- Farm buildings (e.g. barns and milking sheds) are left derelict as farms change their use, e.g. to arable instead of dairy farming.
- There are fewer farmworkers, because of redundancy and increased mechanisation. They abandon their farm cottages and move to urban areas for work.
- Outside farming contractors work just a few days a year – so there is no landscape maintenance (e.g. hedgerow trimming).
- The contractors replace local people. They are usually workers paid on bonus rates, who have no stake in the landscape.

Figure 5.69 *How different landscapes emerge on the rural-urban fringe* ▲

In this section you will learn about:
- how population change and migration have affected the character of rural Cornwall

Every year, over 4 million tourists visit Cornwall for its spectacular coastal scenery and beaches. Figure 5.70 is typical of the image that many of these visitors have of Cornwall. It's a beautiful area – and the UK's top destination for family holidays and short breaks. Yet many of the people who live there all year round feel that Cornwall is a county in crisis – in desperate need of **regeneration**. Low wages, a declining rural economy, and a lack of services and employment opportunities, all contribute to this feeling. Cornwall has no major cities (Figure 5.71), so it has no dominant population centre to attract outside investment to kick-start the local economy.

Figure 5.70 *The Cornish coast – how can this be a place in crisis?* ▲

Cornwall's population trends

In spite of its economic problems, Cornwall has the UK's fastest-growing population. It increased by 0.8% in 2010 – to 540 000 – and by 30% from 1971 to 2010. There are three main trends affecting this population:

- **A natural decrease**. Deaths in Cornwall have exceeded births for decades (Figure 5.72). This is because Cornwall attracts many retired people – creating an age imbalance (Figure 5.73). They come for the more-relaxed and higher quality of life. So, with an ageing population, Cornwall has a higher death rate than the rest of the UK. However, life expectancy there is above the UK average.

- **High inward migration**. As well as retirees, there has also been an inward migration of other age groups, particularly those aged 50-59 who are preparing for retirement. Increasingly, however, many families are also moving to the south-west. One survey from Cornwall County Council in 2005 estimated that half of all students in Cornish secondary schools had not been born in Cornwall. Families seek a better quality of life. Many purchase a flat – a second home – in cities such as London, and move their main home to Cornwall. The main wage earner then commutes at weekends between the second home and Cornwall.

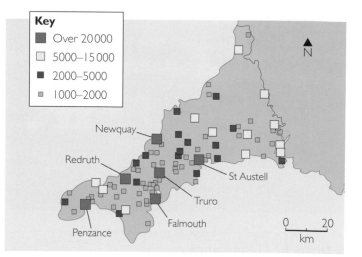

Figure 5.71 *The main centres of population in Cornwall* ▲

Figure 5.72 *Cornwall's natural decrease in population, 1980-2008* ▼

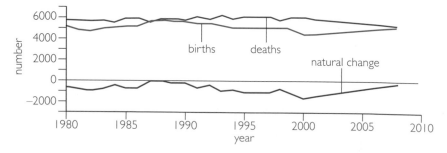

- **Out-migration** by 16-29 year olds. With declining numbers of full-time jobs, and a seasonal tourist economy, over half of those aged 16-18 expect to leave the county. Known as Cornwall's 'brain drain', many young people go away to university and stay away permanently. The opening of Combined Universities of Cornwall in 2004 has helped to reduce this 'brain drain' a bit.

These three trends have combined to increase Cornwall's dependency ratio. The reduced size of the working population, compared with dependent age groups (Figure 5.74), leaves the county with reduced economic potential. This, in turn, affects investment. The emigration of young people (who are high potential earners), in exchange for older age groups on fixed pensions, creates a high proportion of people on low incomes. A quarter of pensioners in rural areas in the UK have low incomes, supplemented by benefits. This results in low spending power in the population, which in turn limits business opportunities. Without extra investment from central government, this leaves local councils with low incomes from council and business taxes. Like Newham in Section 5.7, a negative multiplier exists in Cornwall that is difficult to break (see page 192).

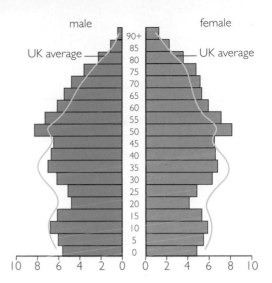

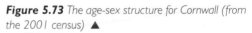

Figure 5.73 The age-sex structure for Cornwall (from the 2001 census) ▲

Figure 5.75 Cornwall's remoteness from the UK's economic core of major population centres ▼

	0-14	15-64	65+	Total of all ages	15-64 (%)
1971	80 400	231 900	66 900	379 200	61
2001	86 200	312 500	100 300	499 000	63
2028	89 800	353 800	177 400	621 000	57

Figure 5.74 Predicted changes to age groups in Cornwall ▲

Economic deprivation in Cornwall

Cornwall has the lowest average annual income per person in England and Wales. The average full-time income in 2008 was £21 522 (25% below the UK average). And this gap is getting wider. Three main reasons explain this – remoteness, decline of traditional jobs, and tourism.

Cornwall's remoteness

Cornwall is geographically isolated. It also suffers from poor infrastructure, i.e. poor roads and communications:

- There are no motorways in Cornwall.
- Rail services are slow. The fastest trains take 4 hours from mid-Cornwall to London, 5 to Birmingham and 7 to Leeds.
- Cornwall has one airport (Newquay), but fares can be expensive.

As a result, Cornwall finds it hard to attract investment, because transport costs are high, even though wages are low. Figure 5.75 shows how far Cornwall is from the UK's main economic 'core', where most goods and services are produced.

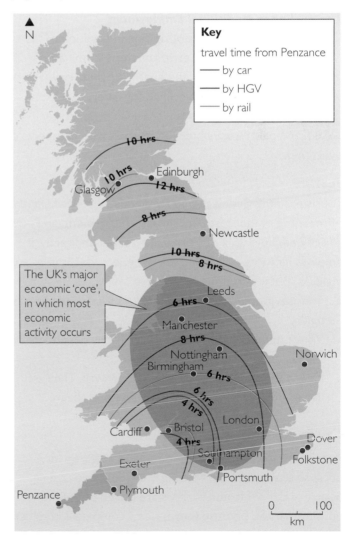

Figure 5.76 A china clay quarry near St Austell, Cornwall ▲

The decline in traditional employment

Cornwall has suffered a major decline in traditional jobs. Until the 1980s, **primary employment** dominated – farming, fishing, tin mining, and quarrying china clay (Figure 5.76). But each sector has declined for different reasons (Figure 5.77). Those lost jobs were year-round, often skilled, and paid reasonable wages (often with prospects of overtime).

Tourism

Tourism is now Cornwall's biggest industry. It has helped to offset some of the losses in primary employment, and now directly employs 25% of Cornwall's people. Indirectly, it employs far more – including shopkeepers, decorators and builders. But it also brings problems:

- Tourism jobs are mainly seasonal, part-time and poorly paid.
- Visitor numbers vary, and they are highly dependent on the weather.
- Only a third of the profits stay in Cornwall. The rest 'leaks' out, e.g. via national pub and hotel chains.

Cornwall's housing problem

Another reason why younger people leave Cornwall is because so much of the housing is unaffordable. Young people and low-income earners are priced out of the housing market, because they have to compete with older migrants into the county, who are often cash-rich. The

Sector	Reasons for its decline
Farming	• Falling farm prices, as supermarkets seek the lowest prices from their suppliers. • Importing food from overseas, where wages and costs are lower. For example, milk from the new EU countries costs 14p a litre to produce, while British farmers can only produce milk for 21p a litre. • Withdrawing EU subsidies, which led to a rapid and accelerating decline.
Fishing	• EU quotas allocated fish supplies to other European countries. • A decline in overall fish stocks, caused by previous over-fishing.
Mining	• The exhaustion of Cornish tin reserves. • A collapse in tin prices, caused by overseas competition. • The strong pound made UK tin more expensive to buy overseas.
Quarrying	• The St Austell area has some of the world's best china clay reserves. However, fewer and larger quarries (using technology rather than people to extract the clay) resulted in cutbacks in the workforce.

Figure 5.77 The decline in Cornwall's primary employment ▲

Joseph Rowntree Foundation lists three of Cornwall's districts in its 40 most unaffordable areas in the UK. In 2010, an average home in Cornwall cost eight times the average Cornish annual income, compared to just six times that across the rest of England and Wales.

Despite this, owner-occupation in Cornwall is high (72% compared to 67% nationally). There is also less social housing (12% compared to 23%). However, the need for social housing in Cornwall is increasing, as more and more local people are unable to buy homes. Two-thirds of those who do not own a home cannot afford one.

The lack of affordable housing has arisen because of the:
- 'right to buy' scheme in the 1980s and 1990s, which allowed those in social housing to buy their homes. This reduced the amount of social housing available.
- purchase of small cottage properties by investors as holiday lets or second homes. 5% of Cornish housing is now used in this way. Although that seems a low total, in some areas the percentage can be much higher – especially in some coastal communities (Figure 5.78). This affects local people looking for somewhere affordable to live. It also influences whether local shops can survive or not.

Figure 5.78 *In coastal communities, such as here in Gorran Haven, almost all traditional cottages are either second homes or holiday lets* ▲

Services and welfare provision

Tourism keeps some services alive in Cornwall, but they are vulnerable. The village store and petrol garage in Gorran, south Cornwall, closed and was redeveloped as housing. Small primary schools are also under threat, and a review of secondary schools concluded that many Cornish schools were too small and uneconomic. Sixth-form provision is mostly provided by Truro College. Few schools in Cornwall have any post-16 provision. Cornwall's size means that some students have to travel three hours to and from Truro every day, and some stay there Monday-Friday. Most hospital care in Cornwall is provided by the largest hospital, in Truro. Access is very difficult for those without cars.

- The decline in rural services is a national issue, e.g. the closure of rural post offices.
- Most rural settlements have neither a general store (78%) nor a village shop (72%).
- 39% of households in rural areas live more than 2 km from an ATM.
- Only 14% of rural parishes have a doctor's surgery.
- 29% of all rural settlements have no bus service.

ageing population The increasing average age of the population, together with an increasing proportion of people of over the age of 65

assisted passages A migration policy used in Australia to encourage young European families to migrate there between the 1950s and 1970s; known as the 'ten pound passage'

birth rate The number of births expressed as a rate per 1000 population in a year

crude rates Measure the basic statistics of any population, such as birth or death rates per 1000

death rate The number of deaths expressed as a rate per 1000 population in a year

demographers People who study population

demographic transition How population characteristics change over time

Demographic Transition Model (DTM) A theory showing how population, food supply, and economic development are linked

dependency ratio The proportion of the population not in work (i.e. children 0-15 and those above 64) who are dependent on those in work. It is normally shown as a percentage

deprivation A low standard of living caused by low income, poor housing and health, and low education qualifications

doubling time The time it takes for a population to double

exponential growth Where growth rates become more and more rapid

fertility rate The average number of children born to a woman over her lifetime, assuming that she survives from birth through to the end of her reproductive life

grey pound The spending power of those who are retired

Gross Domestic Product (GDP) The value of the goods and services produced in a country over a year

host nations Where migrants move to

infant mortality rate The number of deaths of children before their first birthday expressed as a rate per 1000 population

inward migration The process of moving to an area

knowledge economy An economy based on financial, legal, and management and business services, where expertise is 'sold'

life expectancy The expected number of years of life remaining at a given age, usually expressed from birth

median age The middle value of a range of data

migration Movement of people

natural decrease Where population falls as death rates exceed birth rates

negative multiplier Where low spending power caused by low incomes limits economic growth, and may cause decline as demand falls

one-child policy A population policy designed to limit every family to one child

out-migration The process of moving away from an area

population 'explosion' The sustained increase in global population

population density The average number of people per unit area (usually a square kilometre)

population density The number of people per square km

population structure The proportion of people of each sex in each age group

post-production countryside A landscape where leisure and tourism (and not food) earn landowners a living

primary employment Jobs in the production of raw materials or natural products, e.g. farming, fishing, forestry, mining, and quarrying

primary products Raw materials; any goods grown on farms, in forests, or extracted from quarries and mines

Purchasing Power Parity (PPP$) GDP expressed in terms of what per capita income will buy in a country when cost of living is taken into consideration

rate of natural increase The number of people per 1000 by which a population increases or decreases within a year

refined rates Refer to particular changes in a specific population, e.g. whether a local death rate is higher than average

replacement level The number of children needed to maintain a population. This is normally 2.1, to allow for deaths early in life

sheltered accommodation Accommodation designed with the needs of the elderly or less mobile in mind

skills-based migration A migration policy used in Australia to limit migrants to those who are skilled, based on employment, qualifications, age, and English-speaking abilities

soil erosion The removal of fertile soil particles by wind or surface run-off

soil salinity The increase in salt content in soil, beyond which plants cannot grow

source nations Where migrants originate

youthful population Where a high proportion of the population is aged 15 and under

Exam-style questions

1 (a) Study Figure 5.4 (page 161). Briefly explain what the cartoonist is trying to say. (4 marks)

(a) Only a **brief** message is needed.

(b) Explain two reasons for the rapid decrease in global death rates in recent decades. (4 marks)

(b) Focus on two reasons, each explained with two points.

(c) Explain why many low income countries have high fertility rates. (7 marks)

(c) Consider different factors, e.g. poverty, girls' education. Use examples.

(d) Referring to examples, discuss how far low income countries are able to cope with youthful populations. (15 marks)

(d) The command word is 'how far'. Therefore, identify two to three ways in which countries (e.g. Uganda) find it difficult to afford education or health care for children because of debt, and then say how the situation is being improved.

2 (a) Study Figure 5.21 (page 175). Describe the characteristics of the age-sex structure of the UK. (4 marks)

(a) Stick to description only.

(b) Study Figure 5.22 (page 175). Outline one consequence for the UK if the trends shown were to continue. (5 marks)

(b) Look at the trends; decide on one consequence.

(c) Explain why some governments are finding it hard to increase life expectancy in their country. (6 marks)

(c) Consider countries where disease has had a major impact.

(d) Discuss the challenges of an ageing population in one country that you have studied. (15 marks)

(d) Choose the right country – UK? Australia? China? Think about the challenges, e.g. retirement age. Make good use of examples.

3 (a) Study Figure 5.63 (page 196). Describe two features of the data shown. (4 marks)

(b) Explain two reasons for the features you have described. (4 marks)

(c) Explain two other ways in which quality of life can vary across a city. (7 marks)

(d) For any **two** of the following types of area, explain how and why social welfare varies between them:
 - inner city area
 - suburban area
 - rural–urban fringe
 - rural settlement
 (15 marks)

Food supply issues

Working on the land in the Nandi Hills, Kenya

What do you think the woman in the photo is picking?

Do you think she is working on her own farm?

What type of farming is this?

Do food supply issues have anything to do with you?

Have you ever been hungry?

Introduction

There is more than enough food to feed the world's population. In fact, many parts of the world produce far more food than is needed. Why then do so many people go hungry? The problem is really one of distribution, between people and countries. Increasingly, food issues are complex. The world's HICs increasingly purchase food grown in LICs. What impact does this have on the world's poorest farmers?

In this chapter you will learn how food problems can partly be solved by technology, as shown by the Green Revolution of the 1970s and, more recently, the Gene Revolution, as scientists seek ways of growing more food. But who do such developments benefit?

Books, music, and films

Books to read

Stuffed & Starved by Raj Patel

Music to listen to

'Meat is Murder' by The Smiths

'America is not the World' by Morrissey

Films to see

We Feed the World (documentary, 2005)

About the specification

'Food supply issues' is one of the three Human Geography option topics in Unit 1 – you have to study at least one.

This is what you have to study:

- The global patterns of food supply, consumption, and trade. The geopolitics of food.
- Contrasting agricultural food production systems: commercial, subsistence, intensive, extensive, arable, livestock, and mixed farming.
- The management of the food supply, specifically strategies to increase production: the Green Revolution, genetic modification and other high technology approaches, land colonisation, land reform, commercialisation, and appropriate and intermediate technology solutions.
- The management of the food supply, specifically strategies to control the level and nature of food production in the European Union: subsidies, tariffs, intervention pricing, quotas, non-market policies, and environmental stewardship.
- Changes in demand: the growing demand from richer countries for high-value food exports from poorer countries, the all-year demand for seasonal foodstuffs, the increasing demand for organic produce, and moves towards local or regional sourcing of foodstuffs.
- Food supplies in a globalising economy: the role of transnational corporations in food production, processing, and distribution. The environmental aspects of the global trade in foodstuffs.
- The potential for sustainable food supplies.
- Case studies of two contrasting approaches to managing food supply and demand.

In this section you will learn about:
- changing demands for food

Happy Christmas!

When busy British families go to the supermarket for their big Christmas food shop, how many check where their food comes from? Perhaps few realise that much of their Christmas dinner has travelled across the globe to reach them. Figure 6.1 illustrates how a typical British Christmas dinner travels thousands of **food miles** (causing carbon emissions) before arriving on our plates.

The drive for cheaper products – called 'the race to the bottom', for the lowest possible prices – now means that supermarkets seek global suppliers to meet customer demands. For Christmas 2009, British supermarket giant ASDA (part of the US Walmart chain) imported frozen turkeys from Brazil to keep prices low and capture a larger slice of the turkey market. The National Farmers Union in the UK felt that this decision betrayed British farmers.

- British turkey production fell from a peak of 49 million turkeys in 1999, to just 15 million by 2009.
- 'Turkey miles' – the average distance travelled by this food item – increased from a few hundred to nearly 6000 in the same period.
- Low transport and refrigeration costs made it cheaper to transport turkeys over long distances than to produce them at home.

Figure 6.1 *Food miles for a Christmas dinner in London* ▼

Item	Miles travelled	Country	Could it be produced in the UK?
Prawns	5342	Honduras	Yes – Scotland
Avocados	5501	Mexico	No
Smoked salmon	4487	Alaska	Yes – Scotland
Potatoes	463	Scotland (UK)	
Carrots	134	Nottinghamshire (UK)	
Brussels sprouts	168	Lincolnshire (UK)	
Green beans	4228	Kenya	Yes – East Anglia
Mange tout	6312	Peru	Yes – Cornwall
Cranberry sauce	3284	Massachusetts (USA)	No
Goose fat	506	Dordogne (France)	Yes – Norfolk
Turkey	5450	Brazil	Yes – Norfolk
Bacon and pork sausages	596	Denmark	Yes – Suffolk/Humberside
Brazil nuts	6205	Bolivia	No
Stilton cheese	126	Colston Bassett (UK)	
Brandy butter	513	France	Yes
Dried fruit for Christmas pudding	12 427	California	No

Food miles or fair miles?

The journey taken by food from farm to plate is complex, and involves more than just food miles and carbon emissions. The globalised food industry affects livelihoods and environments throughout the world (Figures 6.2 and 6.3).

Cash crops like green beans (from Kenya), mange tout (Peru) and Brazil nuts (Bolivia) all help to sustain the economies of many of the least developed countries – and the lives of many of the world's rural poor. Buying their produce can mean the difference between survival and starvation for many small farmers. The term **fair miles** is used when considering this aspect of the global food trade.

Figure 6.2 *Green beans and other cash crops for export are more valuable to Kenya's economy than tourism* ▲

Figure 6.3 *Huge oil palm plantations are replacing rainforests in places like Malaysia and Indonesia – palm oil is used as a key ingredient in many foods* ▲

Many cash crops for export are grown by large landowners, who buy up the best land – forcing smallholders to farm on less-fertile marginal land, like deforested slopes, where crops can fail and the environment can be damaged.

Did you know?

Most food imported by developed countries has a low carbon footprint, because it's imported by sea. Food imports count for just 10% of the emissions created by the UK's food chain. The energy used to transport food from overseas is often less than that used to produce the same food under heated glass in Europe during the winter.

Changing Chinese diets

The UK is not the only country where eating habits are changing. China's rapid economic growth has led to greater urban populations, with greater spending power. Popular meals there now involve more meat. China's meat consumption has risen by 150% since 1980, while grain consumption has fallen dramatically.

Ten years ago, Zhang swapped his rural home for the grimy suburbs of Beijing. Once a farmer, he no longer grows his own food – he buys it. He was often hungry during his poor rural childhood, but can now afford meat every day.

The trend towards eating more meat is being repeated across China, and it affects global food prices. It also causes hunger for the world's poor, as grain is diverted from feeding people to fattening-up animals.

Figure 6.4 *Adapted from an article by Jonathan Watts in The Guardian in May 2008* ▲

Western suppliers claim that this will affect world markets. 'This is the end of self-sufficiency for China', says James Rice, chief of Tyson Foods (the world's biggest meat producer). '2008 will be the last year that China produces enough corn for itself, and the last that it's self-sufficient in protein. When China goes from being a net exporter to a net importer of anything, it has a big impact on global prices.'

Did you know?

It takes 8 kg of grain to produce 1 kg of beef.

ACTIVITIES

1 Use Figure 6.1 and an outline map of the world to show the origins of a typical British Christmas dinner.

2 a In pairs, discuss and draw up a spider diagram entitled 'Impacts on British farmers of falling supermarket prices'. Then complete a second diagram entitled 'Impacts on British consumers of falling supermarket prices'.

   b How far are the interests of British farmers and consumers the same?

3 In 300 words, explain why the concept of 'fair miles' can be as useful as that of 'food miles' in persuading people to think about what they buy.

4 Briefly explain:

   a how economic development is altering the diets of urban Chinese families

   b how these changes affect global food supplies

   c how they affect people in the UK

Hunger in a world of plenty

In this section you will learn about:
- the global distribution of food and hunger
- how food is produced under different farming systems

Skills

In this section you will:
- interpret patterns and trends from data

Counting calories

Where we live determines what we eat. Globally, the highest **calorie intake** occurs in High Income Countries (HICs). The average American consumes 3826 calories a day, whereas the people of Haiti, Eritrea, Somalia and Burundi consume less than half that (Figure 6.5). Americans consume 815 billion calories *a day* – 25% more than the level recommended by the World Health Organisation (WHO), and enough to feed an extra 80 million people.

As well as eating more than is recommended, richer countries also waste more. Each year, 40-50% of the US harvest is not eaten (30-40% of the UK's food is also thrown away). Enough food is produced globally for everyone to consume as many calories as the Japanese. Yet, in 2009:
- over nine million people died of hunger or malnutrition (one child dies every six seconds from a hunger-related cause)
- over a billion people suffered from a shortage of protein or calories (**under nourishment**)
- two billion more lacked adequate minerals and vitamins (**under nutrition**).

Figure 6.5 *Average calorie consumption per person per day* ▼

The big problem is not the quantity of food available globally, but its distribution and affordability. Advances in farming have more than kept pace with population growth. In 2009, 1.2 billion people worldwide suffered from **obesity**, while at the same time 1 billion were suffering from hunger. Even many of the obese people are low-income earners, with imbalanced diets consisting largely of cheap carbohydrates.

Figures 6.6 and 6.7 show the distribution of hungry people across the world:
- In percentages, the worst affected countries are in sub-Saharan Africa.
- In numbers, nearly two-thirds of hungry people live in Asia (because of the huge populations of many Asian countries, compared to those in Africa).

Did you know?

World agriculture produces an average of 17% more calories per person today than it did 30 years ago, despite a 70% increase in global population.

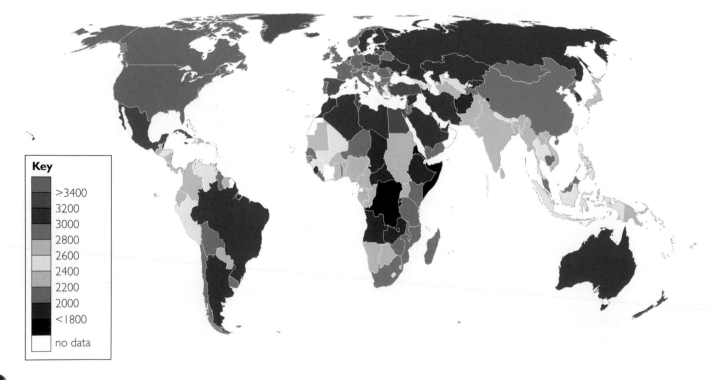

Key
>3400
3200
3000
2800
2600
2400
2200
2000
<1800
no data

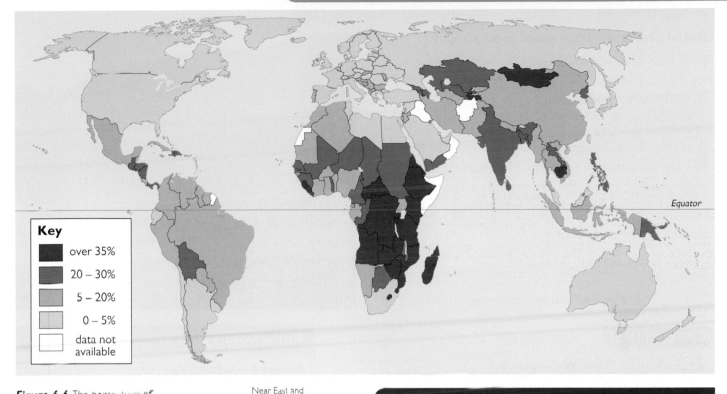

Figure 6.6 *The percentage of people suffering from hunger by country* ▲

Figure 6.7 *The distribution of the 1 billion hungry people by world region (the numbers represent millions)* ▶

Key
- over 35%
- 20 – 30%
- 5 – 20%
- 0 – 5%
- data not available

developed countries
15

Near East and North Africa
42

Latin America and the Caribbean
53

Asia and the Pacific
642

sub-Saharan Africa
265

Missed targets

Today, 20% of the world's population remains short of food. This is rarely because it's unavailable, but because the poorest can't afford to buy it. Wealth determines a person's diet, and the gap between wealthy and poor remains vast.

- In 1996, 900 million people worldwide were classified as suffering from hunger.
- In 2000, the UN's Millennium Development Goals (MDGs) set a target to 'halve the number of **chronically undernourished** people worldwide by 2015'.
- By 2010, the global economic crisis (and two years of food price rises) had derailed this commitment. By 2009, the number of hungry people worldwide had reached 1 billion, making the MDG target for 2015 a mere dream.
- The 2009 UN Progress Report on the MDGs concluded that hunger now affects 17% of the population of Low Income Countries (LICs), compared to 16% in 2006.

The 'haves' and 'have-nots'

Having enough farmland to be self-sufficient is a real asset for a country. But some have to import food, while others produce surpluses. Advances in food processing and storage technology have allowed greater distances between places of production and consumption. As a result, countries can be divided into three broad groups:

- The 'haves' – countries with enough fertile farmland to provide the food they need, plus a surplus (e.g. Europe, the USA, Australia, New Zealand)
- The 'have-nots who can' – countries with limited good farmland, but enough money to buy from the 'haves' (e.g. Singapore, Japan, the Middle East, and increasingly China and Indonesia)
- The 'have-nots who cannot' – countries without enough good farmland or technology to grow sufficient to feed themselves, and who are too poor to import food (most LICs; 50% of the world).

LICs are less and less likely to feed their own people. Increasingly, wealthy landowners in those countries grow cash crops for export, instead of renting land to subsistence farmers. Land that once supported local communities now feeds the world's HICs, and the poor have less chance of feeding themselves. This leads to scarcity in crops like rice and maize – causing higher prices and hunger.

Global shifts – patterns of food supply

There are over a billion farmers in the world. In theory, they produce enough food to feed everyone, but what they farm and how they do it varies hugely.

Subsistence farming

Up to 70% of the populations of LICs are smallholders. Many are **subsistence farmers**. Most subsistence farmers have plots less than five hectares in size (Figure 6.8). Much of their land is farmed **intensively** – maximising yields by using every square metre, maximising labour input, and planting and harvesting all year round (which Uganda's equatorial climate allows).

Figure 6.8 Intensive subsistence farming in Uganda ▶

CASE STUDY

Extensive soy farming in Argentina

Traditionally, Argentina has been one of the world's main cattle countries. But because global soy prices have risen sharply in recent years, it has now switched to **arable farming**. Soy is now its main export. 98% is either exported to Asia to make flour for human consumption, or to Europe as animal feed. Soy is now grown on 80% of Argentina's cultivable lands. Most production is owned by large agri-businesses, which have bought up small independent farms. This has had widespread effects:

- 17 000 dairy farms have closed in Buenos Aires Province alone.
- 500 small towns have been abandoned, as farms have closed and jobs have been lost.
- Milk and vegetables are now imported, because soy is more profitable.
- The price of many foods has risen, because of shortages.

Know your types of farming!

Subsistence farming – growing enough to feed a household, with no surplus

Pastoral farming – grazing animals, e.g. cattle or sheep

Arable farming – growing crops

Mixed farming – growing crops and keeping livestock

Commercial farming – growing produce for sale

Intensive farming – maximising return from the land using high inputs of labour, fertiliser, machinery, and capital

Extensive farming – using low inputs of labour, machinery, and capital

Specialised farming – the intensive production of one crop (also known as monoculture)

CASE STUDY

Land reform in Botswana

Botswana is a landlocked country in southern Africa, much of which is semi-desert. Only 5% is suitable for arable farming. Most Botswanans have traditionally been subsistence livestock farmers – using communal grazing lands open to everyone. Cattle are symbols of wealth and status there, and 40% of the population owns three million cattle – half of which are reared on communal lands. This situation suits the semi-arid climate well, because cattle numbers fall in dry years and rise in wet years.

However, recent **land reform** by the Botswanan government has privatised communal lands – allowing wealthy buyers to create large cattle ranches for commercial livestock farming. The government did this to make the land more productive, because it signed a deal in 2009 to gain access to EU markets for Botswanan beef. As a result, Botswana now has to produce more beef to satisfy the EU demand. And landowners earn more by keeping more cattle. However, this is putting pressure on the environment, because the land has to meet cattle production targets even during droughts.

CASE STUDY

Commercial farming in HICs

By contrast with LICs, less than 3% of the labour force in HICs works in agriculture. Increasingly, farms are growing larger in scale, and are known as **agri-businesses**. They are owned and managed by major companies that organise all inputs, processing, and marketing. They farm **intensively** – maximising yields by investing in inputs like machinery and fertiliser, and increasing profitability by reducing labour costs (Figure 6.9). Most use contractors, who move between farms doing jobs like ploughing or harvesting, using highly technical machinery. As a result, most large farms now employ few permanent workers of their own.

These methods allow countries like Australia and New Zealand to become major food or wine producers – exporting to the USA, EU, China and India. Like farmers in the UK, they seek to produce more every year from the same amount of land. However, artificial nitrogen fertilisers (on which **commercial farming** relies) are becoming more expensive as world oil prices rise, and irrigation water is also becoming more expensive.

Figure 6.9 Agri-businesses work by maximising investment in machinery ▲

Internet research

Use the Internet to research and produce a presentation comparing how the 'inputs' in Figure 6.10 determine farming patterns in Botswana and Argentina. Type the following terms into Google to get you started:

- Botswana Ministry of Agriculture
- Land reforms in Botswana
- Soy farming in Argentina
- World Resources Institute (select from country profiles under food and agriculture)
- CIA World Factbook (select country profiles)

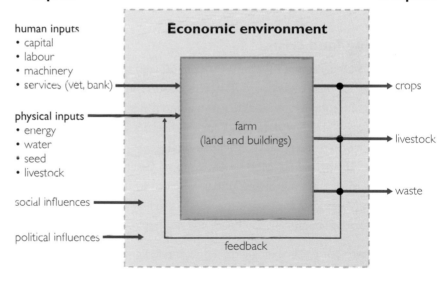

Figure 6.10 Farming as a system ▲

ACTIVITIES

1 Study Figures 6.5-6.7.
   a Identify those countries and regions with the highest and lowest daily calorie intakes.
   b Suggest reasons for the patterns you identified.
   c Is there a perfect match between patterns of hunger and daily calorie intake? Explain your answer.

2 Define the following terms in your own words: under nourishment, under nutrition, subsistence farming, commercial farming, intensive and extensive farming, agri-business.

3 Make three copies of Figure 6.10 – one each for 'Soy farming in Argentina', 'Cattle farming in Botswana', and 'Commercial farming in HICs'. Then annotate your copies to explain how food is produced in each system.

In this section you will learn about:
- changing global demands for food
- the emerging threats to global food security
- who controls food supply globally

Malthus – right or wrong?

In 1798, Thomas Malthus (see pages 160 and 162) predicted that population growth (demand) would outstrip the Earth's ability to feed everyone (supply) – Figure 6.11. Malthus thought that the inevitable result of this imbalance would be 'natural checks', such as famine, disease – and even war – which would bring population levels down again.

However, the increasing use of technology in food production (e.g. chemical fertilizers, pesticides and herbicides, and the development of GM crops) seemed to prove Malthus wrong. But, at the beginning of the twenty-first century, Malthus's theory has risen again. Some people now believe that growing levels of hunger and poverty, particularly in LICs, support Malthus's ideas. So, was Malthus right? Is global population growth the reason for hunger and poverty? Or are there other factors in play?

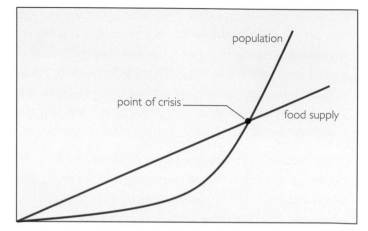

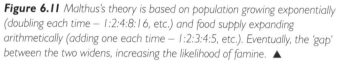

Figure 6.11 *Malthus's theory is based on population growing exponentially (doubling each time – 1:2:4:8:16, etc.) and food supply expanding arithmetically (adding one each time – 1:2:3:4:5, etc.). Eventually, the 'gap' between the two widens, increasing the likelihood of famine.* ▲

An emerging food crisis

In 2007, the US Department of Agriculture announced that world grain stocks were at a record low. The USA is the world's main grain supplier, so people listen when it says that there is a problem. The data speak for themselves – in 1990, the world had a reserve stockpile of 115 days' of grain, but by 2007 this had fallen to just 53 days. Grain output had failed to meet a growing global demand, so the risk of shortages was real. Not since the 1960s had fears about global **food security** hit the headlines like this.

Part of the reason for the growing global demand for grain is its increasing use as livestock feed. This is because, as incomes rise, the world's population is now consuming more and more meat, especially in rapidly developing countries like India and China (see page 211). This is known as the **nutrition transition**. The other main explanation given for the increasing demand has been basic population growth. The equivalent of the entire US population is added to the world every six years, so this has an obvious impact on demand.

Carrying capacity

Theoretically, the natural environment limits the amount of food that can be produced. The fertility of the soil, local climate, and landscape, all present opportunities and challenges for agriculture. Farmers are free to intensify their use of the land to increase yields. But the land's natural **carrying capacity** will be exceeded if each farmer takes more than nature can provide – without replacing what is lost.

If farmers add fertilizers, and install irrigation and drainage systems, the land's carrying capacity can be increased. For a while everyone gains, so intensification continues until the increased yields no longer cover the costs of the extra inputs. From then on, the law of diminishing returns occurs, i.e. crop yields fall as the natural environment degrades through over-use. This is an example of the world 'living beyond its environmental means'.

Living beyond our environmental means?

People in wealthy countries have become used to getting any food they want, from anywhere in the world, at any time of year. Global supply lines mean that geographical and seasonal boundaries can be crossed – strawberries at Christmas? No problem! Increasingly, many LICs – like Kenya (page 210) – are using scarce land and water resources to grow crops for export, to support their economies, rather than subsistence crops to feed their people. Meanwhile, even though it exports many foodstuffs, the UK now depends on the rest of the world to live beyond its environmental means. The veins on Figure 6.12 show food flowing into the UK from every corner of the globe – grapes from South Africa, apples from Chile, palm oil from Indonesia, etc.

Figure 6.12 *The veins of the global food economy* ▲

The impacts on marginal lands

In many cases, demands for food from HICs have led to a process known as **marginalisation** within LICs. Subsistence farmers are forced onto land that will not support their needs. For instance, throughout the semi-arid Sahel region (Figure 6.13), many countries have devoted their best farmland to growing cash crops for export to richer countries. As a result, poor subsistence farmers – who are just trying to grow enough food to feed their families – are forced to use land that would normally be left alone. Fragile soils and a lack of reliable rains mean that crops often fail, or do not provide enough food.

Some LICs are now going even further. Instead of growing cash crops themselves, they are leasing their land directly to foreign companies and governments:

- Karuturi, an Indian farming company, has leased over 300 000 hectares of 'under-used' land in Ethiopia. It is part of an Ethiopian government project to allow 3 million hectares of land to fall into overseas control (Figure 6.14). By doing this, the Ethiopian government hopes that large-scale commercial farming can be introduced into Ethiopia to replace subsistence farming.
- The oil-rich (but farmland-poor) Arab state of Abu Dhabi has leased 30 000 hectares of farmland in Sudan, plus land in Uzbekistan and Senegal, to grow crops for direct export back to Abu Dhabi.

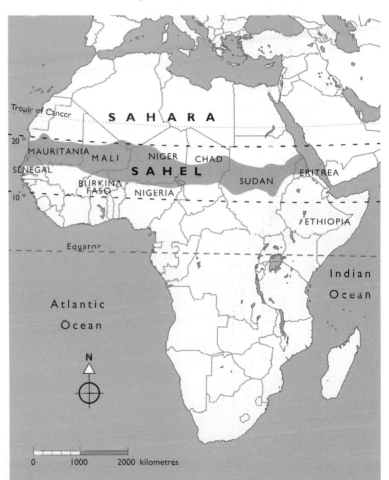

Figure 6.13 *The Sahel* ▲

Figure 6.14 *The Ethiopian government has invited international companies to introduce large-scale commercial agriculture to the Gambella region* ▶

Growing food to pay off debt

In the mid-twentieth century, the fear of food shortages in LICs led international governments and aid agencies to promote agricultural reforms there. The LICs looked to the West – and to a 'technological fix' – to deal with their food problems. Many adopted western techniques that were oil-dependent, e.g. tractors instead of ox ploughs.

To pay for these new techniques, many LICs used precious farmland to grow cash crops for export (such as coffee, beans and cotton), instead of food for local consumption. The income from these cash crops paid for imported fertilizers and machinery. Many LICs also took out cheap loans from international banks (at 5% interest on average) to help pay for the new machinery and fuel.

However, when global interest rates soared to 15% or more in the 1980s, the loan repayments became unaffordable and left many LICs crippled with debt. Global oil prices also rose – handing LICs a further 'double whammy' of increased transport and fertilizer costs.

Many LICs felt that the only way to manage their debts, and to cover their increasing costs, was to grow even more cash crops for export. So yet more precious farmland was taken over for commercial production – and poorer farmers were forced from the best land onto marginal land. As a result, many LICs became even less able to feed their own people than they had been decades before. Bizarrely, many also became more and more reliant on food aid from the USA and the EU to survive.

CASE STUDY

Tanzania's cotton crops

To help deal with its debts and develop its economy, Tanzania turned more land over to grow cotton instead of food. By 2010, it had 500 000 cotton farmers and produced 120 000 tonnes of cotton a year. It is now Africa's fifth largest cotton producer. But this makes it vulnerable to changes in global prices. It suffered badly when cotton prices fell by 28% in the aftermath of the global financial crisis of 2007 onwards. Its cotton farmers could not earn enough money to make a living, and there was no available land for them to grow their own food instead.

Figure 6.15 *The cotton fields of Tanzania – land taken out of domestic production to pay off debts* ▲

Economic development through trade

Since 1945, the world's leading nations have been trying to free-up supplies of goods, including food – a process known as **trade liberalisation**. Their ultimate goal is a world based on 'free trade', where all goods can be traded easily and in greater quantities. Theoretically, this should help to ease global poverty, because LICs would gain larger markets for their produce.

However, it has taken a long time to reach the present trade agreements – from the General Agreement on Tariffs and Trade (GATT) in 1948, to the creation of the World Trade Organisation in the 1980s. Battles are still being fought between individual geographical regions (such as the EU), countries like the USA, and major food corporations. The LICs often have little say in any trade agreements that could help them to increase their foreign earnings and enhance their economic development.

In 2005, ActionAid produced a report showing that the world's food trade is in the hands of major TNCs (transnational corporations). The report said that:

- Nestlé, Monsanto, Unilever, Tesco, Walmart, Bayer, and Cargill have all expanded hugely in size, power and influence because of policies promoted by the USA, UK and other G8 countries.
- the annual turnover of any one of these companies is often larger than the GDP of countries with whom they trade.
- many TNCs squeeze local companies in LICs to drive down prices, and also impose tough quality standards that poor farmers can rarely meet.
- many TNCs lobby (or attempt to influence) politicians in their home countries to set trade rules to suit themselves.

Did you know?

- Just five companies control 75% of the world's traded grain.
- Nestle's profits are bigger than Ghana's whole economy.

Company Name	Nationality	Example brand names include
Nestle	Swiss	Nescafe, Evian, Carnation, Moca, Buitoni, Haagen Dazs, After Eight
Cargill Inc.	American	
Kraft Foods Inc.	American	
Unilever	Anglo-Dutch	
Tyson Foods Inc.	American	
General Mills, Inc	American	
Groupe Danone	French	
Kellogg	American	
ConAgra Foods	American	

Figure 6.16 *The world's main food companies* ▲

Internet research

Compile a factfile based on current international reports into global food, using these websites to help:

- World Trade Organization www.wto.org/ and follow the links to trade topics
- Food and Agriculture Organization of the United Nations http://www.fao.org/
- World Resources Institute http://earthtrends.wri.org/
- Global Issues: http://www.globalissues.org and follow the links to Food and Agriculture

ACTIVITIES

1 Define the following terms: nutrition transition, carrying capacity, marginalisation, trade liberalisation.
2 In pairs, discuss and complete a table to show 'Ways in which Malthus was right' and 'Ways in which Malthus was wrong' about global food supply.
3 Draw a spider diagram to show why the decision to expand the production of cash crops is often a mixed blessing for LICs.

4 How might the TNCs listed in Figure 6.16 argue that they bring advantages to LICs when they trade with them?
5 Make a copy of Figure 6.16. In small groups, research common brand names for each of the companies shown, and complete your table.

In this section you will learn about:
- how 2008 changed global food security and affected the poor
- how Haiti and Cuba manage their food supply and demand

Warnings for the future

2008 was a turbulent year. Some normally fairly stable parts of the world experienced discontent and violence – and it was all related to food supply (Figure 6.17). In some countries:

- food rationing was introduced
- there were demonstrations and protests about rising food prices and growing hunger
- elected leaders were thrown out of power.

Even more seriously, there were many deaths in major food riots in Cameroon and Haiti (page 223).

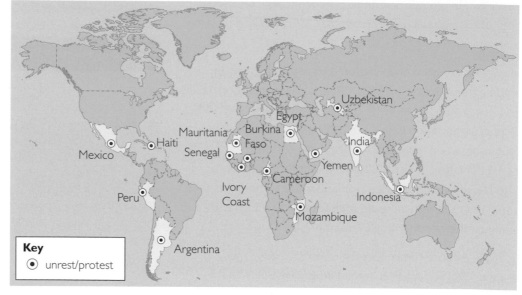

Figure 6.17 *Global food crisis flashpoints by April 2008, and increasing grain prices, 2005-8* ▲ ▶

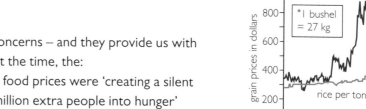

The events of 2008 were all linked by food security concerns – and they provide us with some early warnings of the world's possible future. At the time, the:

- World Food Programme (WFP) claimed that high food prices were 'creating a silent tsunami that threatens to plunge more than 100 million extra people into hunger'
- FAO reported a 45% increase in food prices over a nine-month period
- World Bank recorded food price rises of 83% between 2005 and 2008 – raising fears of real and lasting food insecurity worldwide.

But, in reality, the seeds of this insecurity were sown much earlier.

Figure 6.18 *Subsidised government rice being sold to the poor in Manila under army supervision* ▼

The challenge of food insecurity

Over many decades, rapid global population growth has increased the demand for food – putting upward pressure on prices. People in LICs are always hardest hit by rising food prices, because they spend most of their income on food and water (Figure 6.18). Food rarely accounts for more than a third of the income of a person living in an HIC. The devastating effects of hunger on the most vulnerable are often the result of decisions made elsewhere in the world, especially trade and economic decisions.

Ten reasons why 2008 was a year of food crises

In 2008, ten important reasons combined to make the global food security situation much worse:

1 Rapid economic growth and rising living standards in south and south-east Asia increased their demand for food.

2 Changing diets, particularly in Asia, diverted grain from human consumption to feed livestock. Like China (page 211), India's meat consumption rose sharply between 1993 and 2008.

3 The amount of fertile arable land had declined, because of drought, salinity, and climate change across Africa, Indonesia, Australia, and South America.

4 Unstable weather patterns in 2007 destroyed crops. Grain harvests in 2007-8 were down by 60% in Australia (due to drought), and 10% in China and the UK (floods).

5 The World Bank's free-trade/open-market policies had altered farming patterns and encouraged LICs to produce cash crops like coffee, instead of grain and staple foods.

6 Oil prices rose by 600% between 2002 and 2008, due to increased demand. This forced up the prices of fertilizer, food processing and transport, which then raised food prices.

7 The increasing use of crops like corn to make bio-fuel, instead of food, raised corn prices and reduced the amount of land available to grow food crops.

8 Civil wars forced people in Sudan, Ethiopia and Eritrea into marginal areas, where crops often failed.

9 Commodity traders in London and other financial centres bought up stocks of crucial foodstuffs (like grain) on the 'futures market', and held on to them until prices rose and they could sell them at a profit.

10 Share markets collapsed during 2007-08, so speculators and investment banks had to look elsewhere to make money – and food crops were an easy target.

Food insecurities in the twenty-first century

A dangerous political situation has emerged, where individual governments are acting to ensure that they have food for their own people, at the expense of others. Trade agreements between nations are popular, but the direct leasing of land in other countries (see pages 217 and 228) is new:

- Libya (which imports 90% of its grain) has leased 250 000 acres of land in Ukraine to grow wheat, in exchange for oil.

- Egypt is also trying to secure land in the Ukraine, in exchange for natural gas.

- China has adopted the policy of 'farming abroad'. It wants 2.5 million acres in the Philippines (10% of the countryside), and is trying to forge agreements with African countries, Australia, Brazil and Russia.

Environmental insecurities

Deteriorating environmental conditions – falling water tables, eroding soils, and climate change – are all making food security worse:

- Food production depends on fertile soils and reliable climatic conditions for a stable growing season.

- Water shortages lead directly to food shortages. The main grain producers – China, India and the USA – all extract water faster than it can be replenished naturally.

- Saudi Arabia has decided to stop growing cereals completely by 2016, in order to preserve its water table. It will now need to import 14 million tonnes of wheat, rice, corn and barley a year instead.

Haiti suffers from regular environmental stresses (pages 222-224).

Haiti on the brink

In 2007, many Haitians were resorting to eating 'cookies' made from clay, salt and shortening (Figure 6.19). Something had obviously gone badly wrong with Haiti's food security. But what?

In 1981, Haiti imported just 18000 tons of rice, but by 2010 this had risen to 400000 tons. Less than a quarter of the rice consumed in Haiti is now grown there. But with a population of 9.2 million – and an annual population growth rate of 1.84% – Haiti's demand for food has been increasing rapidly. Unfortunately, huge debts (and the need to grow export crops of coffee, mangoes and tobacco to repay them) mean that Haiti now struggles to afford enough imported rice to feed everyone (Figure 6.20).

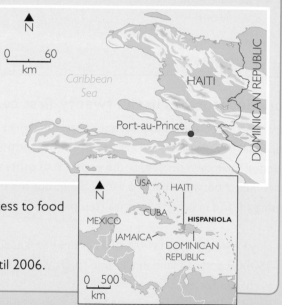

Figure 6.19 Dried 'mud cookies' being collected for sale in the capital, Port-au-Prince, in November 2007 ▲

Figure 6.20 Haiti factfile ▼

- Haiti is the poorest country in the Western Hemisphere, with an average annual GDP per person of US$1300 in 2009.
- 80% of Haitians live below the poverty line, and 54% live in abject poverty.
- 66% of the population are farmers – mostly subsistence smallholders.
- Food production in Haiti fell by 30% between 1991 and 2002.
- In 2004, an FAO report said that 60% of rural Haitians went without food on a regular basis – and 20% did not have access to food at all.
- There is a high infant mortality rate (60 for every 1000 births)
- Haiti did not have its first democratically elected government until 2006.

Why does Haiti have so little food? A number of factors combine to answer that question. Some of them are long-term and date back to Haiti's colonial era – it was first ruled by Spain, and then by France. Other factors are more recent. Some are human and some are natural.

Human factors

Colonialism has a lot to answer for. A combination of colonial trade and inappropriate farming techniques led to the island's major asset – its fertile soil – being washed away. Haiti's extensive hillside forests were cleared for farming, but heavy rainfall from frequent tropical storms caused major soil erosion on the steep slopes (Figure 6.21).

Figure 6.21 Haiti's deforested landscape today ▼

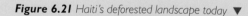

- Haiti's landscape was first put at risk by the Spanish decision to clear many forests and convert the land into sugar plantations.
- The French, who took over from the Spanish, then developed lucrative plantations of coffee, mangoes and tobacco.
- The Spanish and French colonial rulers imported over half a million slaves to Haiti. The black population, although in the majority, has always been the poorest.
- After independence in 1804 – and in huge debt to France – there was a gradual process of shifting poor Haitians to the hillsides, to allow wealthy farmers access to the fertile valleys.
- The poorest farmers were never able to grow enough food to sell, and had to resort to cutting more trees for fuelwood to supplement their incomes.
- A growing charcoal industry also demanded more forest clearance. By 2010, just 4% of Haiti's forests remained.
- In many places, the once-fertile soil has been eroded down to the bedrock, and food production has fallen as a result.

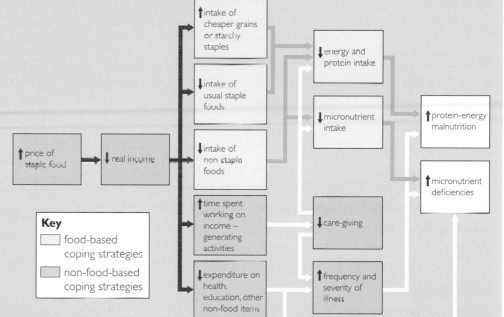

Figure 6.22 *The impacts of food price rises on a typical Haitian family in 2008* ▲

Haiti's role in today's globalised economy

Huge debts have forced Haiti to support itself by growing cash crops – particularly mangoes – rather than rice. It seemed to make economic sense for Haiti to use the profits from selling mangoes abroad to pay for imported rice instead. But massive rises in the price of rice in 2008 (Figure 6.17 graph) changed all that. There was food in Haiti, but it was too expensive for most Haitians to afford. In 2008, Haitian farmers simply could not sell enough mangoes to pay for the expensive imported rice (Figure 6.22).

Ironically, if Haitian farmers had had the opportunity to grow food directly for themselves, they would not have been so badly hit by the 2008 global price rises, which in turn might have prevented the food riots (Figure 6.23).

Figure 6.23 *There were serious food riots in Haiti in 2008* ▼

Food riots in Haiti

There is real desperation on the streets of Port-au-Prince. The food crisis has already claimed lives – with four people dead in food riots – not to mention the many nameless, faceless Haitian children whose deaths have either been caused or hastened by malnutrition. The democratically elected Haitian government has been another victim – brought down by the food riots. Hard-won and long-awaited democratic reforms in this troubled country may also be a casualty of the food crisis.

Natural factors

Cruel weather

Haiti's eroding soils are continuing to undermine the country's food production. The trend towards growing cash crops for export – rather than food for Haitians to eat – is making the problem worse. But, in addition to all that, Haiti's crops are often devastated by torrential rain and flooding. This is because Haiti is located right in the main path of tropical hurricanes and major storms during the hurricane season. For instance:

- In 1998, Hurricane Georges killed 400 and destroyed 80% of Haiti's crops.
- In 2004, tropical storms Jeanne and Gordon killed thousands.
- In 2008, Hurricanes Fay, Gustav, Hanna and Ike killed 793, destroyed or damaged 100 000 homes, and wiped out 70% of Haiti's crops. Children died from hunger.

Nature delivers these hazardous events, but it is the political and economic structure of Haiti that has caused Haitians themselves to turn these events into disasters. The process of clearing land for cultivation, and cutting down trees for fuel and charcoal, has meant rapid deforestation. In 1980, over 25% of Haiti's forests were still standing. By 2008, just 4% was left. Then, when the tropical storms and heavy rains occur, they wash the fertile soils off the steep slopes and further reduce Haiti's capacity to feed itself.

Cruel geology

Being located close to a tectonic plate boundary adds to Haiti's vulnerability. On 12 January 2010, an earthquake with a magnitude of 7.0 on the Richter Scale struck Haiti. Its epicentre was just outside the capital city of Port-au-Prince. Over 230 000 people died – the same number as those killed in the 2004 tsunami in Indonesia. Survivors faced major food and water shortages. With much of Haiti's infrastructure destroyed, including the main port and roads to the interior, it was difficult for food aid to get to Haiti and then to be distributed (Figure 6.24).

In other parts of the world, far more intense earthquakes have much less impact. But poor building quality in the widespread shantytowns of Port-au-Prince caused the collapse of many buildings, which was the biggest cause of death during the earthquake.

Combined factors

When nature wreaks havoc, it's easy to explain away the accompanying food shortages as the consequence of physical geography. But, in Haiti's case – as with so many other areas of the world – it is the economic situation that really multiplies the miseries. Failing economies, rising inflation, unequal trade patterns, and political incompetence have left Haiti vulnerable to each of nature's events. As long as the population remains impoverished, it is difficult to see how their food insecurity will ever be resolved.

Figure 6.24 *Haitians queue for food after the earthquake in January 2010* ▼

Across the Caribbean to Cuba

Just 90 miles of water separate 'capitalist' Haiti from communist Cuba (Figure 6.20). Both countries share the same risks from natural hazards, but far greater earthquakes have left Cuba far less affected than Haiti in 2010.

Also, unlike Haitians, Cubans are well fed and healthy. Since the 1980s, Cuba has aimed for food self-sufficiency and meeting the FAO's recommended nutrition levels. Cuba's centrally planned 'communist' approach has brought new land into use, developed new technologies to boost output, and made good use of the labour force.

Despite the fact that 80% of Cubans live in urban areas, they are still encouraged to grow food at home. Open spaces have been cultivated, along with patio gardens. Intensive gardens maximize yields of vegetables, and all crops are 'shared' locally. By 2003, about 300000 patios were in use, and 18000 hectares of urban agriculture had helped Cuba to exceed the FAO nutrition levels.

Applying socialist principles of co-operatives and collective responsibility may not be to everyone's liking, but it has ensured food security for Cuba. The State organises – the people produce. Organic farming dominates, because Cuba cannot afford oil-based fertilizers or pesticides, and some crops like potatoes exceed European levels of output.

Figure 6.25 *Cuba's communist government uses food rationing to achieve social equality – the picture shows a typical monthly ration* ▲

Figure 6.26 *Every spare urban space is used to grow food in Cuba* ▼

1 Why are food shortages often likely to cause riots?

2 **a** In pairs, discuss and classify the ten reasons for the 2008 food crises on page 221, using a copy of the table below.

	Short-term	Medium- or long-term
Economic factors		
Social factors		
Environmental factors		

b Decide how many of the reasons were due to (**i**) the global economic 'market', (**ii**) government actions. How many could have been predicted and acted upon in advance?

3 Which do you think is the most significant reason to help explain Haiti's food problems? Justify your answer.

4 In pairs, discuss and list the possible advantages and disadvantages of allowing (**a**) the free economic market, and (**b**) governments, to decide what food should be produced, and how much.

5 Draw a simple sketch of Figure 6.26. Label the inputs and outputs of this approach to food production.

In this section you will learn about:
- the role of transnational corporations in the global food chain

Big players in the global food chain

Invisible giants

Behind the many different brand names for food that are used worldwide, there are a small number of invisible corporate giants. Between them, these giants have a stranglehold over a huge proportion of the global food trade:

- 66% of the global pesticides market, 25% of the seed market – and virtually all genetically modified seeds – are controlled by Astra-Zeneca, DuPont, Monsanto, Novartis and Aventis
- 60% of the McDonald's meals eaten worldwide contain wheat supplied by Cargill.
- 75% of the world's grain trade is controlled by Cargill and Archer Daniels Midland.
- 70% of the world's coffee trade is controlled by Philip Morris, Proctor and Gamble, Nestle and Sara Lee.
- 80% of the world's cocoa trade is controlled by Cargill, ADM and Philip Morris.
- Unilever sells food in 150 countries, using the brand names Brooke Bond, Bestfoods, Birds Eye, Knorr, Hellman's, Lipton, Ben and Jerry's, Levers and SlimFast.

Gentle giants?

Supplying big companies can be good for small growers, because it ensures a reliable market for crops like soy. However, in the past, small growers in places like the Amazon were encouraged by many of the big companies to clear the forest illegally – to increase the cultivated area and their crop yields. The farmers knew that more land under soy production meant higher incomes. But allegations of environmental destruction can harm a big company's public image.

Cargill has a purpose-built port on the Amazon River at Santarem (Figure 6.27), where 90% of the local farmers take their soy crops. From there, Cargill exports the soy to fatten-up chickens. The chickens are then processed by McDonalds into McNuggets.

The Responsible Soy Project was established by The Nature Conservancy in the northern Amazon region in 2007. Working with The Nature Conservancy, Cargill has now set a condition that farmers can only sell soy to them if they promise to reforest denuded land and comply with the Brazilian Forest Code. Cargill is supporting the training of farmers to do this. It's a bold attempt at sustainable development.

> **Did you know?**
>
> £1 in every £8 spent in the UK is spent in Tesco.

> **What is Cargill?**
>
> Cargill is a privately owned company that operates in 67 countries – and has an annual turnover of US$1200 **billion**. It buys up, processes and distributes agricultural produce – especially grain. It also supplies animal feed, fertilizers and finance to farmers.

Figure 6.27 *Cargill's soy exporting facility at Santarem* ▼

How the giants won

The liberalisation of global trade opened up agriculture to big business. Efficiency and profitability in commercial farming needs access to large amounts of the best land, and also to global markets. Specialisation in single commodities leads to intensification and monoculture – with cheaper and higher yields. Small farmers just can't compete with this, so they see their incomes fall. Many small farmers then give up control of their own land to work for big agri-businesses.

Supermarket fix

Most of our food is now produced and processed by a few major companies. So, it's not surprising that nowadays we also buy most of our food from large supermarket chains, rather than individual specialist shops. With their increasing size and dominance in the marketplace, supermarkets have gained greater purchasing power and are able to dictate prices to suppliers (Figure 6.28).

Criticisms of supermarkets usually focus on the ways in which their food is sourced and sold. Large supermarkets, like Tesco, Carrefour and Walmart, search the world in order to deliver the lowest-priced goods to their customers. In the case of fresh food, this means that produce is often shipped or flown across the world to supply out-of-season demand. However, while supermarkets provide suppliers with a guaranteed market, and consumers with wider choices, the origins of the food – and the production methods used – often remain hidden (Figure 6.29).

> I get 378 Rand [£32.50] every two weeks, which is not the minimum wage. I can't afford school fees for my daughter, or go to school functions or buy school uniforms.
>
> *Tawana Fraser, a casual farm worker*

> A [supermarket] buyer … picks up the phone and says X is offering me apples for £1 a carton cheaper. Meet his price or I'll change supplier.
>
> *A farmer and supermarket supplier*

> They also change prices suddenly – £1.49 is the price, then they put it on sale and make it 99p. So they can sell it in bulk and increase their profits.
>
> *Another apple farmer*

> Supermarkets … have all the power in the world, and we have to cut costs as far as we can. We're really at their mercy.
>
> *A South African wine producer*

> One supermarket wanted us to change their grape packaging from open to sealed bags. The new bags were three times as expensive. And productivity in the pack-house went through the floor, because it took workers 20% to 30% longer to seal those bags. But the price stayed exactly the same. That's the way it goes.
>
> *A grape producer in the Western Cape*

Figure 6.28 *Supermarkets exposed – messages from South Africa* ▲

◄ **Figure 6.29** *Women in India who process cashew nuts for British supermarkets are scarred by the corrosive oil produced by the nuts when they are shelled*

Depleting nature's store

The rush to produce palm oil for the global food chain threatens huge damage to the environment. A lot of Indonesia's forests have already been cleared to make way for oil palm plantations (page 211), and now the race is on to save Indonesia's peatlands as well (Figure 6.30). Palm oil is used as an additive in many everyday foods – and in cosmetics. It's also used directly as a cooking oil. The big food companies, along with the supermarkets, are accused of promoting deforestation in the push for higher profits.

In Riau, Indonesia, the deep peaty soils are being drained to keep pace with the global demand for palm oil. This demand is expected to double in 25 years. The peatlands contain a large bank of carbon, and their conversion into oil palm plantations releases carbon dioxide into the atmosphere. According to Greenpeace, destroying these peatlands will have the same environmental effect as the carbon dioxide emissions from all of the world's coal- and gas-fired power stations for five years.

The new land colonisation

The quest for food security in HICs is encouraging a new era of land acquisition and neo-colonialism, particularly in Africa (see page 217). Millions of hectares of land in LICs are falling into the hands of overseas tenants, or owners, from HICs. The fear of food shortages, rising fuel prices, population growth, and climate change make this land an attractive and cheap way of ensuring food security in HICs.

In 2010, John Vidal conducted a special investigation for *The Observer* newspaper. It revealed that 50 million hectares of land (double the size of the UK) has been leased or bought in 20 African countries – for the purpose of intensive farming (Figure 6.31). The crops are mostly exported back to the HICs that now control the land.

Vidal's report used the example of a Saudi billionaire. He has taken out a 99-year lease on 1000 hectares of land in Awassa, Ethiopia, as part of a 500 000-hectare project to grow wheat, rice and vegetables for sale in Saudi Arabia (Figure 6.32). This project will employ 10 000 local workers and turn traditional smallholdings into an agri-business managed from a distance.

Figure 6.30 *State-aided rural land colonisation involves clearing Indonesia's peatlands and allowing peasant farmers to grow oil palms there instead. But this degrades the natural environment and releases carbon dioxide.* ▲

- The South Korean company Daewoo is planning to lease 1.3 million hectares in Madagascar.
- Kenya's Tana River delta is being reclaimed and cultivated by Qatar.
- 2.8 million hectares have been leased in the Democratic Republic of the Congo and 2 million in Zambia.
- South Korea has invested in Sudan.
- Saudi Arabia, Kuwait and Abu Dhabi now lease land throughout Africa.

Figure 6.31 *Land leased in Bako, Ethiopia, by a foreign company to grow oil palms to meet the increasing global demand for palm oil, rather than the local demand for staple foods* ▼

In Awassa, millions of tomatoes, peppers and other vegetables are being grown in 500-metre rows in computer-controlled conditions. Spanish engineers are building the steel structure, Dutch technology minimises water use from two bore-holes, and 1000 women pick and pack 50 tonnes of food a day. Within 24 hours, the food has been driven 200 miles to Addis Ababa and flown 1000 miles to the shops and restaurants of Dubai, Jeddah and elsewhere in the Middle East.

◀ **Figure 6.32** *Adapted from an article by John Vidal in The Observer in March 2010*

The bio-fuel rush

The food crisis of 2008 was partly responsible for the recent rush to acquire land in LICs. But the EU's target of having 10% of all transport fuels produced from bio-fuel by 2015 has added to the urgency of the land grab. Land in LICs is cheap, and some of the world's biggest agri-businesses, investment banks, and pension funds have been buying it up with future profits in mind. An ActionAid report has suggested that European companies have already bought 3.9 million hectares of land to grow crops for bio-fuels. But 14 million hectares will be needed if the 10% target is to be achieved.

There has been some evidence of local people being evicted from their land without the promise of jobs or compensation. Claims that the land being acquired is unused and therefore 'spare', ignore the ways in which traditional smallholders work. Land might be left fallow to help save it for future years. Huge water demands for the new crops also take a valuable resource away from local communities.

> Farmland in sub-Saharan Africa is giving 25% returns a year, and new technology can treble crop yields in a short period of time.
>
> *Susan Payne, Chief Executive of Emergent Asset Management*

ACTIVITIES

1. In what ways has globalisation increased the distance between the food producer and consumer? Is this appropriate in your opinion?
2. Refer to Fig 6.28 and your own experiences. List the positive and negative consequences of the ways in which supermarkets secure their supplies.
3. Explain whether you think that it's fair for land in LICs to be used to satisfy the demands of HICs.
4. Make two copies the following table and explain the advantages and disadvantages of supermarkets for
   (**a**) food consumers, (**b**) food producers.

	Advantages	Disadvantages
Social		
Economic		
Environmental		

Internet research

In pairs, choose one 'invisible giant' and investigate:

a where it invests overseas

b which factors have encouraged it to invest in LICs.

Present your findings in two parts:

- As an advertisement for the company – showing it in a positive light.
- As a critique of the company's activities overseas – showing the nature of its impacts.

In this section you will learn about:
- how the EU manages food supply for its member states
- how Spain manages its food supply and demand

Does the CAP fit?

When the 'Common Market' (now called the EU) was first formed in 1957, its first priority was to increase food production in Europe – to guarantee food security. It did this by giving farmers quotas (or targets) to meet, subsidies for doing so, and guaranteed prices (Figure 6.33). The six original members (France, Belgium, West Germany, Italy, Luxembourg, and the Netherlands) all agreed to protect their farmers' incomes and secure their collective food supplies. The Common Agricultural Policy (CAP) was designed to achieve this, and it began operating in 1962. Despite a number of reforms over the years to allow for the EU's expanding membership (now 27 states), the CAP still follows the same basic set of core aims (Figure 6.34).

For decades, the CAP was seen as the answer to European food security. It involved a series of economic mechanisms to support the EU's farmers and the European supply chain. They included:

- setting **tariffs** (import taxes) on imported foods. This meant that cheaper foreign foods became artificially expensive, and European shoppers would be encouraged to buy EU-produced foods instead.
- setting **quotas** for the amount of non-EU food that could be imported. This was also to protect the farmers of member states from cheap foreign competition.
- setting an **internal intervention price**. This meant that if the global market price fell below the guaranteed price, the EU would buy up any surplus produce from its farmers to ensure that their incomes were protected.
- offering **direct subsidies** to farmers to grow particular crops. This provided a guaranteed income, based on the amount of land devoted to that particular crop (e.g. oilseed).

Dairy farmers allow millions of litres of milk to go down the drain

Dairy farmers in France, Belgium and Luxembourg are protesting about the low prices they receive for their milk in the European Union (EU). The farmers say that they are being paid no more than 20 to 24 eurocents per litre – whereas it costs them double that to actually produce the milk.

In a recent move to simplify the EU's Common Agricultural Policy (CAP), Brussels has relaxed the milk quota restrictions. The quotas will be abolished completely by 2015.

Figure 6.33 *Adapted from Radio Netherlands News, 18 September 2009* ▲

Figure 6.34 *The core aims of the Common Agricultural Policy* ▼

- To raise productivity.
- To protect farmers' living standards.
- To stabilize agricultural markets.
- To guarantee the food supply.
- To ensure affordable food in the member states.

Surplus in a world of hunger

The CAP certainly increased European food security. However, by the late twentieth century, over-production had become a big problem. Many European farmers were happy to over-produce – the EU would always step in, anyway, and buy up any surplus produce for a guaranteed price. So they were protected against the global market.

This attitude by many farmers created 'food mountains' of surplus produce that could not be sold on into the global market at the prices guaranteed to European farmers. Some of this excess food then had to be destroyed, because the surpluses were too expensive to store. However, since 2005, a series of reforms to the CAP have started to cut overproduction and reduce payments to farmers:

- Lower production quotas and **set-aside policies** have actually taken land out of use.
- For a while, farmers were paid 'not to produce food' and to use the land for other income-generating activities. This brought about the idea of a **post-production countryside** – a landscape where leisure and tourism and environmental management schemes could earn farmers an alternative living (Figure 6.35).
- Today, farmers receive financial support for looking after their animals, the environment and the landscape, and for **diversifying** – moving away from crops already in surplus.
- In future, all quotas and set-aside schemes will be phased out and 'single farm payments' (lump sums, rather than subsidies) will be paid to farmers who 'produce in response to consumer demand'.

Market forces

The biggest criticism of the CAP is that it protects European farmers, while limiting access to Europe for producers in other countries. The World Trade Organisation (WTO) and many large food companies have argued for reduced import tariffs and more free trade. Some other opinions about the CAP are shown in Figure 6.36.

> **Did you know?**
>
> The CAP cost 43.8 billion euros in 2010. That's 31% of the EU's total budget. Even so, 75% of the EU's farmers earn less than £5000 a year.

Figure 6.35 *A post-production countryside – farmland no more!* ▲

Figure 6.36 *Who is in favour of the CAP?* ▼

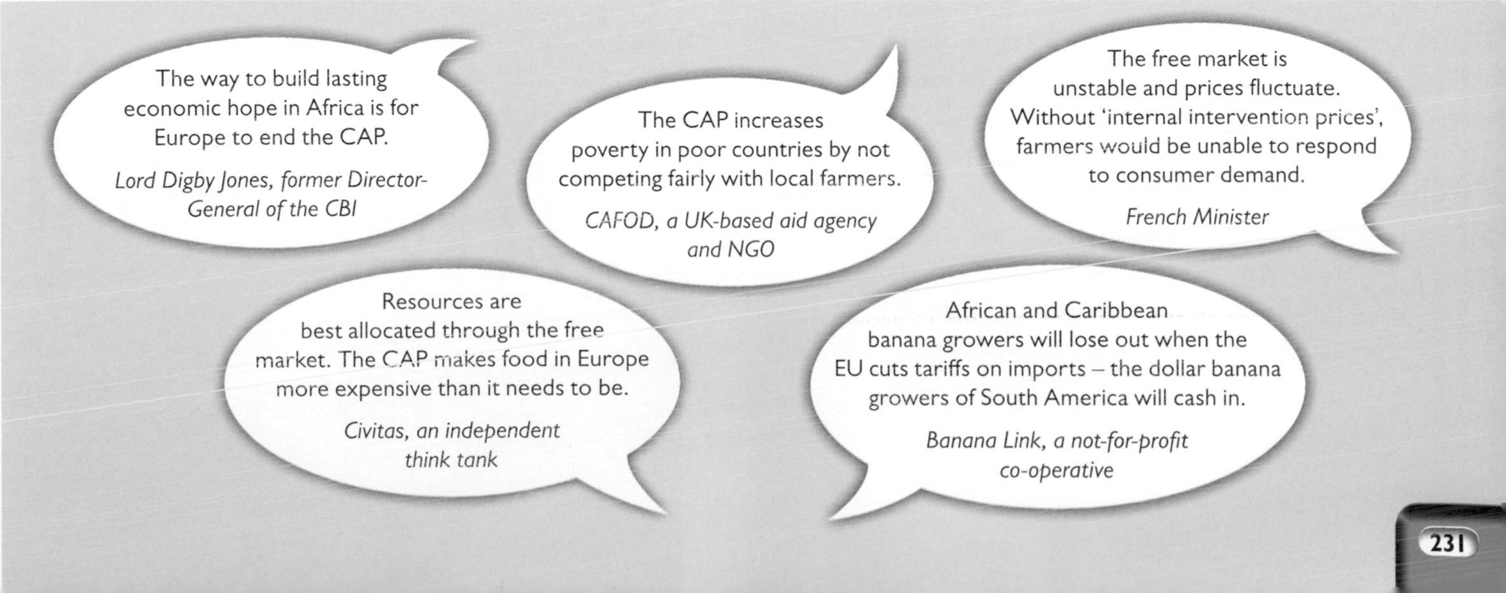

The way to build lasting economic hope in Africa is for Europe to end the CAP.

Lord Digby Jones, former Director-General of the CBI

The CAP increases poverty in poor countries by not competing fairly with local farmers.

CAFOD, a UK-based aid agency and NGO

The free market is unstable and prices fluctuate. Without 'internal intervention prices', farmers would be unable to respond to consumer demand.

French Minister

Resources are best allocated through the free market. The CAP makes food in Europe more expensive than it needs to be.

Civitas, an independent think tank

African and Caribbean banana growers will lose out when the EU cuts tariffs on imports – the dollar banana growers of South America will cash in.

Banana Link, a not-for-profit co-operative

Spain's plastic revolution

How do Spain's farmers manage food production within the EU? Almería (Figure 6.37) used to be one of Spain's poorest regions. Then it discovered the polytunnel! The UK's demand for year-round fruit and vegetables has now transformed the Plain of Dalias and the Alpujarra Hills there. 350 km² of market gardens now grow produce for the UK. The crops grow in bags beneath hectares of plastic (Figure 6.38), as the box on the right explains. Cheap migrant workers from Mali, Colombia, Ukraine, Romania, Poland and Morocco help to keep the production costs low.

This is an intensive agricultural system – and big business. Yields of crops like strawberries and tomatoes have increased by more than 35% since polytunnels were introduced. The EU now consumes almost 1 million tonnes of Spanish polytunnel tomatoes each year! For Almería, this plastic revolution has transformed an impoverished, dry, dustbowl landscape into one of high productivity.

However, this transformation has also brought a surge in demand for water to one of Spain's driest areas. So, as a result, there is now a multi-billion-euro plan to transfer water from the wetter areas of northern Spain to Almería and Murcia in the south. A 1000-km system of pipes and canals, and 118 dams, will link the River Ebro to the thirsty greenhouse economy of the south (Figure 6.37).

It didn't used to be like this

Almería used to be one of the poorest regions in Spain, where farmers could barely subsist. In the 1980s it suffered depopulation, as younger generations moved away from a hard lifestyle with low pay and limited opportunities. Government policies in Almería concentrated on small, family-run businesses and avoided mega-projects. Instead of dams, they opted for small-scale drilled water wells and farms suitable for individual families to run. Plots of land no bigger than a hectare were sold, along with new rural houses and planned villages.

Growing under plastic – how it's done

- The grow bags contain oven-puffed grains of perlite stone to keep soils warm and help drainage.
- Chemical fertilizers are drip fed to each plant by computer.
- 40 kg of pesticides are used per hectare.
- Strawberries are grown on soils that are sterilized using ozone-depleting methyl bromide (to avoid pests).
- There is continuous cropping from October to July for tomatoes, and from January to July for strawberries.

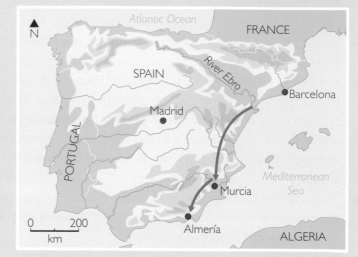

Figure 6.37 ▲

The small coastal plain 30 km southwest of the city of Almería has been intensively developed for agriculture. It accounts for over $1.5 billion in economic activity. The area has a dry, mild, Mediterranean climate and is further sheltered on the north by the Sierra de Gador mountains. With just slightly more than 200 mm of annual precipitation to support crop growth, the area also relies on groundwater fed by small stream aquifers from the mountains to the north.

Figure 6.38 Plastic tunnels in Almería take advantage of the long hours of sunshine and mild breezes to produce salad crops ▶

Then there was a big change. The harsh climate – dry and hot – was harnessed by tapping into underground water supplies. With no risks of frost, plenty of sunshine, and year-round warmth, the stage was set for horticultural success. The coastal breezes helped to ventilate the greenhouses and reduce humidity and crop-damaging condensation. The result of this change is that Almería can now produce 'seasonal crops' all year round in unheated, cheap plastic tunnels.

But not everyone is happy

The landscape has changed (Figure 6.39), and the environmental impacts are serious:

- Trees have been cleared and terraces of plastic now replace the olive groves.
- There is a risk of flooding as deforestation leaves soils exposed when it rains.
- Farmers have taken too much water out of the rivers, and several have dried up. Groundwater levels have also been reduced by 50%.
- Streams are being polluted with plastic and pesticide drums. The over-use of pesticides is raising the cancer risk for workers and locals.
- Small villages have become dependent on single crops.

Figure 6.39 *The concentration of greenhouses and polytunnels in Almería is the largest in the world, and can be seen from space* ▲

ACTIVITIES

1 In pairs, discuss the aims of the EU's Common Agricultural Policy (Figure 6.34) and complete a table showing the advantages and disadvantages of each one.

2 Explain why it has been necessary to reform the CAP in recent years.

3 Complete a SWOT analysis of Almería as a place for intensive farming (a SWOT analysis looks at its **S**trengths, **W**eaknesses, **O**pportunities for the future, and **T**hreats to its future).

4 Should the EU step in to deal with the weaknesses and threats that you have identified, or should it leave Almería for the global market to decide?

Internet research

1 Research and draw up a list of the aims of the World Trade Organization.
- Which of them are similar to the aims of the EU's Common Agricultural Policy (Figure 6.34), and which are different?
- How far do the CAP's aims actually conflict with those of the WTO? Explain your views.

2 Research the relationship between the EU and the WTO in more depth. Then prepare a 500-word argument for, and another one against, the following statement:

'EU agricultural policy favours European consumers at the expense of farmers in LICs.'

Key these phrases into Google to help you get started:
- World Trade Organization
- Common Agricultural Policy
- EUROSTAT farming
- European Commission trade in bananas
- Global issues banana wars

In this section you will learn about:
- technological solutions can increase global food supply
- these solutions may have different effects for different people

The Green Revolution – new crops for old

In the 1960s and 1970s, 'Miracle Rice' and 'Wonder Wheat' were developed as scientifically generated solutions to global food shortages. Many thought that a technological fix had been found which could increase yields of staple foods, feed the world, and prove Malthus wrong. It was called the 'Green Revolution', and was needed because the populations of many MICs and LICs were growing rapidly – and so too was the possibility of famine. Feeding people was essential if social unrest and political revolution were to be avoided.

Figure 6.40 *The Green Revolution in the Punjab* ▲

The Green Revolution began with the development of High Yielding Varieties (HYVs) and Modern Varieties (MVs) of rice, wheat and maize in Mexico and the Philippines. Wheat output in those countries increased by 300% and rice by 600%! Scientists seemed to have discovered how to improve global food supply. They had taken high-yielding strains of each species and crossbred their strongest characteristics to come up with new varieties.

However, the Green Revolution was about more than just seeds. It was part of a 'package'. To be most effective and produce bumper crops, the HYVs and MVs needed:
- precise amounts of water, fertilizer and pesticides
- mechanisation (to improve the speed of sowing and harvesting)
- large fields (to enable machinery to move easily), which meant **land reform** (to merge small plots of land into larger ones)
- suitable infrastructure (e.g. irrigation pipes, fuel for machinery).

But these all came at a price – both social and economic – and ushered in an era of **commercialization** (to encourage farmers to think about cash crops, not subsistence).

Unfortunately, most poor farmers could not afford the 'miracle package'. Many either had to take out loans or sell their plots to the wealthier landowners, who merged them to create larger fields. In terms of yields, the Green Revolution was a great success in parts of India. The Punjab (Figure 6.40) achieved massive increases in yield (up 300% in 10 years). India became largely self-sufficient in food as a result. But the Green Revolution also had serious side effects (Figure 6.41).

Figure 6.41 *The good and bad sides of the Green Revolution* ▼

Good	Bad
• It increased crop yields.	• HYVs were vulnerable to pests and diseases, so chemical pesticides were needed.
• The new varieties had shorter stems, and put more energy into producing their seed heads.	• Fertilizers were also needed, because the new varieties needed high nutrient levels.
• They also grew and ripened more quickly – reducing the growing season and allowing more crops to be planted in a year.	• Some pesticides are dangerous to human health (e.g. a by-product of some pesticides is dioxin, which causes birth defects).
• The Green Revolution produced surpluses for sale and raised incomes.	• The wealth gap widened, as the poor took out loans and fell into debt.
• It created new jobs in food processing, machinery and maintenance, and irrigation.	• Mechanisation increased rural unemployment.
• Small plots of land were merged, field boundaries were removed, and the amount of land available for cultivation increased.	• The new crops needed more water, which lowered water tables and river levels.
• India became a major exporter of rice.	• Irrigation caused **salinisation** and damaged soil fertility.

Did you know?

Many of the trees in the Punjab have now stopped bearing fruit, because heavy use of pesticides has killed so many of the pollinators – the bees and butterflies.

CASE STUDY

The downside of the Green Revolution

Kalahandi District is in the western uplands of Orissa State in eastern India (Figure 6.42). It is home to 1.3 million people – 80% of whom are farmers. Orissa is economically remote from the wealthy Indian cities, and Kalahandi is even more remote. Its forested hills play little part in the rapidly developing Indian economy, and local people are marginalised from the processes of change – the Green Revolution has by-passed Kalahandi.

The livelihoods of local people depend on their daily reactions to circumstances – some within their control and some not. Droughts have persisted there for 30 years. The bare soil on deforested slopes reflects their attempts at survival. In 2007, starvation played a significant role in 250 cholera-related deaths. The BBC visited the district and found people with no food surviving on leaves and the bark of trees (Figure 6.43). They had been unable to afford rice for a year – but how could this be?

Figure 6.42 *Kalahandi – a remote part of an economically isolated Indian state* ▲

Hunger in spite of plenty

There is a bitter irony in Kalahandi. Improved farming techniques, and years of aid programmes, have increased rice yields there. Kalahandi's rice production exceeded local demand between 1998 and 2003 – and actually contributed to India's central food reserves.

However, despite this apparent success, most Kalahandi farmers have reduced nutrition levels. This is because they do not own their own land. Instead, they either rent it from absentee landlords, or work as landless labourers. Rent is high, the tenant farmers are in debt (due to loans for seed and fertilizer), and wages for landless labourers are low. As a result, most local people cannot afford to buy the rice they grow. In 2007, 50 million tonnes of rice rotted in the countryside while farmers starved. The economic system keeps them poor, while the traditional caste system keeps them in their place.

Figure 6.43 *Poor villagers trying to cultivate marginal land in the forest in Orissa state* ▲

Bananas to the rescue

Now, some Kalahandi farmers have begun to diversify into bananas, as a way out of poverty and hunger. 5000 hectares are under extensive banana cultivation, as part of Orissa State's diversification project. A farmer has to spend about 14000 rupees ($340) a year for every hectare of banana cultivation, but can earn around 35000 rupees ($850) net profit.

Now it's the Gene Revolution!

50 years after the Green Revolution, a new revolution is taking place – genetic modification (GM). Imagine rice that can resist disease and drought, or varieties that prevent blindness. That's what scientists promise – higher yields, better health, and starvation a distant memory. So what is there to protest about (Figure 6.44)?

Did you know?

Scientists working on GM crops have introduced a component of car anti-freeze into strawberries – to stop them from freezing!

India now grows over 4% of the world's GM crops (particularly cotton, which has been widely adopted). Some farmers claim huge benefits, because of disease resistance, while others protest about the high costs. Some GM companies force farmers to sign contracts to use only their chemicals, or prevent them from saving seed to use the following year. GM engineers can even stop seeds in some crops from germinating. Many of India's poorest farmers rely on saved seed each year, but now they have to buy new seed every year. Like the Green Revolution, wealthier farmers benefit most.

The GM debate – the case against

- Even though relatively little is known about how genes work in most organisms, GM technology is still being adopted.
- Genetic modification is unpredictable. By inserting genes into foodstuffs from organisms that have never been eaten before, new proteins are being introduced into the food chain. Will they have health side effects?
- Many GM foods contain genes that are resistant to antibiotics. Could these be passed on to bacteria in the guts of humans and animals?
- Herbicide resistance. What happens if seeds are transplanted elsewhere (e.g. by birds) and grow wild as weeds? If GM traits (such as insect resistance) are passed on to wild plants, new 'superweeds' could develop.
- Crop contamination. GM and ordinary crops can cross-pollinate each other, so all farmers could find their crops contaminated. Pollen can travel several kilometres in the wind.

Figure 6.44 *Activists dressed as Indian politicians 'force feeding' volunteers with genetically engineered (or modified) aubergines at a protest in New Delhi against GM (or GE) crops* ▲

What is genetic modification?

GM occurs when scientists take 'genes' from one organism and place them into another. This changes the way the organism develops, creating new types of plants and animals. It allows crossbreeding between species and organisms that would never breed naturally. In the Green Revolution, scientists bred **hybrids** within species, e.g. crossbreeds in grain crops.

Now there are no barriers to prevent scientists taking genes from unrelated organisms (e.g. a rat and a cotton plant). GM allows genes to be crossed between species. A gene from a fish, for example, has been put into a tomato. Scientists cut up DNA and use the parts they want from different organisms – and even make synthetic DNA. The development costs are enormous, so almost all GM research is carried out by TNCs, e.g. Monsanto in the USA. Using patent laws, TNCs 'own' every GM plant grown from their seed. Farmers therefore sign contracts before adopting the new seeds.

Will GM crops save the world?

Can GM crops help to feed the world's population by increasing yields and fighting disease (Figure 6.45)? Many people actually suffer from hunger because they cannot afford to buy food – not because it's unavailable (e.g. Kalahandi). There is more than enough food to feed everyone on the planet, yet nearly two billion people are malnourished. So, do we need GM crops at all, or just a better, fairer and cheaper food distribution system?

On the horizon are bananas that produce human vaccines against infectious diseases such as hepatitis B; fish that mature more quickly; cows resistant to bovine spongiform encephalopathy (mad cow disease); fruit and nut trees that yield years earlier, and plants that produce new plastics with unique properties.

Adapted from the Human Genome Project website

Laxman Sahu had always grown vegetables and rice in Kalahandi. When the Syngenta Foundation established KARRTABYA – an advisory service – he attended farm workshops to learn about hybrids of hot chilli peppers, and bought seeds from KARRTABYA. In 2006-07, he cultivated these. He achieved a bumper crop of chillies. His net profit came close to 100000 rupees. On seeing his success, several other local farmers have now followed this pattern.

Adapted from Syngenta's 'Foundation for Sustainable Agriculture' website (Syngenta is a leading bio-technology agri-business)

Figure 6.45 *The benefits of GM foods, as seen by a range of interests* ▲ ▶

GM crops could help to address some of India's severe nutritional problems. Roughly 50000 children in India go blind every year from vitamin A deficiency, while iron deficiency is a major threat to the health of women. The possibility of engineering iron-rich rice or vitamin A-rich rapeseed oil could be very interesting in this context.

Adapted from Policies Toward GM Crops In India *by Robert Paarlberg*

ACTIVITIES

1 Make two copies of the table below to list the advantages and disadvantages of the Green and Gene Revolutions.

	Advantages	Disadvantages
Social		
Economic		
Environmental		

2 a Read the material about Kalahandi and identify (i) the issues faced by people there, (ii) the needs of its people.
   b To what extent are they likely to have these issues and needs met by the Green and Gene Revolutions? Explain your answer.

3 In your view, can technological solutions ever completely overcome socio-economic problems?

Internet research

Compile a factfile of examples where the Green and Gene Revolutions have been applied. Use the following hints to you get started:
- Google – key in 'Global Policy Forum GM crops'
- Go to the BBC news website – news.bbc.co.uk – then key in 'Green Revolution' and 'GM crops'

Present your findings to your class. When the presentations are complete, make additions and amendments to the tables that you created for Activity 1.

In this section you will learn about:
- appropriate/intermediate technology in developing sustainable food supplies
- the growth of local produce markets and organic produce

Can feeding people be sustainable?

Sustainability is about 'meeting the needs of the present without compromising the ability of future generations to fulfil their needs'. Feeding today's global population is hard enough, so a possible 3 billion extra people by 2050 could be a real problem. Technological fixes – like the Green and Gene Revolutions – can increase food supplies, but they also cause problems (pages 234-237). Sustainable farming means that food should be affordable, and also produced in ways that don't harm communities or the environment.

The days of huge plantations, owned by big corporations, might be ending. Instead, TNCs are increasingly entering into individual contracts with thousands of smallholders in LICs. Small, family-run farms – using **intermediate technology** – could provide the sustainable solution needed.

Africa's smallholders – an intermediate way forward

Exporting coffee, fruits and vegetables has transformed Karatina on the northern slopes of Mount Kenya (Figure 6.46). In the 1980s, the company 'Sunrise Kenya' helped this community by providing technical support and new seeds. 300 smallholders and their families now benefit from increased yields and incomes – earning their way out of poverty. All work is done by hand, or using locally made tools. The women take control of the proceeds and invest them in healthcare, schools and churches. Literacy has improved, because more children now stay on at school. Many go on to work in urban office jobs to help support their families.

Figure 6.46 *A successful smallholder in Kenya picking beans for export to the UK* ▲

When cash crops drive a local economy like this, everyone gains:
- Kenya exports 450 000 tonnes of vegetables, fruits and flowers a year – worth US$1.3 billion in 2009.
- 4.5 million Kenyans are employed growing this produce.
- No subsidies are paid, and minimal use is made of machinery and oil.
- Kenyans rarely use artificial fertilizers, weedkillers or pesticides, so their produce helps to meet the rising demand for organic produce in HICs.
- But what about 'food miles'? The farmers in Figure 6.47 think they are 'fair miles' – and, in any case, the produce is transported in the baggage holds of passenger flights, which makes tourist travel more efficient.

> I earn seven times as much money from growing green beans as I would from maize. The money feeds my family, pays for medicine and schools, and promises a better future. We just use hoes, pangas and manual labour on our two-hectare farm.
>
> *Muru from Karatina, Kenya*

> I could not afford chemicals when I inherited my 2.5 hectares, so I grew pineapples the traditional way. Then 'Blue Skies' offered me micro-loans, training and technical advice about how to grow organic fruit. They now pay me 15% more for my pineapples and provide a ready market. I have now been able to install a clean water supply and proper toilet.
>
> *Kweku Ayuba from Ghana*

◀ ▲ **Figure 6.47** *Adapted from National Geographic, 2010*

Eating locally?

While cash crops can help build sustainable lifestyles for Kenyans, and provide organic food for consumers in HICs, what about the plight of small farmers in the UK? Faced with huge influence from supermarkets, how can they compete with mass-produced food?

Grampound is a small village of about 600 people in mid-Cornwall (Figure 6.48). Within 12 kilometres, there are six major supermarkets. Cornwall is England's lowest wage-earning county, so affordable food is critical. In 2008, the villagers started a 'transition movement'. Its purpose is to look for sustainable ways of feeding people and generating local benefits – including for local farmers.

On the last Saturday of every month, Grampound Village Hall is a busy place, as people arrive for the local produce market (Figure 6.49). A number of stalls are set up – each one run by people from within the village. They include:

- two farms, one selling beef and lamb, another pork, all from animals reared locally
- seasonal fruit and vegetables grown by local producers
- bread and pork pies from local bakeries
- cakes, quiches, and savouries baked by local villagers
- chutneys and jams made by local people, and honey
- free range eggs from a local farm.

Every product is sold at – or often below – supermarket prices. Yet, the producers claim that they make more money from the market than they do selling to large companies. Building on this, a local landowner has agreed to provide the village with 16 allotments to encourage more people to grow their own food.

Figure 6.48 *Grampound in mid-Cornwall* ▲

Figure 6.49
Grampound's monthly produce market ▶

PENHAYES DEXTER BEEF

GRAMPOUND

agri-businesses Large farms or estates owned and managed by major companies that organise the purchase of all inputs, labour, processing, and marketing

arable farming Growing crops

bio-fuel When crops are grown for fuel

calorie intake The number of calories consumed

carbon footprint The amount of carbon taken to produce goods or services

carrying capacity The ability of the environment to produce food and resources sustainably, without limiting access of these to future generations

cash crops Those crops which are grown for cash income

commercial farming Growing produce for sale

Common Agricultural Policy (CAP) A policy set up in 1962 by the EU to raise productivity among farmers

diversification A shift from relying on one product to a wider range

extensive farming Using low inputs of labour, machinery, and capital

fair miles When consumers consider the world's poor in deciding to purchase food from developing countries

food miles The number of miles between source of food and the consumer

food security How secure a country's supplies of food are

free trade Trade without tariffs on goods

Genetic Modification When scientists take 'genes' from one organism and place them into another

Green Revolution The development of High Yielding Varieties (HYVs) and Modern Varieties (MVs) of rice, wheat, and maize in the 1970s

hybrids Breeding different strains of plant to increase yields

intensive farming Maximising return from the land using high inputs, e.g. labour, fertiliser, machinery, and capital

internal intervention price A guaranteed price for farmers, whereby the EU buys surplus produce from farmers to protect their incomes

land reforms Where land is redistributed or re-grouped, usually to make ownership easier (which benefits the wealthy), or distribution fairer (which benefits the poor)

malnutrition Hunger, or insufficient calories

marginalisation The ways subsistence farmers are forced on to land that will not support their needs

mixed farming Growing crops and keeping livestock

nutrition transition Increasing consumption of meat caused by increasing incomes

obesity A condition caused by over-consumption of calories

pastoral farming Grazing animals, e.g. cattle or sheep

post-production countryside A landscape where leisure and tourism (and not food) earn landowners a living

quotas Limits or targets on products (e.g. how much grain can be grown or imported)

salinisation Increasing salt content of soil such that plants are less able to grow

set-aside policies Taking land out of production

specialised farming The intensive production of one crop. Also known as monoculture

subsidies Grants paid to farmers to reduce their costs in producing food

subsistence farming Growing enough to feed a household, but with no surplus

tariffs Taxes on goods imported from overseas

technological fix Solving problems using technology

trade liberalisation The process of freeing up flows of goods traded between countries

under-nourishment A shortage of protein or calories

under-nutrition Lacking adequate minerals and vitamins

World Trade Organisation (WTO) A global decision-making body which aims to expand free trade between countries

EXAMINER'S TIPS

1 (a) Explain the difference between 'malnutrition' and 'under-nutrition'. *(4 marks)*

(a) Focus on the differences.

(b) Study Figure 6.6 (page 213). Describe the distribution of countries where 35% or more of the population suffers from malnutrition. *(4 marks)*

(b) Use evidence from the map, focusing on just the one category of countries.

(c) Explain one factor which may lead to high levels of malnutrition. *(7 marks)*

(c) Consider one factor only, such as drought or conflict.

(d) Using examples, explain why some countries are unable to feed themselves. *(15 marks)*

(d) Use either a HIC which can source its food from further away or a LIC which faces problems of food shortages.

2 (a) Study Figure 6.18 (page 220). Describe the global distribution of those countries suffering food crises in 2008. *(4 marks)*

(a) Look at the general pattern, then use examples.

(b) Describe one factor which can lead to a country becoming increasingly 'food insecure'. *(4 marks)*

(b) Consider factors such as general food insecurity brought about by war, or specific shortages affecting some people, e.g. where exports of cash crops can lead to food shortages.

(c) Explain how countries can become less food insecure. *(7 marks)*

(c) Think of examples where countries have increased their food supply.

(d) Referring to one country you have studied, assess the impacts of variations in food supply. *(15 marks)*

(d) Remember – impacts can be social, economic, or environmental.

3 (a) Study Figure 6.44 (page 236). Explain why these people might be demonstrating. *(4 marks)*

(b) Outline briefly one recent application of technology to food production. *(4 marks)*

(c) Explain why improvements in food technology do not always benefit everyone. *(7 marks)*

(d) Discuss the benefits and problems that arise when countries source food globally. *(15 marks)*

Energy issues

The oil rig has landed

Is oil a renewable resource?

What effect can getting oil have on the environment?

What effect does using oil have on the environment?

Do you think the world's demand for oil is increasing or decreasing?

How much oil do you use?

Introduction

How important is energy? Imagine a world without oil – the basis of most transport fuels and a raw material for clothing fibres, plastics, and artificial fertilisers. Whoever possesses oil controls a portion of the world's most vital raw material. What happens if they turn off the pipeline taps? What happens when – and this is the fear of so many people – the oil runs out?

In this chapter you will learn how far the world can continue using up finite energy reserves, or whether sustainable supplies might replace them. Perhaps consumption can be reduced. Perhaps energy conservation can become the policy of the future. Perhaps homes, workplaces, and transport might have to be designed for sustainability.

Books, music, and films

Books to read
Powerdown: options and actions for a post-carbon world by Richard Heinberg (2004)

Music to listen to
'Temper' by System of a Down
'The Price of Oil' by Billy Bragg
'Black Gold' by Soul Asylum

Films to see
There will be blood (2007) by Paul Thomas Anderson
Three Kings (1999)
Jarhead (2005) about the Desert Storm conflict in 1991
The Age of Stupid (2009) by Franny Armstrong

About the specification

'Energy issues' is one of the three Human Geography option topics in Unit 1 – you have to study at least one.

This is what you have to study:
- The different types of energy: renewable (flow) resources, non-renewable (stock) resources, and primary and secondary energy. The primary energy mix, looked at in a national context.
- The global patterns of energy supply, consumption, and trade. Recent changes in these patterns.
- The geopolitics of energy – which means conflict and co-operation in world affairs. The role of transnational corporations in world energy production and distribution.
- The environmental impact of energy production, including fuel-wood gathering and nuclear power and its management. The use of fossil fuels, including acid rain and the potential exhaustion of fossil fuels.
- The potential for sustainable energy supply and consumption. Renewable energy: biomass, solar power, wind energy, wave energy, and tidal energy. Appropriate technology for sustainable development.
- Energy conservation – such as designing homes, workplaces, and transport for sustainability.
- Case studies at the national scale of two contrasting approaches to managing energy supply and demand.

In this section you will learn about:
* some of the energy dilemmas facing the world

Hard choices in Copenhagen

In December 2009, the world's leaders gathered at Copenhagen in Denmark to discuss the issue of global climate change. They were hoping to reach an agreement to reduce carbon emissions and our dependence on fossil fuels. But they were also aware that 1.5 billion people (nearly a quarter of the world's population) still have no proper access to electricity. Dealing with both of these problems presents a real dilemma for the world.

Fossil fuels drove the economic development of the High Income Countries (HICs) in the nineteenth and twentieth centuries. Limiting their use now might deny Low Income Countries (LICs) the chance to develop economically and catch up. However, despite this possibility, the International Energy Authority (IEA) said that: 'The time has come to make the hard choices needed to combat climate change and enhance global **energy security**'. The IEA also said that reducing greenhouse gases in the atmosphere to 450 parts per million by 2030 – whilst maintaining global energy supplies – would require: 17 new nuclear power stations, 17 000 new wind turbines, and two hydroelectric power schemes on the scale of China's Three Gorges Dam (see page 250) *every year*. This would mean an investment of US$20 trillion over 20 years. So, there is no quick and easy answer to the problem of energy supply now and in the future.

> **Energy security** means access to reliable and affordable sources of energy. Countries like Russia are very energy secure, because they have surplus supplies. The USA and the UK are energy insecure, because they have to import a lot of their energy sources.

The world at night

Figure 7.1 shows the light made by humans across the planet. The concentrations of light roughly correspond with the heavy local use of fossil fuels to produce electricity. However, the brightness doesn't always correspond to population size – many dark areas have high densities of people. For instance, the island country of Madagascar – with a population of over 21 million – uses very little energy, because it's an LIC. By contrast, Manhattan island in New York city is home to 1.6 million comparatively wealthy people – living in an HIC – who each consume more energy in a year than an average Madagascan will in their entire lifetime! So, Figure 7.1 correlates well with levels of economic development and helps to reveal the scale of the global 'energy gap' between HICs and LICs.

Figure 7.1 The world at night ▼

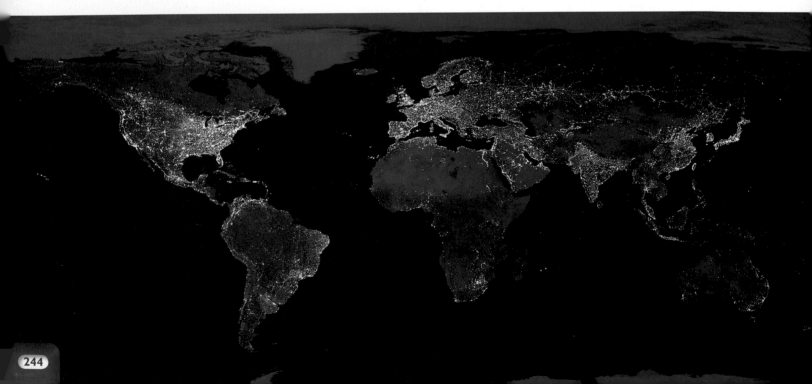

City lights

Over half of the world's population now lives in cities – and their energy consumption continues to increase. Cities consume 75% of the world's energy, and produce 80% of its greenhouse gas emissions. The small area of London that forms the City of London – also known as the Square Mile (Figure 7.2) – generates 1.7 million tonnes of carbon a year. Its energy demands are so high because it depends on a financial and service economy that is based in well-lit climate-controlled offices, using a lot of electronic data-processing systems. The well-being of millions of people across the UK now relies on this urban service economy.

London's energy demands are being met through an intricate web of national and international supply lines, which extend as far as Russia and involve the cooperation of several transnational corporations (TNCs).

Rural shadows

By contrast with the City of London, many rural villages in Peru are only now being connected to a proper electricity supply. Government projects are bringing solar power to 500 000 people in remote areas not reached by the national electricity grid. The new electricity supply powers lights, radios and televisions – and extends the working day.

The solar power also allows families to use electrical machinery and tools. The resulting increased productivity in processing things like cereals, meat and cocoa helps to boost rural incomes and raise the standard of living. Renewable energy is being used to bring about sustainable development and a brighter future.

Figure 7.2 The Square Mile – a city that never sleeps ▲

Figure 7.3 Solar power in rural Peru ▼

ACTIVITIES

1 a Describe the energy use pattern shown in Figure 7.1.
   b How far is this photo a useful indicator of the energy dilemmas facing the world?
2 How have fears about global climate change forced people to reconsider the ways in which their electricity is produced?
3 Define the term 'energy security' in your own words.
4 Why could some countries become more energy insecure as they develop economically?

In this section you will learn about:
- factors determining a country's energy mix

Skills

In this section you will:
- interpret a compound line graph

Falling stocks and rising flows

Energy resources can be classified into two types: **stock resources** and **flow resources**. For example, the amount of oil on the planet today is the stock. The flow is the incoming energy source that replenishes the stock. In the case of oil there is no flow today, because the oil was formed millions of years ago. The stock is falling and not being replaced; it is non-renewable. The same can be said of the other fossil fuels – coal and natural gas (Figures 7.4 and 7.5).

As the non-renewable energy resources are used up, the world is increasingly turning to renewable energy resources instead (Figures 7.6 and 7.7). Renewable energy resources are continuously replenished, because their flow is constant. For example, solar power has a constant flow of energy emitted from a big stock – the sun. Wave and wind power depend on the flow of moving air, which is also constant.

Stock and flow energy resources

Figure 7.4 *An opencast coal mine* ▲

Figure 7.5 *Oil rigs in the Gulf of Mexico* ▲

Figure 7.6 *A hydroelectric power (HEP) scheme* ▲

Figure 7.7 *A huge solar power plant in California* ▲

Primary and secondary energy sources

Coal, oil and natural gas are **primary energy sources**. Burning them generates power that can be used directly to move vehicles and machines. This is primary energy use. But the heat generated by burning these resources can also be used to power a turbine that drives a generator to produce electricity. Electricity is a **secondary energy source**.

Electricity has shaped the modern world. We now use a number of different sources to produce more and more of it. But as the primary sources change from stock to flow resources, the environmental impacts are also changing (Figure 7.8). Each energy choice has different impacts, e.g. greenhouse gas emissions or the visual impact on the landscape, and people have different opinions about which option is best.

Getting the mix right

Every country wants a secure energy supply, but for most countries ensuring energy security is not that simple – unless they have big energy reserves, like oil in Saudi Arabia and natural gas in Russia. The increasing global demand for energy means that stock resources are being used up more and more quickly, and threats of shortages are starting to emerge. As countries such as the LICs develop – and their standards of living rise – they want access to more energy. But if the energy sources are limited (or cannot be acquired easily, quickly or cheaply), the price charged for them rises. Some resources then become too expensive, so countries look for cheaper and more accessible alternatives. The different primary energy sources used to satisfy the increasing demand for energy make up the **energy mix**.

Figure 7.8 *Different sources, different impacts* ▲

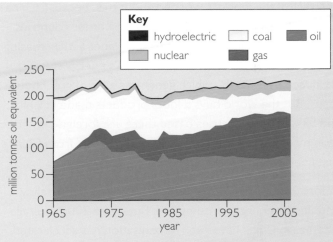

Key
- ■ hydroelectric
- □ coal
- ▨ oil
- ▨ nuclear
- ▨ gas

Figure 7.9 *The UK's changing energy mix, 1965-2006* ▲

The UK's energy mix

The UK's energy mix has changed over the last century. Up to the 1960s, the UK was dependent on domestic supplies of coal. Then, in the 1980s-90s, the energy mix shifted to more use of oil and natural gas (with the discovery and exploitation of reserves under the North Sea). However, the North Sea stocks of oil and gas are now starting to run out. This has forced the UK to import more of its primary energy sources (e.g. natural gas from Russia), and to broaden its energy mix with the development of wind farms and other renewable sources (as well as more nuclear power). The increasing reliance on imported energy sources affects the UK's energy security, and this has now become a political issue.

The privatisation of the UK's energy supply industry, in the 1980s, also means that overseas companies – like France's EDF and Germany's E.ON – now play a big part in deciding which energy sources are used to meet demand in the UK. They buy primary energy supplies on international markets, rather than just using UK supplies. There are enough un-mined coal stocks in the UK to last another 150 years, but current technology and environmental policy make mining them unrealistic and economically unviable. However, should a cleaner way of burning coal be developed, the UK's energy mix could change again.

ACTIVITIES

1 Define the following terms: stock, flow, non-renewable and renewable resources, primary and secondary energy sources, energy mix.

2 How might the following influence a country's energy mix?
- a resource availability
- b accessibility and affordability
- c government policy
- d technology
- e environmental concerns
- f demand
- g market
- h ownership

3 Study Figure 7.9.
- a Describe the trends shown in the compound graph.
- b Calculate the percentage of energy provided by each primary energy source in 1965, 1975, 1985, 1995 and 2005. Which was the most important energy source in each of those years?
- c Which energy sources increased and decreased their share the most over the 40 years?
- d How far have the actual amounts of each energy source changed over the time shown?

In this section you will learn about:
- how China is managing its energy mix and causing the rest of the world to worry

Dominant China

Record economic growth rates since 2000 (Figure 7.10) have seen China emerge as the world's second-largest energy consumer (behind the USA) – and the largest producer of greenhouse gases. China's economy has been doubling in size every eight years. As a result, its share of total global energy demand rose from 10% in 2001 to 16% in 2010.

The rest of the world is worried about this, because China's growing hunger for energy is threatening other countries' energy supplies. It's also polluting the planet and increasing global warming, because most of China's energy is produced using coal – which generates a lot of greenhouse gases. But the world buys a lot of products made in China, and 53% of the industries based there are foreign-owned, so in many ways the rest of the world benefits from China's sustained economic growth.

Before 2000, China's economic growth rate had already doubled its energy consumption. But, since 2000, average incomes in China have risen, urbanisation has accelerated, and Chinese industry has become more energy intensive. Energy demand there has risen by a further 86%, and China is expected to account for nearly 40% of the world's total energy consumption by 2030.

The energy demand per person in China is still low, because of its huge population, but as average wealth continues to increase there, the energy demand per person will rise as well. Large amounts of imported coal, oil and gas will be needed to meet China's growing energy demands – leading to increasing energy insecurity there.

China's hunger for fuel

China controls 3% of the world's oil reserves. It was self-sufficient in oil until 1993, but since then has needed to import it to fuel its rapid economic growth (Figure 7.11).

Figure 7.12 shows how China's energy demand is projected to increase up to 2030. However, this steep rise in energy consumption is not just fuelled by basic economic growth and the demands of new industry, but also by rapid urbanisation and growing car ownership.

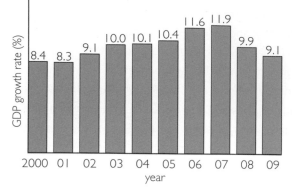

Figure 7.10 *China's economic growth rate, 2000-2009* ▲

Figure 7.11 *China's oil production and consumption, 1986-2006* ▼

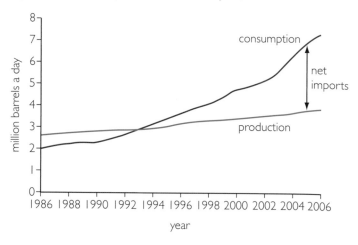

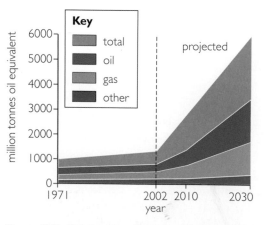

Figure 7.12 *China's projected growth in energy demand by 2030* ▲

Rural-to-urban migration in China is the highest ever recorded – an average of 8.5 million people a year. Most of these urban migrants head for the industrial centres by the coast, where energy consumption is significantly higher than in rural areas.

China's roads used to be filled with bicycles, but now cyclists battle with increasing numbers of cars (Figure 7.13). This is because one of the social effects of China's rapid economic growth and increasing wealth is the desire to own a car:

Figure 7.13 *1000 new cars arrive on Beijing's streets every day* ▲

- Chinese car ownership is expected to jump from 16 cars per 1000 people in 2002, to 267 cars per 1000 people by 2030.
- In 2009, 13.6 million cars were sold in China. That's 3.5 million more than in the USA that year.
- By 2020, China is expected to have 140 million private cars on the road – even more than the USA.

At the moment, China only uses 10% of its energy for transport (Figure 7.14), but it will need a lot more oil to fuel its anticipated growth in car ownership. Much of that oil will have to be imported – China already imports about 60% of its oil needs. This raises serious issues about China's long-term energy security (see page 251).

Energy demand by economic sector

China is still largely an agricultural society. 60% of the population lives in rural communities, and 45% of the workforce is employed in agriculture. However, agriculture contributes only 13% of China's GDP, and only uses a fraction of its energy compared with industry (which accounts for 71% of China's total energy use). Figure 7.14 shows how China's energy use by sector compares with a number of other countries.

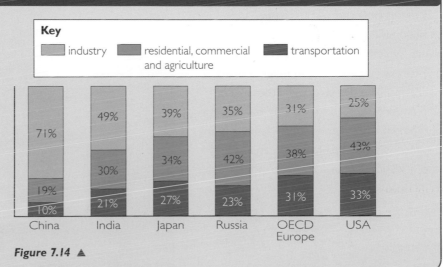

Key
- industry
- residential, commercial and agriculture
- transportation

	China	India	Japan	Russia	OECD Europe	USA
industry	71%	49%	39%	35%	31%	25%
residential, commercial and agriculture	19%	30%	34%	42%	38%	43%
transportation	10%	21%	27%	23%	31%	33%

Figure 7.14 ▲

Fueling the dragon

Maintaining China's economic growth depends on secure energy supplies. China's electricity comes from a number of primary energy sources (Figure 7.15).

Coal

China is the world's biggest producer and consumer of coal – it generates nearly 75% of China's electricity. In fact, as Figure 7.15 shows, China's reliance on coal is expected to increase rather than decrease in the foreseeable future.

The huge and growing demand for electricity in China means that the country is building an average of three new coal-fired power stations a week! In 2006, China added 102 gigawatts of generating capacity to its electricity grid. That's as much as the entire generating capacity of France.

Coal is a cheap source of fuel, but it's also a dirty form of energy. Increasing pollution problems mean that China now needs to build new, cleaner, coal-fired power stations. It's also given the go-ahead for several huge new projects to turn 'dirty' coal into 'clean' gas. This is an expensive but necessary route to take if China is to overcome the serious pollution problems shown in Figure 7.16.

HEP

Hydroelectricity (HEP) accounts for 16% of China's energy production. Major HEP projects are part of the country's long-term energy security strategy.

The turbines of the giant Three Gorges Dam on the Yangtze River (Figure 7.17) will generate 2.5 gigawatts of electricity at maximum output – making it the biggest HEP scheme in the world. That is equivalent to a third of the UK's total energy capacity.

But that is only the start. China wants to build HEP dams on all of its major rivers, which has caused concerns for environmental groups and neighbouring countries. The disastrous earthquake in Sichuan province, in May 2008, damaged a number of large dams there – leaving many dangerously close to collapse, and calling into question China's 'big dam' policy.

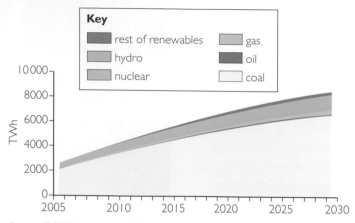

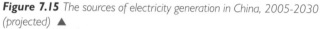

Figure 7.15 *The sources of electricity generation in China, 2005-2030 (projected)* ▲

Figure 7.16 *Living in the shadow of a coal-fired power station* ▲

Figure 7.17 *The Three Gorges Dam on the Yangtze River* ▶

Future fears for energy security

Oil and natural gas

Production at China's largest oil fields has now peaked, and exploration has begun for new fields in the far west and offshore (Figure 7.18). But there are problems with this. Large oil deposits in the Tarim Basin have failed to attract any investment, because of their remote location and difficult geology. Deepwater exploration in the South China Sea is affected by the danger of territorial disputes with neighbouring countries like Vietnam and the Philippines. Natural gas is a cleaner fuel, but it's proved costly and difficult to build pipelines from the gas fields in western China.

Coal

China's coal reserves are largely located in the north and far west, while the industry representing 71% of China's demand for energy is mainly located much further east and south (Figure 7.18). It's difficult to mine, move and burn coal quickly, cleanly and in sufficient quantity to meet China's escalating demand. For example, a surge in electricity demand in 2002, led to coal shortages, power cuts, spikes in demand for oil, and a rapid deterioration in air quality due to pollution from the coal-fired power stations.

Securing the future

China's foreign policy can be described as 'resource-based', with **energy dependency** becoming a national security issue (Figure 7.19).

> **Energy dependency** is measured by assessing the level of energy imports as a proportion of total energy consumption. The higher the proportion of energy imported, the more energy dependent the country is on others.

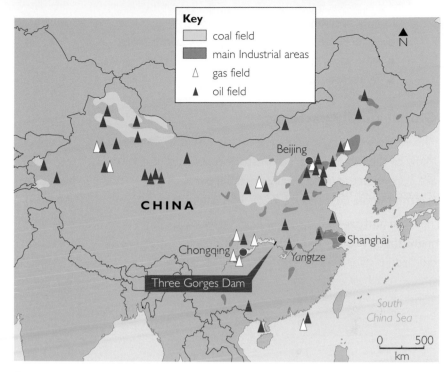

Figure 7.18 *China's energy resources are often far away from the centres of demand* ▲

Energy efficiency	Low
Diversification of energy sources	Poor – China still relies on coal for nearly 75% of its electricity generation (Figure 7.20)
Diversification of energy suppliers	Good – China considers Central Asia to be the most secure source for its future energy supplies, but it's also investing in – and importing energy from – Iran, Sudan, Canada and Australia
Control over the means of supply	Poor – with 90% foreign suppliers
Military protection for energy supplies	Good – China has a powerful navy. It is: • expanding naval access to Bangladesh • investing in the port of Gwadar in Pakistan • upgrading its military airstrip on Hainan island in the South China Sea • forging closer ties with Myanmar (Burma) • strengthening links with Iran, Venezuela and Russia

Figure 7.19 *China's energy security* ▲

	2005	2010
Coal	76.5	74.7
Oil	12.6	11.3
Natural gas	3.2	5.0
Nuclear	0.9	1.0
Hydroelectricity	6.7	7.5
Other renewable energy	0.1	0.5

◄ **Figure 7.20**
China's changing energy mix as a result of government policies and management (the numbers are percentages)

ACTIVITIES

1 In pairs, draw a spider diagram to show the reasons for China's growing energy insecurity.
2 What steps is China taking to overcome its energy insecurity?
3 Compare the map of China's energy reserves with an atlas map showing China's population distribution. Describe the differences in terms of energy poverty, surpluses and imbalances.
4 What are the short- and long-term problems associated with China's increasing energy demands?

In this section you will learn about:
- where the main energy supplies come from
- who uses the most energy
- fears for the future

Global supply patterns

The world's energy sources are distributed unevenly. Many of the largest reserves of stock resources (e.g. oil) are concentrated in politically unstable parts of the world, like the Middle East (Figure 7.21). This can lead to potential disruption in supply. The distribution of flow resources (e.g. wind and solar) also varies globally, because some places are windier and/or sunnier than others (Figures 7.22 and 7.23).

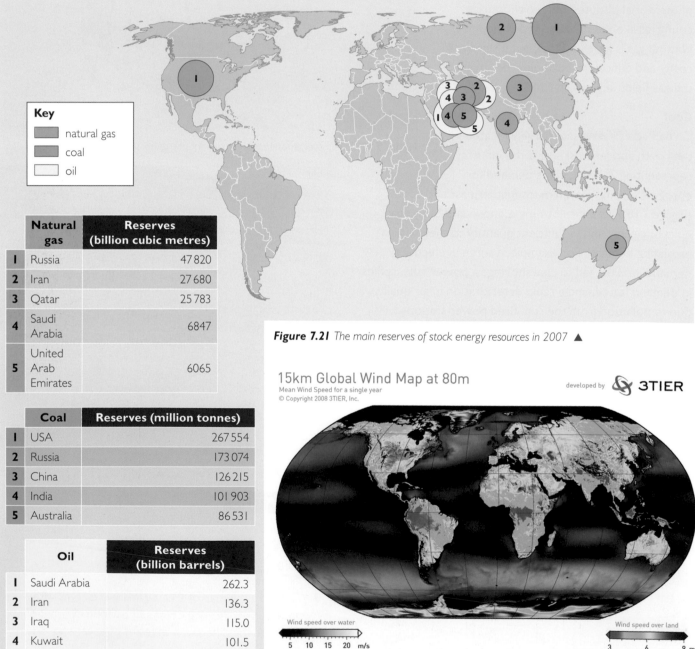

Key
- natural gas
- coal
- oil

Natural gas	Reserves (billion cubic metres)
1 Russia	47 820
2 Iran	27 680
3 Qatar	25 783
4 Saudi Arabia	6847
5 United Arab Emirates	6065

Coal	Reserves (million tonnes)
1 USA	267 554
2 Russia	173 074
3 China	126 215
4 India	101 903
5 Australia	86 531

Oil	Reserves (billion barrels)
1 Saudi Arabia	262.3
2 Iran	136.3
3 Iraq	115.0
4 Kuwait	101.5
5 United Arab Emirates	97.8

Figure 7.21 *The main reserves of stock energy resources in 2007* ▲

15km Global Wind Map at 80m
Mean Wind Speed for a single year
© Copyright 2008 3TIER, Inc.

developed by 3TIER

Wind speed over water
5 10 15 20 m/s

Wind speed over land
3 6 9 m/s

Figure 7.22 *Global mean wind speeds over land and sea at 80 metres above the surface* ▲

Averaged Solar Radiation 1990-2004

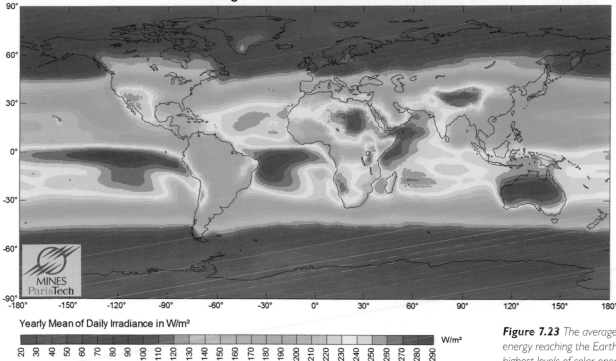

Yearly Mean of Daily Irradiance in W/m²

Figure 7.23 The average amounts of solar energy reaching the Earth, 1990-2004. The highest levels of solar energy are shown in red, followed by orange, yellow, green, blue, purple and finally pink. ▲

Global consumption patterns

The global energy mix still depends largely on fossil fuels like coal. As you can see in Figure 7.24, fossil fuels currently make up 85% of world energy consumption. The overall demand for energy is expected to continue increasing until at least 2050, but energy consumption varies dramatically between regions (Figure 7.25). As they continue to develop economically, China and India are expected to consume an increasing share of the world's primary energy resources, e.g. coal, in the coming decades (Figure 7.26).

Figure 7.24 The global energy mix still depends on fossil fuels ▶

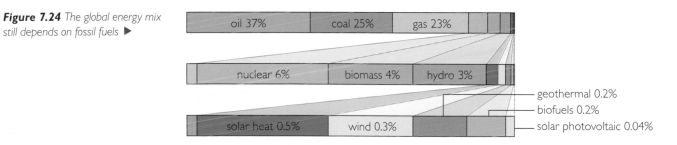

oil 37% | coal 25% | gas 23%

nuclear 6% | biomass 4% | hydro 3%

geothermal 0.2%
biofuels 0.2%

solar heat 0.5% | wind 0.3%

solar photovoltaic 0.04%

Figure 7.25 Predicted world primary energy consumption by 2050 ▼

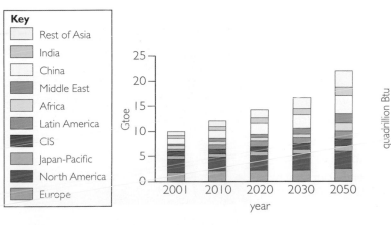

Key
- Rest of Asia
- India
- China
- Middle East
- Africa
- Latin America
- CIS
- Japan-Pacific
- North America
- Europe

Figure 7.26 World coal consumption, 1990-2030 ▼

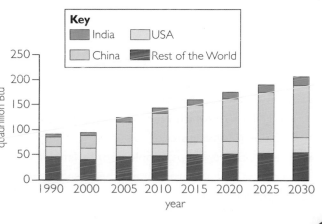

Key
- India
- USA
- China
- Rest of the World

Potential energy gap – is there enough?

By 2030:

- the global demand for energy is predicted to have grown by 50% (with demand from LICs and MICs surging by 85%)
- electricity generation is expected to double (with natural gas and coal providing the primary sources)
- nuclear power will probably have increased its share of the energy mix, as countries turn to recognised and reliable energy sources to plug potential gaps in their supply.

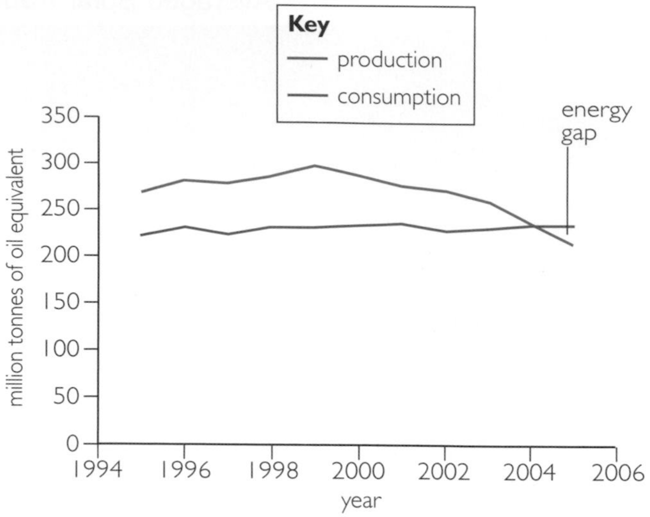

The UK faces a serious energy shortage as old power stations come to the end of their lives and stock resources of North Sea oil and gas are used up. As Figure 7.27 shows:

- energy production in the UK peaked in 1999 – since then, primary energy production has been falling at just over 5% a year
- in 2005, the UK became a net importer of energy, as consumption finally overtook production.

Figure 7.27 *The growing energy gap in the UK* ▲

The idea of **peak oil** (or gas) refers to the point at which global oil production reaches its maximum level. After that point, production will fall into sustained decline. The date of peak oil is not certain, because new finds and changing technology can alter calculations – but it will definitely happen. In fact, some experts believe that peak oil will be reached as early as 2020. After the peak, many forecasters expect global oil production to fall by 3% a year:

- With growing demand, this means that the gap between the amount of oil we want and the amount of oil we get will grow by more than 4% a year.
- Within 10 years, we could have only about half the amount of oil required to sustain economic growth.
- This could lead to spikes in oil prices and deep recessions.

Energy poverty

While some countries have an energy surplus, the unequal distribution of energy sources means that other countries or areas suffer from **energy poverty**:

- 25% of the world's population still have no access to electricity (80% of these people live in rural areas).
- 2.4 billion people rely on traditional biomass for cooking and heating (Figure 7.28).

Not only that, but:

- the use of traditional biomass is killing people – 2.5 million women and children die each year from lung conditions caused by smoke from traditional cooking stoves
- energy poverty keeps people poor by limiting women's ability to gain an education and earn money – because they have to spend hours each day collecting fuel.

Figure 7.28 *Women carrying dried animal dung and wood to use as fuel in sub-Saharan Africa* ▼

Tar sands – a dirty future?

One country that takes the threat of rising oil prices and falling supplies seriously is Canada. Some of the world's largest reserves of tar sands are found there. (Tar sands – also known as oil sands – are naturally occurring mixtures of sand or clay and a very dense viscous form of oil, called bitumen.)

Alberta, in western Canada, has three major deposits of tar sands. Together they cover an area larger than England. Extracting the oil is expensive and difficult, but, by 2030, these tar sands could supply 16% of North America's oil needs.

But there is an environmental price to pay to get at the oil. The tar sands are extracted by opencast mining (Figure 7.29):

- This process is very energy intensive. It takes the equivalent of one barrel of oil to produce three barrels of crude oil from tar sands.

- The energy intensive production process also produces a lot of greenhouse gases.
- It takes 2-5 barrels of water to produce every barrel of oil.
- Two tonnes of mined tar sands are needed to produce one barrel of oil. This leaves a lot of waste and destruction behind.
- So far, 470 km² of woodlands have been removed to allow the opencast mining, and lakes of toxic wastewater cover 130 km².

Figure 7.29 *The Alberta landscape has been scarred by huge opencast mines, toxic waste ponds and hundreds of miles of pipes* ▶

Internet research

Research the energy resources of two countries at different levels of economic development.

a What types of energy resources do they have access to?

b Where are their main reserves found?

c How large are the reserves?

d Are both countries energy secure?

e What are the energy related issues in your chosen countries?

ACTIVITIES

1 Study Figure 7.21.
   a Which 'energy superpowers' currently control the world's energy resources?
   b What problems could this cause?

2 How serious are the concepts of 'peak oil' and 'an energy gap' for:
   a the UK
   b LICs
   c the world at large?

In this section you will learn about:
● how world energy production and distribution are influenced by individual countries and companies

The East Siberia-Pacific Ocean oil pipeline

The East Siberia-Pacific Ocean (ESPO) oil pipeline will be 2600 miles long when completed in 2016 (Figure 7.30). This pipeline will offer Russia a new **energy pathway** into Pacific markets.

The Russian, Chinese and Japanese governments are all interested in oil for a number of **geopolitical** reasons. China and Japan have been competing for access to Russia's oil – and the ESPO pipeline project – in order to secure their future energy supplies.

> **Geopolitics** is the study of the ways in which political decisions and processes affect the use of space and resources. It's the relationship between geography, economics and politics.
>
> An **energy pathway** refers to the flow of energy from the producer to the consumer. The ESPO oil pipeline will allow Russia's oil to flow more easily to China and Japan, plus any other Pacific countries that want to import Russian oil (including the USA).

Inconvenient neighbours

Russia and China share a 4300 km long border, so a good relationship between them is vital. However, they are suspicious of each other – with Russia seeing China as a rival power and potential threat. Therefore, Russia has been reluctant to commit itself too heavily in terms of energy supply to China.

For the reasons outlined in Figure 7.31 opposite, the Japanese government was keen for the ESPO pipeline not to end in China (as originally proposed), but for it to be extended to the Pacific coast (bringing the oil closer to Japan). Therefore, it offered to finance a large proportion of the ESPO project ($7 billion).

The Japanese financing will enable the Russians to build the most expensive pipeline in history, while restricting Chinese access to Russia's oil and helping to rebuild the relationship between Russia and Japan. Russia will also be able to export its oil more widely to other countries in the Pacific region. However, a spur will still be built off the main ESPO pipeline to run into China (Figure 7.30) and, as a 'sweetener', Russia has promised to increase oil exports to China by rail to 300 000 barrels a day.

> **Did you know?**
>
> The ESPO pipeline had to be re-routed to avoid the last remaining habitat of the endangered Amur leopard (on the Chinese border, near Vladivostok). Only 30-40 Amur leopards still survive in the wild.

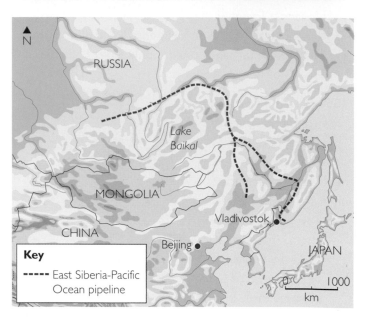

Figure 7.30 *The route of the ESPO oil pipeline* ▲

> **Internet research**
>
> Use the following websites to build up a factfile of key political, economic and environmental issues related to the construction of major international oil/ gas pipelines:
>
> http://maps.grida.no/go/graphic/ major-oil-pipeline-projects
> http://news.nationalgeographic.com/ news/2007/01/070104-gas- pipeline.html

Figure 7.31 *China versus Japan* ▼

China

China's needs for energy, and its relationship with Russia, are complex:

- China needs more energy to support its rapid economic growth (pages 248–251).
- Reliable energy supplies are also vital for maintaining the control of the Chinese Communist Party and the internal security of the country.
- Most of China's current external oil supplies travel by oil tanker through the Strait of Malacca (between Indonesian Sumatra and Malaysia). This Strait is narrow (only 2.8 km wide at its narrowest point), crowded (50 000 ships use it every year), and subject to frequent attacks by pirates. So China wants to develop new energy pathways to widen its supply options (see the graph) and increase its energy security.
- China and Russia have a joint political interest over the issue of US military presence in Central Asia and the Middle East, and also US policies relating to the promotion of democracy and some regime change there. Beijing is concerned that US strategies aim to block or prevent China's expanding economic, political and energy ties.

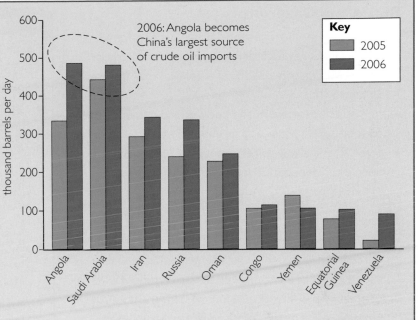

The main sources of China's oil imports in 2005 and 2006 ▲

Japan

Japan also wants some of Russia's oil:

- Japan has almost no oil reserves of its own.
- It is the world's third largest oil consumer, after the USA and China.
- In 2007, Japan imported over three-quarters of its oil from the Middle East (see the graph). Access to the ESPO pipeline would reduce Japan's oil dependence on the Middle East by 10–15%.
- Japan wants to engage with Russia and increase its economic and political influence, which declined in the 1980s and 1990s as China and Russia strengthened their ties.

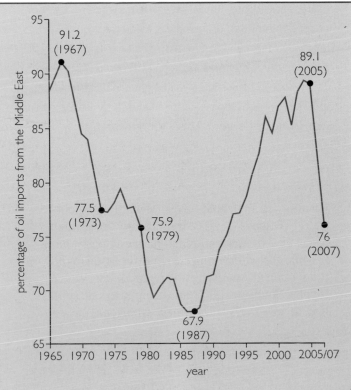

Japan's reliance on Middle Eastern oil ▶

Russia and Gazprom

Russia is re-emerging as a global player. Its economic power lies in its key natural resources – particularly oil and gas – with energy becoming a political tool (Figure 7.32). Energy has:

- helped to re-assert Russia's power and influence over former Soviet states and neighbours
- given Russia a way to restore its international position and regain geopolitical importance.

Critics say that Russia has been using the supply of gas as a weapon:

- In November/December 2004, as part of a peaceful democratic revolution, Ukraine (a former Soviet republic) replaced its pro-Russian government with one led by pro-Western reformers. In January 2006, Ukraine found its gas (supplied by Gazprom – see right) cut off, after Russia decided to quadruple the price and Ukraine's new government refused to pay.
- In March 2008, Gazprom again cut gas supplies to Ukraine over 'disputed debts' (Figure 7.33). However, Ukraine was seeking to join NATO and the EU at the time, which angered Russia.

Alexei Pushkov, a professor of international relations and Russian TV presenter, says that it's a misconception if people think Russia is using gas as an economic weapon. 'Gazprom is an instrument of foreign policy, like American oil companies are instruments of American foreign policy. If Russia was using gas to blackmail foreign countries, we wouldn't have special deals with European nations who are eager to have long-term deals with Gazprom'.

Russia's stranglehold over dwindling global energy resources was dramatically confirmed when new figures showed that the country has become the world's biggest exporter of oil. With production in August hitting record levels, Russia toppled Saudi Arabia from the number one spot. It is already the world's largest exporter of gas.

Figure 7.32 *Adapted from an article in The Guardian in September 2009* ▲

Gazprom factfile

Gazprom has rapidly become one of the world's most powerful companies. Based in Moscow, it is the world's largest gas supply company. Gazprom:

- controls about a third of the world's gas reserves.
- accounts for 92% of Russia's gas production.
- provides 25% of the EU's natural gas. In 2007, it provided 4% of Britain's gas supplies (which is expected to rise to 10% within five years). Over 80% of the gas exports to Western Europe cross Ukraine (**a transit state**).
- is the sole gas supplier to Bosnia-Herzegovina, Estonia, Finland, Latvia, Lithuania, Macedonia, Moldova, and Slovakia.
- is the world's third largest corporation (after Exxon Mobil and General Electric). Gazprom's annual earnings (in 2009) were £61.95 billion.
- began in 1992, when the former Ministry of the Gas Industry was reorganised (the government of the Russian Federation still owns 50.002% of the shares in Gazprom, and Gazprom retains very close links to the State).
- employed 432 000 people in 2006.

http://www.website...

Gazprom cuts Ukraine's gas supply

Gazprom, Russia's gas monopoly, has cut supplies to Ukraine. State-owned Gazprom said it would cut gas shipments to Ukraine by 25%, but Ukraine's national gas firm, Naftogas, has claimed that the reduction is 35%. The dispute centres on a US$1.5 billion (£770 million) debt that Gazprom says it is owed, and Ukrainian officials say has been paid.

Gazprom has offered assurances that the gas supply to the rest of Europe will not be affected. A previous row saw Russia cut the gas supply to Ukraine in 2006, which then hit gas exports to Western Europe that pass through Ukraine.

Figure 7.33 *Adapted from an article on the BBC News website on 3 March 2008* ▲

Gazprom. The soaring concrete office block thrusts towards the sky in downtown Moscow, like a rocket with boosters strapped to its side. After passing through security and handing over your passport to a stern-looking official, you cross a bleak windswept plaza to the tower of power. Once inside, your papers are scrutinised again and your bags put through X-ray machines. It is hard not to think of Fritz Lang's science-fiction film 'Metropolis', with its giant towers filled with toiling workers.

Adapted from an article in *The Independent* on 3 September 2007

Figure 7.35 *Gazprom's headquarters building in Moscow* ▶

Europe's energy security

The cutting off of gas supplies to Ukraine by Gazprom alarmed many European countries, because they depend on Russia for so much of their gas. Was Europe right to be worried about its energy security?
Yes and no.

Yes – because of the amount of gas Russia supplies to Europe (and the fact that most of it comes via three pipelines through Ukraine). All of the gas supply valves lie within Ukraine. When Gazprom shut down the pipeline in 2006, the flow of gas to the rest of Europe fell by 40% in some areas. In the winter of 2009-10, Gazprom cut supplies to Poland, Hungary, Bulgaria and Romania in a new dispute.

Figure 7.36 *Europe depends on this – one of the gas pipelines running through Ukraine from Russia, which was turned off in March 2008 in the second dispute between Russia and Ukraine* ▶

No – for the following reasons:

- Gazprom cannot cut off supplies within Russia (where demand is growing), and its export markets to Western Europe are too valuable to lose. So it's former Soviet states, such as Ukraine and Belarus, which are likely to lose out.
- Even during the Cold War, the supply of Russian gas was stable.
- Gazprom is now helping to secure Europe's energy supplies, with the construction of new pipelines bypassing Ukraine and Belarus (Figure 7.37):
  o The Nord Stream pipeline will be a new energy pathway for Russian gas to Europe. It will run for 1200 km along the bed of the Baltic Sea – with no transit countries involved – thus reducing any possible political interference with energy supplies.

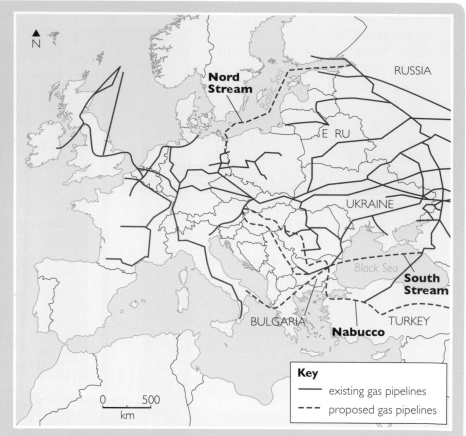

Figure 7.37 *The main gas pipelines across Europe* ▲

  o The South Stream pipeline will run under the Black Sea from the Russian coast to the Bulgarian coast.
- In an effort to enhance its energy security, the EU is planning its own pipeline. The Nabucco pipeline will bring gas from Central Asia and the Caspian Sea across Turkey into the EU. But it may only be able to supply about 5% of Europe's needs.
- The South Caucasus pipeline (opened 2006) will also bring gas from Azerbaijan to Europe via Turkey.
- The EU (including the UK) is looking at alternative energy sources, possibly involving more nuclear power, but this could take years.
- In spring 2010, Gazprom renegotiated its gas prices and contracts with European countries. As the continuing recession hit demand, these countries had reduced their gas consumption and got cheaper supplies elsewhere. Gazprom was worried about this loss of business and income.

What about OPEC?

The Organisation of Petroleum Exporting Countries (OPEC) is a permanent intergovernmental organisation. Its members are oil producing and exporting countries, like Saudi Arabia. For nearly all of them, oil is their main – or only – export, and is vital for their development and social and economic well-being.

OPEC's objective is to 'coordinate petroleum policies to ensure fair and stable prices for producers, an efficient and regular supply to consumers, and a fair return on capital for those investing in the industry'.

OPEC is a powerful player in the global energy supply business. It sets oil production quotas for member countries in response to economic growth rates and demand-and-supply conditions. To maintain stable prices, OPEC boosts supplies when demand rises and reduces them when demand falls. Over three-quarters of the world's proven oil reserves are controlled by OPEC members, so it's a powerful organisation.

2008 was a bad year for oil prices. They surged upwards and the global economy shuddered as the dollar slumped and share prices fell. In response, the USA and other major oil consumers urged producers like Saudi Arabia to boost oil supplies to bring the price down (Figure 7.38).

Figure 7.38 Adapted from an article in The Independent on 16 June 2008 ▶

Top ten oil producers in 2007	Million barrels a day	Top ten oil consumers in 2007	Million barrels a day
Saudi Arabia	10.2	USA	20.7
Russia	9.9	China	7.6
USA	8.5	Japan	5.0
Iran	4.0	Russia	2.9
China	3.9	India	2.7
Mexico	3.5	Germany	2.5
Canada	3.4	Brazil	2.4
United Arab Emirates	2.9	Canada	2.4
Venezuela	2.7	Saudi Arabia	2.3
Kuwait	2.6	South Korea	2.2

Figure 7.39 *Who produces and consumes the world's oil?* ▲

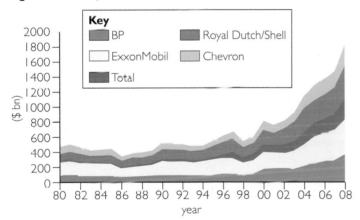

Key
- BP
- Royal Dutch/Shell
- ExxonMobil
- Chevron
- Total

Figure 7.40 *The performance of the big five oil companies, 1980-2008* ▲

Internet research

1. Conduct some research into Gazprom's plans for the future, by referring to www.gazprom.com. Find out more about their pipelines, oil and gas field developments, and new energy sources.
2. Choose either Gazprom, or one of the big five energy companies named in Figure 7.40. Use the relevant company website to develop a geographical profile of a TNC. Show where they operate, what their future plans are, and how they relate to the governments of the areas they work in.

Saudi oil output to rise

At the request of consumer nations, Saudi Arabia has agreed to pump an extra half a million barrels of oil a day from July 2008. The price of oil has risen to $135 a barrel – sparking protests around the world. Saudi Arabia has been under pressure from the USA to increase production – as American petrol prices hit record highs – but the Saudis have argued that the high oil prices are not just caused by excess demand and not enough supply, but also by 'speculators' in the oil markets.

Saudi Arabia is worried that the current high oil price will dampen growth in the industrialised West and lower demand – hurting Saudi Arabia's economy.

In the summer of 2008, protests related to the high price of oil swept across Europe and Asia, like this protest in the Philippines

ACTIVITIES

1. Why are China and Japan so keen to have better access to Russia's oil?
2. In what ways do Gazprom's close links with the Russian government pose a threat to other countries dependent on Russian gas?
3. Hold a class debate on the following: 'Russia uses its gas as an economic weapon in the same way that OPEC uses oil'.
4. Study Figure 7.39. What does it say about the USA's energy security? Does this surprise you? Why?
5. Study Figure 7.40. Comment on the relative performances of the big five oil companies between 1988, 1998 and 2008. Which company consistently performed the best?

In this section you will learn about:
- the negative impacts of energy production on the environment

Skills
In this section you will:
- use atlas maps to plot wind directions and their influence on patterns of energy-related pollution

Energy at any cost?

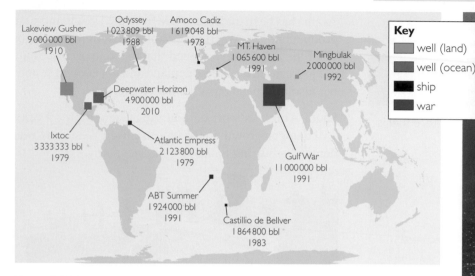

Figure 7.41 *Major oil spills across the world have caused serious environmental problems. The Deepwater Horizon oil spill in 2010 (pictured) caused huge environmental damage to the Gulf of Mexico. It took BP three months to finally stop the oil spewing out of the Macondo Well below the Deepwater Horizon oil platform.* ▲

The fuelwood crisis

Juliette, in Burkina Faso, spends up to three hours a day collecting fuelwood. Twice a week, her husband 'cycles' a bike-load of wood to the market to sell. They earn around £10 a month, and have 'free' firewood for their own cooking. In Burkina Faso, like so many African countries, gathering firewood is a way of life for women. In rural areas, no dead wood is wasted, and – until recently – very little live wood was cut.

For generations, firewood has provided the main source of cooking fuel, heating and lighting in many rural areas of the world. It's cheap, accessible and renewable. However, development and the use of more secondary energy (electricity), has now changed the quantity and type of fuel sources used to support modern lifestyles in HICs and MICs.

But, in LICs, development has not always brought about a big switch to electricity use. Today, vast areas of precious forest are being lost in LICs, as more and more timber is used as fuel. For instance, more than 90% of the wood cut in Burkina Faso is still used for cooking.

Figure 7.42 *Taking fuelwood to market in Uganda* ▲

Did you know?

40% of the world's trees are removed for fuelwood. Gathering dead wood in rural areas for local use is sustainable, but the commercial collection of live wood for urban areas is now outstripping new growth.

Poor people living on the margins know that trees provide their fuel, and they will not normally cut live wood – because that's unsustainable and undermines their future fuel supplies. So why are many forests in LICs being 'harvested' – with deforestation and desertification the inevitable results? Figure 7.43 provides an explanation.

Urbanisation is the real problem. Demands for fuel in urban areas are much greater than in rural areas, because:

- urban populations are larger and are growing faster
- eating habits in urban areas are changing to three warm meals a day (still only one in rural areas), so more fuel is needed to heat the food
- charcoal is the preferred fuel source in urban areas – it's lighter to carry than wood, but contains only 40% of the original wood's energy, so more is needed for the same result (Figure 7.44)
- the demand for charcoal also means that larger lumps of wood are needed to create it – not twigs! (Charcoal is made by heating wood in the absence of oxygen to carbonize it and drive out the water.)

The sale of fuelwood is a relatively recent phenomenon in rural areas. People there used to meet their fuel needs by using dead twigs and small branches collected close to their homes. However, population growth – and the realisation that there is money to be made by selling wood to urban people – has encouraged large-scale woodcutting of live wood. Juliette's role is being repeated right across Africa. She collects the wood that is eventually trucked to the city. Her husband also makes and sells sacks of charcoal.

Charcoal production is illegal in most countries, because it accelerates the removal of the forests and governments are attempting to control the rate of fuelwood cutting. The Ministry of the Environment in Burkina Faso sells licences to limit tree cutting. Juliette's husband now has to cycle to the market three times a week to maintain their income, and she has to cut live trees to pay for the licence fees.

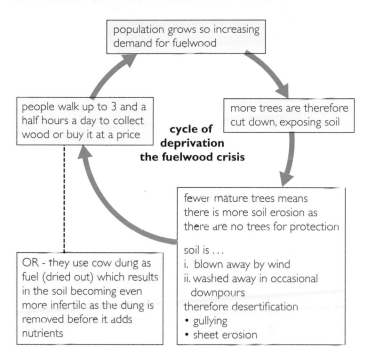

Figure 7.43 *The vicious cycle of fuel deprivation* ▲

Figure 7.44 *Wasting primary energy sources like there's no tomorrow – a charcoal seller in Khartoum, Sudan (a country in the Sahel region, like Burkina Faso, at risk of deforestation and desertification)* ▲

Burning issues

While the forests burn, and primary energy sources are wasted, it's also clear that fossil fuels are not going to last forever. If the world maintains a 'business as usual' approach (i.e. that it will continue to rely on fossil fuels), then it will have to face up to rising prices as fuel stocks dwindle – as well as the increasing threat of climate change. The International Energy Agency's *World Energy Outlook 2010* report leads with a telling statement:

'The message is simple and stark: if the world continues on the basis of today's energy and climate policies, the consequences of climate change will be severe. Energy is at the heart of the problem – and so must form the core of the solution.'

What is acid rain?

Acid rain occurs when emissions of sulphur dioxide and nitrogen oxides react in the atmosphere with water, oxygen and oxidants to form various acidic compounds, including sulphuric acid and nitric acid. These then fall to the earth as dry particles, gas, or rain, snow and fog (Figure 7.45).

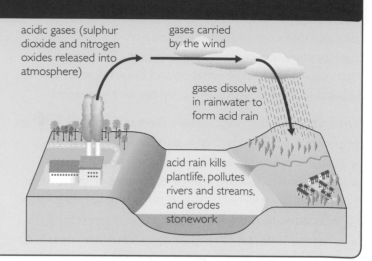

Figure 7.45 *The formation of acid rain* ▶

Acid rain returns

Acid rain held 'celebrity status' in the 1980s and 90s. Talk of corroded buildings, dying forests and dead lakes drove home the message that burning fossil fuels (which give off sulphur dioxide and nitrogen oxides) has a serious environmental impact. It was the first big **trans-boundary pollution** event.

Most industrialised nations took action to address the acid rain problem. In the UK, the coal industry took the blame as the main culprit. Since then, coal-fired power stations have been phased out, or filters have been fitted at great expense, to try to clean up Europe's air. The 1979 Geneva Convention set emission targets – and international cooperation, Clean Air Acts and improved technology have reduced the acid rain problem in both Europe and the USA.

But the global threat is far from over. Asia is predicted to be the new acid rain hotspot (Figure 7.46). Energy use has surged in China, India, South Korea and Thailand. Much of that energy is being generated using coal, which releases a lot of sulphur dioxide into the atmosphere. The effects of the resultant acid rain are already widespread:

- Indian crop yields close to the power plants are down by 49%.
- Sichuan and Guizhou provinces in China are experiencing acid rain over 66% of their farmland.
- The growth rates of pine and oak trees across South Korea are declining.
- Water supplies are more acidic, which has implications for human health as well as for fish and wildlife.
- Air quality is predicted to be at dangerous levels across large parts of Asia by 2020 (Figure 7.47)

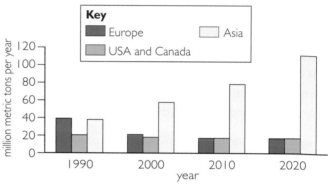

Figure 7.46 *Growing sulphur dioxide emissions in Asia* ▲

Figure 7.47 *Predicted sulphur dioxide concentrations across Asia by 2020. The World Health Organization recommends that concentration levels for acceptable air quality should be no higher than 15-20, so large areas of China – in particular – face air-quality levels that would be dangerous to health.* ▼

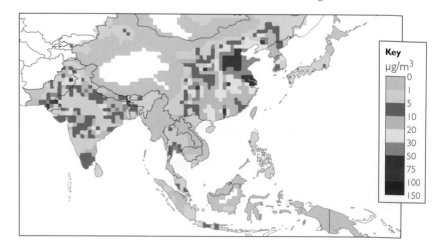

The nuclear alternative

On 26 April 1986, the world awoke to the news that a huge explosion at a nuclear power plant in Chernobyl, Ukraine, had released 100 times more radiation into the atmosphere than the atomic bombs at Hiroshima and Nagasaki in 1945. This radioactive fallout spread right around the world (Figure 7.48).

Fears surrounding nuclear power continue to haunt the public. Yet this extremely expensive and potentially hazardous way of generating electricity is increasingly being seen as the solution to different countries' energy security problems (Figure 7.49). Figure 7.50 gives the alternative view.

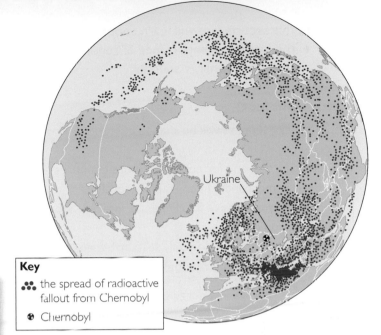

Key

⬡ the spread of radioactive fallout from Chernobyl

⊕ Chernobyl

Figure 7.48 *Silent but deadly clouds of radioactivity* ▲

China embraces nuclear future

China intends to spend $50 billion to build 32 nuclear power stations by 2020. Some analysts say that it will build 300 more by 2050, which will generate about the same power as all the nuclear power stations currently operating in the world today. By comparison, the USA currently has just over 100 operating nuclear power stations.

China's plans are being greeted with both optimism and concern. Nuclear power stations release few greenhouse gases. However, their safety and the issue of what to do with the radioactive waste are problems which the Chinese government still has to solve.

Figure 7.49 *Adapted from an article in* The Washington Post, *29 May 2007* ▲

ACTIVITIES

1 Explain why and where there is an emerging fuelwood crisis.
2 What is meant by 'trans-boundary' pollution?
3 Why has acid rain re-emerged as an environmental problem?
4 Use an atlas to:
   a plot the major industrial areas on an outline map of Asia.
   b draw on the prevailing wind patterns over Asia.
5 How strong is the link between your annotated map of Asia and the sulphur dioxide patterns shown in Figure 7.47?

Greenpeace China gives the following reasons why Asia should not develop nuclear power:

- The costs associated with safety, security, insurance, liability in case of accident or attack, waste management, construction and decommisioning are rising rapidly for nuclear power. But the costs of solar and wind power are falling.
- Nuclear waste disposal is an unsolved problem. The waste remains radioactive for up to 10000 years, and no safe containment solution has been worked out yet.
- Nuclear technology is also used in nuclear weapons production, so there is a risk of nuclear weapons proliferation (one of the reasons why some people are worried about Iran and North Korea).
- Mining, extracting, processing and transporting nuclear fuel produces carbon dioxide emissions at every stage.
- Any investment in nuclear power is money denied to developing and promoting renewable energy and energy efficiency.

Figure 7.50 *Nuclear power – yes or no?* ▲

Internet research

a Find out how European countries dealt with acid rain 20 years ago.
b In pairs, look up the history and ongoing environmental problems associated with nuclear power – the BBC websites on the right will help.

http://news.bbc.co.uk/1/shared/spl/hi/guides/456900/456957/html/nn3page1.stm

http://news.bbc.co.uk/1/shared/spl/hi/guides/456900/456932/html/nn1page1.stm

c Find out why managing spent nuclear fuel also causes environmental difficulties.

Sustainable energy supply

In this section you will learn about:
- the increasing importance of renewable and sustainable energy sources

Future perfect?

Future energy supplies will be shaped by three driving forces:
- Increasing demand for energy, as more countries – especially India and China – rapidly develop.
- Increasingly expensive fossil fuels, as available stocks begin to run out.
- The search for technologically sustainable energy sources, driven by fears about global climate change.

It's clear that renewable energy sources will play an increasing role in meeting global demands for energy (Figures 7.51 and 7.52). But, what are the renewable options? Are they readily available and can they plug any future energy gaps?

Defining renewable energy

It makes sense to use energy sources that are renewable. So why have they not been used before? Well, in fact, they have. Windmills have long been used to drive pumps and provide energy for milling. And waterwheels were central to early industrialisation. Fossil fuels just took over from these traditional sources because they could provide more energy in the form of electricity, more cheaply and over larger areas.

The main renewable energy sources today are:
- biomass – energy from organic materials like wood, plants, animal waste, general waste
- solar power – energy direct from the sun can be used to provide heat and to generate electricity
- wind energy – moving air turns a propeller-driven generator to produce electricity (Figure 7.53)
- wave energy and tidal energy – moving water flows through a barrage, releasing energy
- hydroelectric power (HEP) – the vertical release of water turns the electricity generator

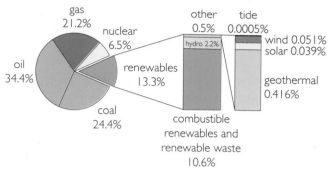
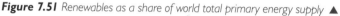

Figure 7.51 *Renewables as a share of world total primary energy supply* ▲

Figure 7.52 *Average annual global growth rates of various energy sources* ▼

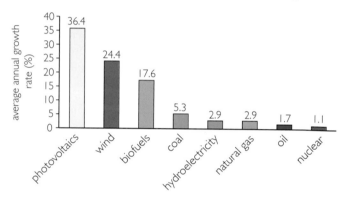

Figure 7.53 *The offshore wind farm at Thanet in Kent* ▼

CASE STUDY

Wind power

The British government anticipates that an extra 4000 onshore and 3000 offshore wind turbines will be built as part of a £100 billion boost to the renewable energy strategy. From June 2010, plans to increase clean energy production by effectively introducing a 'carbon tax' on 'dirty fuels', means that energy companies in the UK are likely to turn increasingly to wind power. Agreements are already in place for massive wind farms in the North Sea – and the UK is set to become the world leader in offshore wind power. An underwater 'supergrid' will be constructed to transmit electricity to the mainland.

However, some wind power schemes are less ambitious. Delabole's wind farm in north Cornwall is a good example. In 2010, Delabole's ten old wind turbines were replaced by four more-efficient ones – generating 9.2 MW of electricity for 7800 local homes.

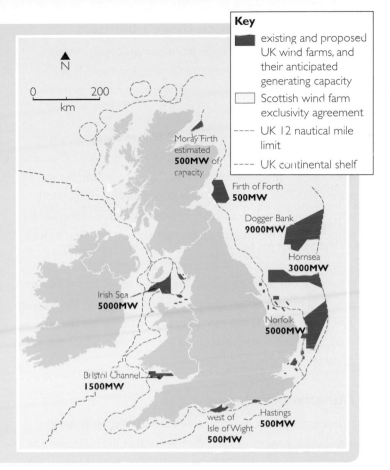

Key

- existing and proposed UK wind farms, and their anticipated generating capacity
- Scottish wind farm exclusivity agreement
- ---- UK 12 nautical mile limit
- ---- UK continental shelf

0   200 km

Moray Firth estimated **500MW** of capacity

Firth of Forth **500MW**

Dogger Bank **9000MW**

Hornsea **3000MW**

Irish Sea **5000MW**

Norfolk **5000MW**

Bristol Channel **1500MW**

west of Isle of Wight **500MW**

Hastings **500MW**

Figure 7.54 *The UK's existing and planned offshore wind farms* ▶

CASE STUDY

Wave and tidal power

Once again, the UK is well positioned to take advantage of its geography:

- In January 2009, the Scottish government granted permission for the world's largest wave-energy project. Harnessing power from the Atlantic waves will generate enough electricity for 1500 homes in the Western Isles.
- Cornwall could also become a leader in wave technology. A project named the 'WaveHub' is being considered off the southwest coast. It could produce electricity for 7500 homes and save 300 000 tonnes of carbon emissions every year.
- Eight of the UK's river estuaries also offer prime sites for major tidal energy projects. Daily tides move vast amounts of water, and it has been suggested that a fifth of the UK's energy could be produced in this way. The Dee, Solway, Humber and Severn Estuaries are likely sites. Only 20 viable sites have been identified in the whole world.

Figure 7.55 *The tidal power scheme on the Rance Estuary, in France, was opened in 1966 - and is still the only one in the world. Although it can only generate power when the tide is flowing in or out, that is entirely predictable and reliable.* ▶

Where there's muck there's biomass

In October 2010, gas made from human sewage was pumped into the national grid for the first time. It will be used in homes for cooking and heating, just like normal natural gas.

Sewage from Thames Water's 14 million customers is treated at Didcot in Oxfordshire. Biogas is given off when the sludge – made up of solid waste – is warmed up in giant vats. This process is known as anaerobic digestion, and involves the breakdown of biodegradable materials by bacteria. The resultant gas, biomethane, is purified before being piped into the grid.

It takes around 20 days from flushing a toilet to the biogas reaching homes. The £2.5 million project currently supplies 200 homes, and is the first of many similar projects. It's thought that biomethane could account for 15% of the UK's domestic gas supply by 2020.

Adnams brewing up a gas

Adnams Brewery in Southwold, Suffolk, has started a pioneering energy system that generates biomethane gas from brewery and local food waste. The bio-energy plant generates enough biogas to heat more than 230 family homes for a year. Microbes accelerate the decomposition of the waste – yielding biomethane and liquid fertiliser.

The new anaerobic digestion facility pumped gas into the National Grid for the first time on 9 October 2010. Adnams' own delivery lorries will also be converted to run on the gas, and the liquid fertiliser will be used to fertilise the fields where the brewery's barley is grown.

Eddie Stobart announces renewable energy venture

In March 2010, the Stobart haulage company announced a 20-year contract to deliver 750 000 tonnes of biomass fuel a year to ten potential sites across the UK. For example, Stobart's trucks will take wood chippings to the Steven's Croft biomass power station at Lockerbie in Scotland, which opened in 2008 and provides electricity for 70 000 homes.

Figure 7.56 ▲

Adnams Bio-energy will have the capacity to break down 12 500 tonnes of organic waste each year – enough to power a family car for 4 million miles. The brewery itself uses solar panels, photovoltaic cells and harvests its own water – so it's already highly sustainable.

◄ ▲ **Figure 7.57** Installing the anaerobic digestion tanks (right), and the completed Adnams bio-energy facility (left)

Small-scale appropriate technology for sustainable development: micro-hydro

96% of Kenyans have no access to grid electricity. Mbuiru village is a typical Kenyan village – 200 km north of Nairobi. The Tungu-Kabri Micro-hydro Power Project (funded by the United Nations Development Programme, and developed by Practical Action East Africa and the Kenyan Ministry of Energy) harnesses the energy of falling water to create electricity. The project is cheap, sustainable and small-scale. It generates enough electricity to benefit about 1000 people – providing light, saving time, and allowing people to run small businesses. It is an example of appropriate technology working for the needs of people by making use of local renewable resources.

Figure 7.58 *Local villagers were involved in building the micro-hydro project* ▲

Internet research

Research a variety of sources and case studies of renewable energy sources and compile a SWOT analysis to show the strengths, weaknesses, opportunities and threats posed by each source.

These websites will help:

- Renewable energy sources: http://home.clara.net/darvill/altenerg
- Wind in the UK: http://www.bwea.com/ref/faq.html and http://www.bwea.com/ukwed/offshore.asp
- Post Copenhagen Summit action in the UK: http://www.actoncopenhagen.decc.gov.uk/en/ukaction/business/case-studies/woking-council
- CHP schemes: http://www.cowi.com/menu/news/newsarchive/utilities/Documents/Case%20stories%20from%20central%20Copenhagen.pdf
- Solar in the Mohave Desert: http://www.inhabitat.com/2008/04/10/mojave-desert-solar-power-fields/
- Wave and tidal power: http://news.bbc.co.uk/1/hi/8564662.stm
- Biomass: http://www.biomassenergycentre.org.uk/portal/page?_pageid=76,15049&_dad=portal&_schema=PORTAL

ACTIVITIES

1. Define the main types of renewable energy sources.
2. Explain why renewable energy sources are likely to take on a greater role in providing the world's electricity.
3. Define biomass energy and identify the different sources that can be used to produce biogas.
4. Do you think that biomass energy will contribute increasing amounts to the UK's energy mix? Why has it not been used in the past on a large scale?

In this section you will learn about:
- the increasing importance of reducing energy consumption

Making the future work

Some of the old coal-fired power stations are only 40% efficient, and all over the world lots of energy is wasted. So it's not just a case of finding new ways to produce more and more electricity. **Demand management** techniques can be used to help reduce energy consumption and demand in the first place. These techniques normally involve charging energy inefficient consumers more and reducing waste. Conserving energy and reducing the amounts we actually use could reduce the size of any future energy gap.

Decentralising energy generation and developing local energy generators, along with energy efficiency measures, can be very effective, as the two case studies below show.

CASE STUDY

Copenhagen's combined heat and power

Copenhagen's combined heat and power (CHP) system supplies 97% of the city with clean, reliable and affordable heating – as well as 15% of Denmark's total heating needs. The system was set up in 1984 as a partnership between local councils and energy companies.

The CHP system uses a combination of:
- waste heat from electricity generation (normally released into the sea or rivers, but now taken through pipes to people's homes)
- surplus heat from incinerating waste
- geothermal energy
- biofuels (wood pellets and straw)
- and small amounts of natural gas, oil and coal.

By 2005, household fuel bills in Copenhagen were 1400 euros lower than if oil had been used for heating. And, by 2009, carbon dioxide emissions were down to 0.6 million tons – compared with 3.5 million in 1995. Sulphur dioxide emissions have also fallen by 35% over the same period. It's a clean, cheap and efficient system, that wastes only 6% of the energy generated. In other words, it captures most of what used to be wasted. A sophisticated computer-controlled system also redirects excess heat from one area to another which needs more. The Danish government supports the scheme by offering tax incentives to energy companies.

CASE STUDY

Low-carbon homes

Woking Borough Council in Surrey has reduced carbon dioxide emissions by 82% and energy consumption by 52%. It has developed a network of 60 local generators, near to where the electricity is actually used. They are used to power, heat and cool council buildings and social housing – as well as town centre businesses. The sustainable installations include the use of solar power (PVs) to provide background heat and CHP. The schemes cost the council £12 million to set up, and the first 5 years saw energy bill savings of £5.4 million.

In 2008, Woking introduced the Low-Carbon Homes Programme to encourage its residents to minimise their carbon dioxide emissions and water consumption. The first phase of the programme was to convert a detached house as an example, to show homeowners what steps they could take for themselves to conserve energy and reduce their carbon footprint (Figure 7.59).

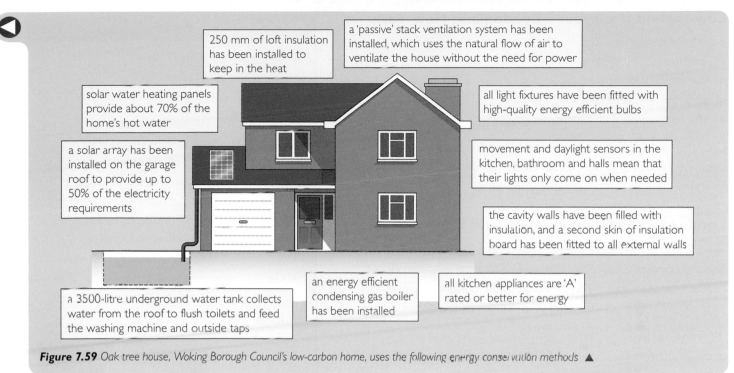

250 mm of loft insulation has been installed to keep in the heat

a 'passive' stack ventilation system has been installed, which uses the natural flow of air to ventilate the house without the need for power

solar water heating panels provide about 70% of the home's hot water

all light fixtures have been fitted with high-quality energy efficient bulbs

a solar array has been installed on the garage roof to provide up to 50% of the electricity requirements

movement and daylight sensors in the kitchen, bathroom and halls mean that their lights only come on when needed

the cavity walls have been filled with insulation, and a second skin of insulation board has been fitted to all external walls

a 3500-litre underground water tank collects water from the roof to flush toilets and feed the washing machine and outside taps

an energy efficient condensing gas boiler has been installed

all kitchen appliances are 'A' rated or better for energy

Figure 7.59 *Oak tree house, Woking Borough Council's low-carbon home, uses the following energy conservation methods* ▲

A brighter energy future?

Turning to renewables and reducing pollution-generating fossil fuel dependency could lead to more lights at night and a brighter future for all. Current trends are shown in Figures 7.60 and 7.61. Green taxation or reward schemes may be needed to complement nationwide strategies and encourage local action. Individual homes with solar panels, wind turbines and CHP systems are developing fast – and also offer technological fixes to the potential energy gap. A shift away from private transport to public or mass transport infrastructures will also change the shapes of our cities.

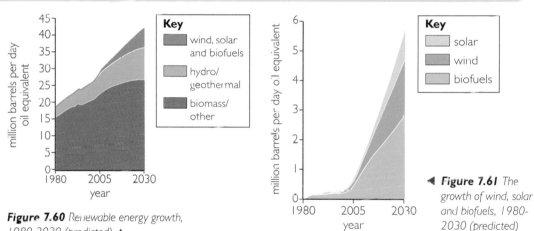

Figure 7.60 *Renewable energy growth, 1980-2030 (predicted)* ▲

◄ **Figure 7.61** *The growth of wind, solar and biofuels, 1980-2030 (predicted)*

Managing supply and demand

Most countries are managing their energy supply to some extent. You have already seen in this chapter how China and the UK are managing their supplies by, for example, changing their energy mixes and considering more use of nuclear power and renewables like HEP and wind power.

But countries find it much harder to manage demand. In the UK, official building regulations aim to ensure the energy efficiency of new buildings by encouraging the use of things like double-glazing and cavity wall insulation as standard. Government grants are also available for the installation of loft and cavity wall insulation, and solar panels, in existing buildings. Electrical appliances carry energy-use ratings to encourage consumers to think about their energy demands. Some power companies have also offered their customers free electricity monitors, so that they can keep tabs on their day to day energy use. The companies then offer rewards if the electricity use falls by a certain amount. But, so far, measures like these have only had a small impact on the UK's demand for energy.

Low-carbon transport – London and New York

Traffic congestion in cities means that:

- transport is inefficient
- fuel is wasted
- pollution is increased
- the environment and the economy suffer.

Charging individual road users to enter some areas has the combined effect of reducing fuel consumption and improving urban lifestyles.

London's congestion charging and Boris's bicycles

Since 2003, drivers have been charged £8 a day to drive in the Central London Congestion Zone. The scheme was introduced because London had the UK's worst traffic congestion. The cost to the economy of drivers spending half their time in queues was estimated to be £2-4 million a week.

In theory, the congestion charge should encourage people to make more use of public transport, which is cheaper and more fuel-efficient. The income from the congestion charge has been used to replace or upgrade every vehicle in the fleet of London buses. The positive effects of the congestion-charging scheme have been dramatic:

- Traffic in the zone has fallen by 21%.
- Bus passenger numbers have increased by 45%.
- Cyclists have increased by 43%.
- Road traffic accident rates are down by 5%.
- Carbon dioxide and nitrous oxide emissions have fallen by 12%.
- There has been 60% less disruption to bus services.
- Retail sales in the area have increased.

New York – a failed attempt

Other cities have considered copying London's strategy. In 2007, New York City Mayor, Michael Bloomberg, outlined plans to reduce emissions and fuel consumption by:

- increasing electricity bills by $2.50 a month – to pay for improved power stations. This would cost every household $30 a year until 2015, but save them $240 a year after that.
- congestion charging in southern Manhattan – with the aim of raising $380 million in the first year to fund public transport and pay for a new transport authority to organize public transport schemes.

However, these ideas faced political opposition and, in 2008, the state government shelved them. Many argued that the 'tax was regressive and would penalise the city's poor'.

Boris's bicycles

'*London is being gradually liberated from the tyranny of the combustion engine.*'

(*Guardian*, October 2010)

Ten weeks after London's Mayor – Boris Johnson – launched a city wide cycle hire scheme, the millionth ride was achieved. 6000 bicycles have been provided so far, along with docking stations and 12 'cycle super highways' – at a cost of £140 million. Barclays Bank has provided £25 million of sponsorship, and the scheme is set to expand to 10 000 bikes. This bicycle hire scheme is designed for short journeys, and the first 30 minutes of bike hire are free.

Figure 7.62 *Boris Johnson promoting London's bicycle hire scheme* ▲

Figure 7.63 *Traffic congestion in Manhattan* ▲

Workplace efficiency

Many parts of the world are starting to reconsider the ways in which architects design buildings. The new German parliament building (Figure 7.64) uses renewable energies for 80% of its fuel.

In 2001, The UK Climate Change Levy was introduced to 'encourage improved energy efficiency and reduce greenhouse gas emissions' (UK Carbon Trust). All businesses have to pay this tax, but employers do receive tax incentives if they adopt energy efficiency schemes and other good practices.

Businesses have responded to the levy in a number of ways:

- Turning everything off when it's not in use
- Installing movement-sensitive lights
- Having regulated air conditioning and heating (that's only on when work spaces are actually being used)
- Recycling and re-using materials to reduce waste
- Using SMART metering systems
- Completing energy audits
- Installing solar panels, heat transfer systems, and insulation
- Training all staff and reviewing the energy situation regularly

ACTIVITIES

1 Make a list of the broad issues faced by countries as they decide on their 'energy futures'.
2 Define the terms: demand management, combined heat and power systems, congestion charging.
3 Discuss with a partner the difficulties that individuals might face in reducing their own energy demand.

The Reichstag's energy system is based on a mixture of solar energy, geothermal power, combined heat and power, biofuel generators, and innovative ventilation. Special insulation limits heat loss. More than 80% of the electricity it uses is generated internally. A geothermal installation cools the building in summer and provides heat in winter. Annual carbon dioxide emissions have been reduced from 7000 to less than 1000 tons.

Figure 7.64 *The German Reichstag (parliament) building* ▲

Internet research

In groups, devise a case for a multi-million pound research project into one of the following:

a Energy conservation and building regulations
b A CHP scheme for every town over 100000 people
c Alternative energy sources

Use the following websites to help.

British government:
- http://www.environment-agency.gov.uk/business/topics/pollution/98263.aspx

- http://www.decc.gov.uk/en/content/cms/what_we_do/lc_uk/crc/crc.aspx

Somar International:
http://www.energysavingnews.com

Warwickshire County Council:
http://www.actonenergy.org.uk/switch-it-off-campaign

Transport for London:
http://www.tfl.gov.uk/roadusers/cycling/15025.aspx

bio-fuel Crops grown for fuel

Combined Heat and Power (CHP) Power stations that produce electricity and which also use their surplus heat either by recycling it, or by piping it off to local homes so that it is not wasted

energy dependency The level of energy imports as a proportion of total energy consumption. The higher the proportion of energy imports, the more energy dependent the country is on others

energy efficiency The balance between the amount of energy used and that lost, e.g. up chimneys of power stations

energy gap The difference between energy supply and demand

energy mix Primary energy sources (e.g. coal, wood, oil, gas) used to meet demand

energy pathway Refers to the flow of energy between producer and consumer

energy poverty Having less energy than is required to meet demand

energy security Means having access to reliable and affordable sources of energy

energy deficit Means having insufficient energy to meet demand

energy surplus Having more energy available than is required to meet demand

flow resource Infinite, continuous energy sources which can be constantly renewed

fossil fuel Fuel produced over millions of years from the breakdown of organic material, e.g. wood into coal

geopolitics The study of the ways in which political decisions and processes affect the way space and resources are used. It is the relationship between geography, economics, and politics

hydro-electricity Energy generated by moving flows of water

low carbon economy An economy which seeks to use low amounts of carbon-based energy

low carbon homes Homes which are deliberately intended to use minimal energy

peak oil The theoretical year in which oil is or was produced at its maximum and which can never be repeated as stocks decline

primary energy source Resources which can be burned to generate energy direct, e.g. coal, oil, and natural gas.

privatisation The sale of government assets to private shareholders, so that a company is run for profit rather than as a government service

recyclable resource Reprocessed resources, e.g. plutonium and heat capture systems

renewable energy sources Those whose flow is continuous, e.g. solar or wind power

secondary energy source Energy sources generated using another fuel, e.g. electricity

solar power Energy generated from the sun

stock resource Finite energy sources whose use means they will eventually run out

tar sands Also known as oil sands, these are naturally-formed mixtures of sand or clay and a dense viscous form of petroleum called bitumen

trans-boundary pollution Pollution which crosses national boundaries, e.g. acid rain caused by burning fossil fuels which emit sulphur dioxide and nitrogen oxides

transit state A country or state through which energy flows on its way from producer to consumer

wave power Energy generated from waves

Exam-style questions

1 (a) Study Figure 7.9 (page 247). Outline the key changes in the UK energy mix 1965-2006. *(4 marks)*

(a) Focus on a few important changes only.

(b) Suggest two reasons for the changes you have outlined. *(4 marks)*

(b) Don't write too much – 4 marks, achieved by brief explanation of two points.

(c) Explain the challenges that lie ahead for the UK if these changes in energy mix were to continue. *(7 marks)*

(c) Think about the future and what could happen if current trends are allowed to continue.

(d) Referring to one country you have studied, explain how its energy mix is creating challenges. *(15 marks)*

(d) You must name a country – so which one? Russia? China? The UK?

2 (a) Study Figure 7.18 (page 251). Describe the distribution of China's energy resources. *(4 marks)*

(a) Stick to description only.

(b) Explain one challenge created by the distance between China's energy resources and the distribution of its major cities. *(4 marks)*

(b) Consider factors such as distance and where the demand is.

(c) Outline two environmental issues created by the energy industries in a named country that you have studied. *(7 marks)*

(c) Consider environmental issues such as air pollution..

(d) Using named examples, explain the options open to countries which don't have sufficient energy resources of their own. *(15 marks)*

(d) Use case study information. What can countries do? Buy some? Develop new sources?

3 (a) Study Figure 7.39 (page 261). Outline two features of the pattern shown in the data. *(4 marks)*

(b) Explain one issue that can arise when energy resources are concentrated in the hands of a few providers. *(4 marks)*

(c) Explain why some countries are facing a 'fuel-wood crisis'. *(7 marks)*

(d) Using examples, assess the view that the world's energy supplies will have to come from renewable resources in future. *(15 marks)*

Health issues

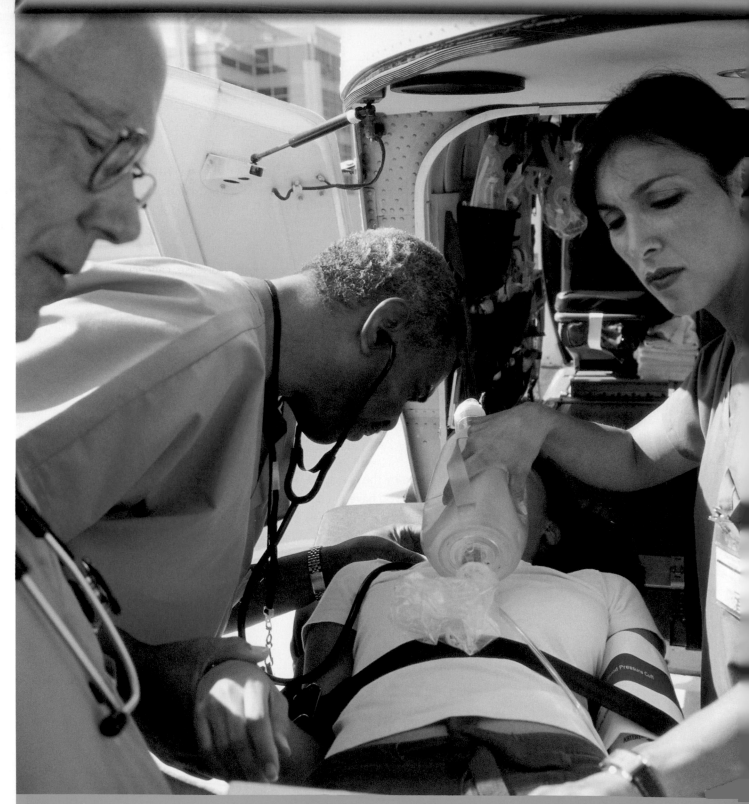

Someone's life hangs in the balance

Where do you think this photo was taken?

The person on the stretcher – unlucky? Or lucky?

Is this the sort of care you would expect to get?

How much would you pay for this sort of care?

What proportion of the world's population would get care like this?

Introduction

Globally, many diseases are found more in some regions rather than others. Environmental factors, such as climate, influence the distribution of some diseases. Malaria would be an example. Human behaviour influences the distribution of others. Examples include patterns of sexual behaviour and HIV/AIDS, smoking and lung cancer, and diet and heart disease and cancer.

In this chapter, you will learn about different health challenges, such as HIV/AIDS and cancer. You will also see how inequalities of wealth – globally or locally – can significantly affect human health. In all cases, success in meeting health challenges depends on human systems – for example, who makes decisions, or how health programmes are funded.

Books, music, and films

Books to read

The Spirit Level: Why More Equal Societies Almost Always Do Better by Richard Wilkinson

Music to listen to

'Smokers Outside the Hospital Door' by The Editors

Films to see

Zombieland
I Am Legend
Doomsday

About the specification

'Health issues' is one of the three Human Geography option topics in Unit 1 – you have to study at least one.

This is what you have to study:

- The global patterns of health, morbidity, and mortality. The role and importance of health in world affairs.
- The study of one infectious disease (e.g. malaria, HIV/AIDS), including its global distribution and its impact on health, economic development, and lifestyle.
- The study of one non-communicable disease (e.g. coronary disease, cancer) including its global distribution and its impact on health, economic development, and lifestyle.
- Food and health – causes and patterns of malnutrition, periodic famine, and obesity.
- Contrasting health care approaches in countries at different stages of development.
- Health matters or issues in a globalising world economy – transnational corporations and pharmaceutical research, production, and distribution; tobacco transnationals.
- Regional variations in health and morbidity in the UK.
- Factors affecting regional variations in health and morbidity – age structure, income and occupation type, education, environment, and pollution.
- Age, gender, and wealth and their influence on access to facilities for exercise, health care, and good nutrition.
- A local case study on the implications of the above for the provision of health care systems.

In this section you will learn about:
- how geography and health issues are inter-related
- how health issues are never far from the news

Figure 8.1 is not an image of a third-world country plagued by war, famine, or natural disaster – but the USA – the world's richest country. However, despite this wealth, nearly 46 million Americans (15.3% of the population) have no health insurance at all, and millions more are under-insured. The USA does not have a system of free health care for everyone, like the NHS, so health insurance is crucial for anything more than the most basic of health needs. Full health care in the USA is not considered a basic right, but is the responsibility of individuals to provide for themselves – and many Americans, like those in Figure 8.1, just cannot afford to do this.

Figure 8.1 *The health charity Remote Area Medical (RAM) at work in the USA. RAM offers free health care treatment to Americans who do not have health insurance and cannot afford the treatment they need. The charity is staffed by volunteer doctors, dentists and other health care professionals.* ▲

Health care reform – USA vs. UK

In 2010, the USA and the UK both embarked on changes to their health care systems. These changes caused a lot of heated debate in both countries. The issue of health and health care is always an emotional one, because nobody likes to see themselves or their loved ones in pain. Any health care debate will generally raise questions about:
- whether there should be a basic right to health care
- who should provide that care
- and perhaps, most importantly, who should pay for it?

As you can see in Figure 8.2, the USA spends nearly twice as much on health care (as a percentage of GDP) as the UK. However, over 52% of that health care is privately funded, whereas in the UK most health care is publicly (or State) funded, through the NHS. So there are big differences in the approach taken to health care in these two HICs.

	USA	UK
Expenditure on health care (% of GDP)	16	8.4
Expenditure on health care (per person in US$)	7290	2992
Expenditure from the private sector (%)	52.8	12.9
Life expectancy at birth (years)	78	79
Death rate (per 1000 population)	8.4	10
Infant mortality rate (per 1000 live births)	6.2	4.8

Figure 8.2 *How the USA and UK compare* ▲

President Obama's health care plan

On 21 March 2010, the US House of Representatives voted to pass a health reform bill that would enable more than 32 million extra Americans to gain access to health insurance. The passing of this bill was a hotly contested issue in America. President Obama had to tour the country drumming up support for his plan. Attempts by previous Democratic Presidents, such as Bill Clinton, to enact health care reform had all failed. Even in 2010, President Obama's health reform bill was passed by just seven votes! Not a single Republican representative voted to support the bill. The political divide in America had never been so stark.

Figure 8.3 *President Obama signing the historic Health Insurance Reform Bill in the White House* ▶

A summary of US health care reforms in 2010

- The cost of the reforms was expected to be $940 billion over 10 years.
- 32 million extra Americans, who had no health insurance at all, now had access to it. But that still left millions of Americans without health insurance.
- Health insurers could no longer deny coverage to those with pre-existing medical conditions, or remove coverage if someone became long-term sick.
- The self-employed, who had no health insurance provided for them by employers, could buy health insurance through new insurance exchanges.
- Low-income individuals and families who wanted to purchase their own health insurance were now eligible for government subsidies.
- A new 'individual mandate' was introduced, which meant that those not covered by Medicaid (government-funded healthcare for those on very low incomes), or Medicare (government-funded health care for the over-65s), must have health insurance or pay an annual fine.

The Coalition's NHS reforms

Only two months after coming to power, David Cameron's new Coalition government announced that the National Health Service (NHS) was to have a major shake-up – one of the most radical in its history. The government's White Paper, announced on 12 July 2010 (Figure 8.4), proposed that:

- £80 billion of NHS funding would be transferred to General Practitioners (GPs), to enable them to purchase health care for their patients directly
- primary care trusts and strategic health authorities – the bodies that currently organise local health care – would be abolished by 2013.

The aim of the reforms was to make the NHS more responsive to patients' needs, and to give GPs the power to select appropriate health care for their patients. The government also hoped that the changes would save money – the NHS had to save up to £20 billion by 2014.

Figure 8.4 *Secretary of State for Health Andrew Lansley announces the Coalition government's White Paper on health in Parliament* ▲

Health issues around the world

On any given day, health issues make the news all over the world. A search of international newspapers on the Internet – on 26 July 2010 – showed a range of different health issues making the news (Figures 8.5-8.8).

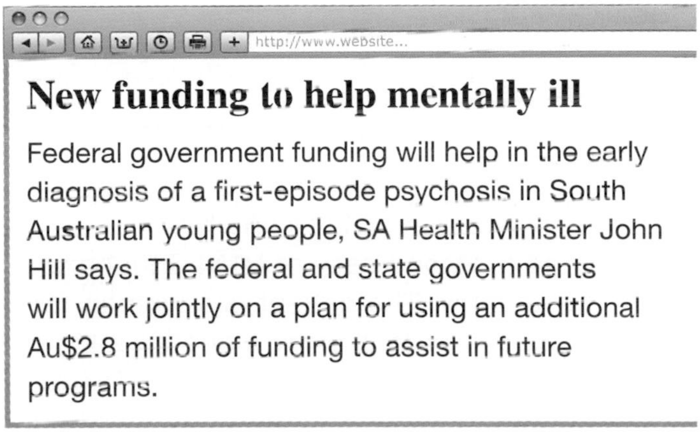

Figure 8.5 *Australia – adapted from a report on* The Sydney Morning Herald *website* ▶

http://www.website...

New funding to help mentally ill

Federal government funding will help in the early diagnosis of a first-episode psychosis in South Australian young people, SA Health Minister John Hill says. The federal and state governments will work jointly on a plan for using an additional Au$2.8 million of funding to assist in future programs.

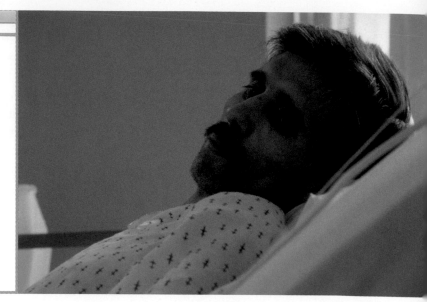

Vaccine hopes can't mask the extent of Russia's AIDS crisis

HIV infection in Russia remains at epidemic levels, and along with the rest of eastern Europe this is the only part of the world where infection rates are still rising. So while last week's news of a pilot vaccine developed in Novosibirsk is welcome, experts fear a relapse in official attitudes to a disease still often believed to be a scourge of undesirables or a foreign conspiracy.

Figure 8.6 *Russia – adapted from an article on* The Moscow News *website* ▲

Doctors use iPhone and iPad to provide treatments

Health care and fitness applications for sophisticated mobile devices like Apple's popular iPhone, and the recently released iPad, are booming – and Japan's medical fraternity are getting with the program. Jikei University Hospital has launched an iPhone application which is designed to assist with the diagnosis of stroke symptoms in patients. It is also partnering with Fujifilm Corp. to develop a system for other hospitals to use the application.

Figure 8.7 *Japan – adapted from an article on the* Daily Yomuiri Online *website* ▲

Figure 8.8 *Argentina – adapted from a report on the* Buenos Aires Herald *website* ▼

Second case of H1N1 virus in Argentina

Health Minister Graciela Ocaña has confirmed a second case of H1N1 Influenza A in the country. The patient is a woman who arrived in Argentina by plane on 9 May from the USA.

Internet research

Find out more about:
a how health care is provided in the UK and the USA
b how each health care system is funded
c the advantages and disadvantages of each country's system.

Good sources of information include BBC News Health: http://www.bbc.co.uk/news/health/

ACTIVITIES

1 Use the information in this section to produce a mind map showing why geographers study health.
2 Why do you think that health issues feature prominently in the news?
3 Study the information in this section about health care in the USA and the UK.
  a Identify the questions being asked about health care in both countries.
  b In what ways are those questions (i) similar and (ii) different?
  c What are the advantages and disadvantages of (i) the government providing health care, and (ii) individuals having their own health care insurance?
4 In pairs, discuss why health care provision is often an area that politicians disagree about. Feed back your ideas in class.

In this section you will learn about:
- the meaning of the terms morbidity and mortality
- the global prevalence of malaria, and how it affects both morbidity and mortality rates
- how and why we measure morbidity and mortality rates

Skills

In this section you will:
- produce choropleth maps to show morbidity and mortality statistics
- compare choropleth maps to identify patterns and trends

What are morbidity and mortality?

Morbidity refers to ill health. It includes any diseased state, disability, or condition of poor health – due to any cause. The term might be used to refer to the existence of any form of ill health, or to the degree to which a health condition affects a patient.

Mortality refers to deaths. It is the condition of being mortal, or susceptible to death.

Malaria

Cheryl's not feeling too well!

In July 2010, popular X Factor judge and Girls Aloud star Cheryl Cole (Figure 8.9) fainted at a photo shoot in London. After undergoing medical tests, Cheryl was diagnosed with malaria and admitted to hospital for treatment. She had been feeling unwell for several days, but had ignored it because she thought she was just tired from her busy work schedule. The early symptoms of malaria include headaches, aching muscles, and weakness or a lack of energy, so it can often be confused with exhaustion or flu.

Malaria is transmitted through a bite from an infected mosquito, and it only takes one bite to contract the disease. Cheryl Cole had recently returned from a holiday in Tanzania, on the east African coast, where malaria is rife. She had been taking anti-malarial tablets, but some forms of malaria are now resistant to the drugs. About 400 cases of malaria are reported in the UK each year – mainly from people who have been travelling abroad, like Cheryl.

Malaria in Tanzania

Every year, malaria infects about 18 million Tanzanians – nearly half the population of 41 million. It is a **chronic disease** that can debilitate its victims for weeks at a time. Even though most Tanzanians survive their encounters with malaria, they can suffer up to four attacks a year. These attacks cause wider problems, because they keep children out of school and farmers away from their fields. They also prevent workers from earning much-needed income to support their families (Figure 8.10).

Figure 8.9 *Cheryl Cole returning to the X-Factor after recovering from malaria* ▲

Figure 8.10 *A boy sick with malaria in hospital in Tanzania, watched over by his anxious parents* ▼

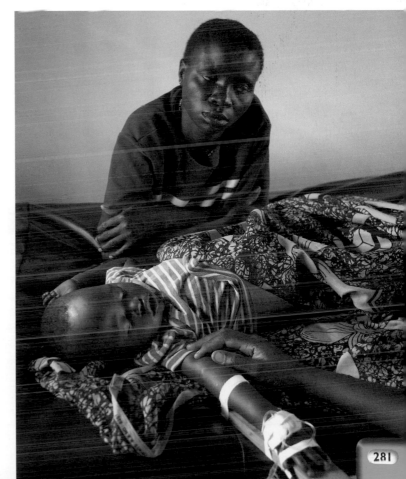

281

Is malaria a global disease?

Malaria is a **parasitic infection** – spread by mosquitoes that carry the disease. The World Health Organisation (WHO) estimates that:

- about 3.3 billion people (half of the world's population) are at risk of contracting malaria
- about 247 million new cases occur every year
- malaria is present in 108 different countries.

Malaria is prevalent in Asia, Latin America, and – to a lesser extent – in the Middle East. Some parts of Europe are also affected. But, of all the people living with malaria, 92% live in Africa (Figures 8.11 and 8.12). So is malaria really a 'global' disease?

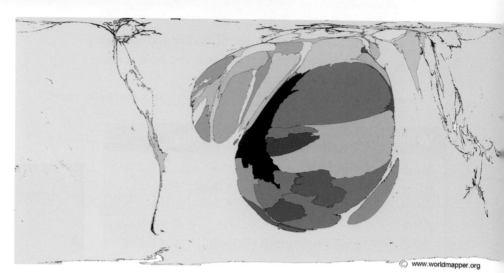

© www.worldmapper.org

Figure 8.11 *A Worldmapper map showing the distribution of malaria cases worldwide. Each country has been drawn in proportion to the number of malaria cases there, so Africa is huge and North America and Europe are tiny.* ▲

Measuring morbidity

Morbidity statistics are widely used by epidemiologists (people who study patterns of disease) to analyse ill health in human populations. There are two main measures of the **morbidity rate**:

- The **prevalence rate** gives an indication of the number of people in a given population who are suffering from a particular condition at any one time.
- The **incidence rate** shows how many people in a population develop a particular condition in a given period of time (usually a year).

Morbidity statistics are collected through a variety of sources, including official statistics on **contagious diseases** and other **notifiable illnesses**, hospital in-patient records, records of claims for sickness benefits, and local or national interview surveys that gather self-report data on ill health.

Rank	Country	Number of cases
1	Nigeria	57 506 430
2	Congo (Dem. Rep. of the)	23 619 960
3	Ethiopia	12 405 124
4	Tanzania	11 539 867
5	Kenya	11 341 750
6	India	10 649 554
7	Uganda	10 626 930
8	Mozambique	7 432 539
9	Ghana	7 282 377
10	Cote d'Ivoire	7 028 990

Figure 8.12 *The top ten countries with the highest incidence rate of malaria in 2006 – all but one were in Africa* ▲

What is a notifiable disease?

A notifiable disease is any disease that is required by law to be reported to the government. In the UK, under the Health Protection (Notification) Regulations 2010, GPs must report any cases of the following diseases to the Health Protection Agency:

Acute encephalitis	Enteric fever (typhoid or paratyphoid fever)	Legionnaires' disease	SARS
Acute meningitis	Food poisoning	Leprosy	Smallpox
Acute poliomyelitis	Haemolytic uraemic syndrome (HUS)	Malaria	Tetanus
Acute infectious hepatitis	Infectious bloody diarrhoea	Measles	Tuberculosis
Anthrax	Invasive group A streptococcal disease and scarlet fever	Meningococcal septicaemia	Typhus
Botulism		Mumps	Viral haemorrhagic fever (VHF)
Brucellosis		Plague	Whooping cough
Cholera		Rabies	Yellow fever
Diphtheria		Rubella	

Malaria can kill!

Meanwhile ... Alexander's not feeling well either

Alexander Patris, a young boy of two, is sitting on his mother's lap coughing. Anna Patris pats her young son on the back – whispering words of comfort into his ear. They are in the malaria ward of a hospital in Arusha, Tanzania (Figure 8.13). Each time Alexander coughs, Mrs Patris prays that he doesn't die from malaria – the same disease that killed two of her other young children.

Tanzania has the highest malaria death rate in the whole of sub-Saharan Africa. A Tanzanian dies of malaria every 5 minutes – particularly pregnant women and children under 5. Of the 18 million Tanzanians who contract malaria every year, 100 000 will die and 80% of those will be children aged 5 or younger.

Is malaria a global killer?

In 2008, the WHO estimated that there were about 247 million new cases of malaria worldwide. These cases caused nearly one million deaths – mostly among children living in Africa (Figures 8.14 and 8.15). In 2003, 94% of all deaths from malaria were in Africa, so is malaria really a 'global' killer?

- A child dies of malaria in Africa every 45 seconds.
- The disease accounts for 20% of all childhood deaths there.

Figure 8.13 *Mothers at a hospital in Tanzania have brought their children to be checked for malaria* ▲

Rank	Country	Number of deaths
1	Nigeria	225 424
2	Congo (Dem. Rep. of)	96 113
3	Uganda	43 490
4	Ethiopia	40 963
5	Tanzania	38 730
6	Sudan	31 975
7	Niger	31 501
8	Kenya	27 049
9	Burkina Faso	25 625
10	Ghana	25 075

Figure 8.14 *The top ten countries with the highest malaria death rates in 2006 – all were in Africa* ▲

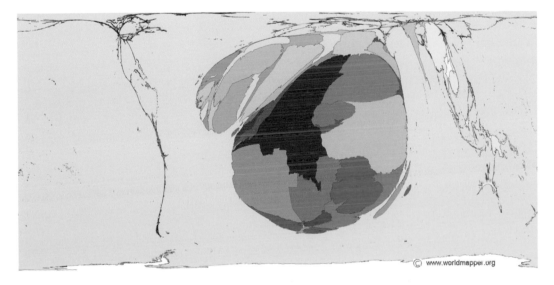

◄ **Figure 8.15** *Another Worldmapper map – this time showing the distribution of deaths from malaria worldwide*

Measuring mortality

Mortality statistics are used by epidemiologists to analyse the impacts of different causes of morbidity (or ill health) on a given population. The **mortality rate** is a measure of the number of deaths (in general, or due to a specific cause) in a population.

- The **crude mortality rate** is measured using the formula:

$$\text{Crude mortality (death) rate} = \frac{\text{Number of deaths during the year}}{\text{Total mid-year population}} \times K$$

K is a constant (usually 100, 1000, or 100 000), which is used to turn the rate into a whole number (per 1000, for instance).

- The **infant mortality rate** is the number of deaths occurring in children under the age of one year.
- The **case mortality rate** is the number of people dying from a disease, divided by the number of those diagnosed with the disease.

ACTIVITIES

1 Prepare a factfile for malaria: What is it? How do you get it? What causes it? What are its symptoms? How does it affect people? How is it treated? How can it be reduced?

2 Compare what it's like to suffer from malaria in the UK and in Tanzania. Explain the differences.

3 Why do you think that governments like to keep a record of:
   a the morbidity rates of certain diseases?
   b mortality statistics and causes of death?

4 What important information do you think should be recorded by a **coroner** when somebody dies? Explain your answer.

5 Use the data in Figure 8.12 (and look at the skills box on page 189).
   a Produce a choropleth map showing the top ten countries with the highest incidence rates of malaria in 2006. Use the darkest colour for the highest incidence rates.
   b Describe the pattern that your map shows.

6 Now use the data in Figure 8.14.
   a Trace an overlay for your first choropleth map and produce a second map showing the top ten countries with the highest malaria death rates in 2006.
   b Describe the pattern that your map shows.

7 Compare your two choropleth maps. How close is the correlation between the countries with the highest incidence rates of malaria and those with the highest mortality rates?

Internet research

Try to explain some of your findings from Activities 5-7, by conducting Internet research into some of the countries listed in Figures 8.12 and 8.14. Type phrases such as 'malaria in Nigeria' into Google to get you started.

In this section you will learn about:

- the global distribution of the HIV virus
- the causes of HIV and AIDS, and how the infection is spread
- the impacts of HIV/AIDS in Russia and India

What is HIV/AIDS?

The human immunodeficiency virus (HIV) is a **retrovirus** that eventually leads to acquired immune deficiency syndrome (AIDS). The HIV slowly invades white blood cells (the body's main defence against infection) and stops them working effectively (Figure 8.16). This process can take many years (often with few outward symptoms), until the body's immune system becomes so weak that other 'opportunistic' infections and diseases develop, such as pneumonia. It is this stage that is referred to as AIDS.

How is it spread?

HIV/AIDS is a **behavioural disease**, spread through human activity. The most common ways in which people become infected with HIV are:

- by having unprotected sexual intercourse with an infected partner (Figure 8.17)
- by injecting drugs using a needle or syringe that has already been used by someone who is infected with HIV/AIDS
- as a baby, from an infected mother – during pregnancy, labour or delivery, or through breastfeeding.

In HICs and MICs, certain groups have been particularly badly affected by HIV/AIDS. They include injecting drug users, sex workers, and homosexual men. In some people's minds, HIV and AIDS are closely linked with these groups, which can lead to increased stigma and prejudice against them.

The global distribution and spread of HIV/AIDS

HIV/AIDS occurs all over the world (Figure 8.18). However, some places are more affected than others. The worst-affected region is sub-Saharan Africa, where over a quarter of the adults in some countries are HIV positive (Figure 8.18). But, according to the United Nations, HIV is actually spreading fastest in Eastern Europe and Central Asia, where the number of people living with HIV increased by 67% between 2001 and 2008.

- An estimated 39 million people worldwide are now living with HIV.
- An extra 2.7 million people, on average, become infected with HIV every year. But there is some evidence that the number of new cases worldwide might be starting to stabilise.
- In 2005, the HIV/AIDS **pandemic** killed more than 2.8 million people.

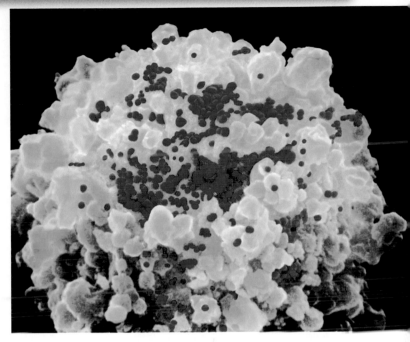

Figure 8.16 The HIV virus (red) attacking a human cell ▲

Figure 8.17 Nadja Benaissa is a singer with the chart-topping German girl band No Angels. She is also HIV positive. In August 2010, Nadja was given a two-year suspended prison sentence for having unprotected sex with three men at different times – without telling them that she was HIV positive. One of the men later tested positive for HIV. ▲

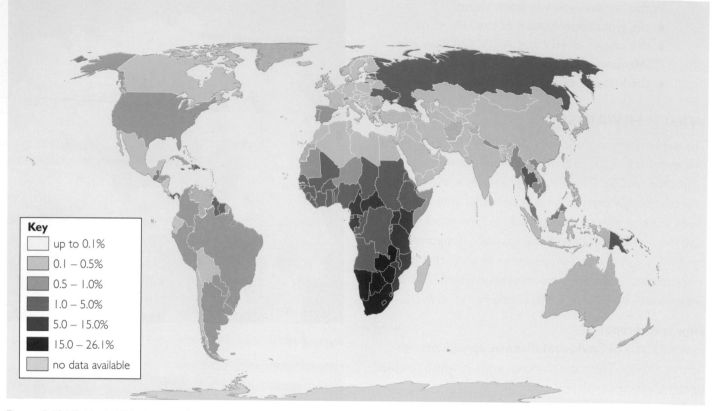

Figure 8.18 *Worldwide HIV infection rates for adults (aged 15-49) in 2007* ▲

Key
- up to 0.1%
- 0.1 – 0.5%
- 0.5 – 1.0%
- 1.0 – 5.0%
- 5.0 – 15.0%
- 15.0 – 26.1%
- no data available

The economic impacts of HIV/AIDS

In 1997, the World Bank published a report called 'Confronting AIDS', which summarised the impact that AIDS could have on the economic output of different countries around the world. The report concluded that the economic impact of AIDS 'although varying across countries, will generally be small relative to other factors'. It gave several reasons for this optimistic conclusion:

- AIDS could, in theory, reduce economic growth through the loss of many workers – because AIDS generally affects working-age adults. However, evidence shows that AIDS has had little impact on the economies of many countries, because unskilled workers are easily replaced, while more-skilled workers tend to have much lower infection rates to begin with.

- AIDS could also have an economic impact through reduced investment, because many households might cut back their spending in order to pay for medical care and funerals. But the World Bank pointed out that in many countries these costs are borne by the State, so they have little impact on economic growth at a national level.

- Finally, history shows that GDP per person tends to be unaffected by epidemics. The Black Death, which killed 30-50% of Europe's population in the fourteenth century, is thought to have had little impact on GDP growth per person.

However, even if on a broader level the economic effects of HIV/AIDS are considered to be minimal, the impact of the disease on families at an individual level must not be overlooked. In many LICs, the infection of an adult within a family can have devastating impacts on the whole family, both socially and economically.

The spread of HIV/AIDS in Russia

Russia has one of the fastest-growing HIV infection rates in the world:

- The official number of HIV cases diagnosed and registered with health officials now totals 370 000. But that is well below the UNAIDS estimate of 940 000, because the official figure only counts those Russians who have been in direct contact with Russia's HIV-reporting system (Figures 8.19 and 8.21).
- Some experts believe that the actual number of HIV infections in Russia could be as high as 3 million, if all unregistered cases are included.
- One Russian official, who subsequently lost her job as a result, claimed that up to half of all Russian citizens could be infected with HIV by 2020.
- 77% of Russians infected with HIV are men – 60% of them between the ages of 17 and 25.
- An estimated 80-90% of all HIV infections in Russia have been caused by intravenous drug use, which is widespread and rising (Figure 8.20).

The impacts of HIV/AIDS in Russia

Macroeconomic Impacts

Research in 1998 attempted to measure the economic impacts of HIV/AIDS in Russia. At that time, it was estimated that there had been about 1.6 million deaths of working-age men in Russia due to AIDS. It was also estimated that those deaths equated to a loss of between 1.8% and 2.7% of Russia's GDP.

By 2020, it is thought that Russia's GDP could be anything between 1.2% and 10.7% lower than if there had been no AIDS cases in the country. 80% of all those Russians currently infected with HIV are under the age of 30, so they are prime workers whose loss will have an economic effect.

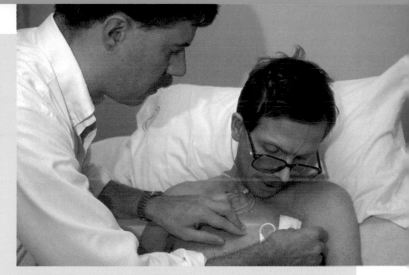

Figure 8.19 *A doctor treating an officially recognised AIDS patient in a Russian hospital* ▲ ▼

Figure 8.20 *A nurse giving HIV safety advice to a drug addict in Siberia, Russia* ▼

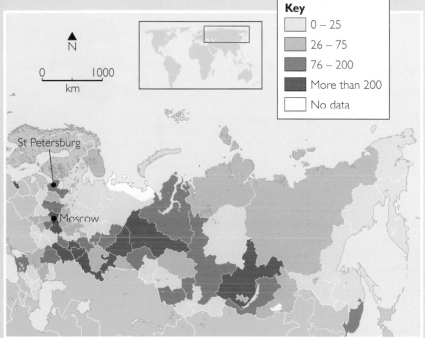

Key

	0 – 25
	26 – 75
	76 – 200
	More than 200
	No data

St Petersburg

Moscow

Figure 8.21 *HIV prevalence per 100 000 people in Russia in 2002, by administrative region (according to official Russian Ministry of Health statistics)* ▶

Microeconomic impacts

On businesses: Businesses will be impacted through higher levels of absenteeism, increasing staff turnover, and reduced productivity. It is expected that the impact will be felt most strongly in small and medium-sized companies, where the loss of a single worker will be more difficult to cope with than in a large industrial firm.

On households: It is likely that most AIDS patients will be adults in their thirties. If an adult becomes sick, the household will lose that person's income, and will also incur the extra costs associated with health care. Therefore, a household with an AIDS case will be more vulnerable to poverty.

By contrast, a healthy household may well benefit from increased employment opportunities and higher wages as a result of AIDS deaths in the workforce. Therefore, there could be an increase in *inequality*.

On the State: In middle-income countries – such as Russia – much of the burden of extra health care costs falls on the State. The proportion of Russian GDP spent on health care rose from 3% in 1990 to 7% in 1998. However, by 2007, it had fallen back to 5.6%

The biggest impact that AIDS is expected to have on Russia is due to the **multiplier effect**. A sick person not only drops out of the workforce, but also causes their partner to stop work – or their children to drop out of school – to care for them. Very soon this can lead to a downward spiral of poverty and deprivation for a family, as well as problems for the wider economy.

CASE STUDY

The spread of HIV/AIDS in India

In 2008, UNAIDS estimated that there were 2.4 million people living with HIV in India. This equates to a relatively low prevalence rate of 0.3%. However, because India's population is so large (over 1.16 billion people in 2010), the country has the third largest number of people living with HIV in the world. An increase in the Indian HIV prevalence rate of just 0.1% would increase the number of people living with HIV by over half a million!

The HIV epidemic arrived later in India than it did in many other countries. It was not until 1986 that the first cases of HIV were diagnosed in India – among a small group of sex workers in Chennai. However, HIV infection rates rose rapidly throughout the 1990s, and today the epidemic affects all sectors of Indian society – not just the groups with which it was originally associated (such as sex workers and truck drivers).

Figure 8.22 *Villagers in south India receiving HIV/AIDS health education* ▲

A 2006 study by the National AIDS Control Organisation (NACO), suggested that HIV infection rates had begun to fall in southern India – the region hardest hit by HIV/AIDS. Despite this evidence, the impact of HIV/AIDS in India will have a devastating effect on the lives of millions of Indians for many years to come. The epidemic there may yet match the severity of the one in southern Africa.

The impacts of HIV/AIDS in India

In 2005, research by the University of Princeton – focusing on the impacts of HIV/AIDS in India – found that there had been significant impacts on households where an active adult had died because of HIV/AIDS:

- Households that had suffered HIV/AIDS deaths reported reduced savings and reduced consumer spending. Many had also been forced to sell key family assets, such as livestock, to get extra money.
- Many families also reported that they had experienced discrimination from their neighbours. A few of them had had to send their children away to distant relatives, or withdraw them from school.
- Children had suffered from discrimination at school if their parents had died from AIDS.
- Because AIDS orphans often didn't receive a proper education, a large proportion of them had had to start work early.

- Members of extended families often provided AIDS orphans with shelter, which caused further economic hardship for the family.

A key finding of the report was that the impacts on both households and children were much more negative in those households which were socially and economically disadvantaged to begin with.

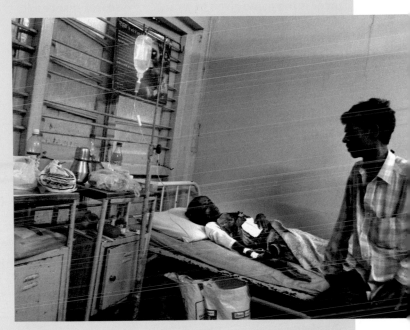

Figure 8.23 *A relative looking after an HIV/AIDS patient at a residential clinic in Mumbai* ▲

ACTIVITIES

1 Prepare a factfile for HIV/AIDS: What is it? How do you get it? What causes it? What are its symptoms? How does it affect people? How is it treated? How can it be reduced?

2 Study Figure 8.18.
   a Describe the geographical distribution of those infected with HIV.
   b Suggest some of the factors that may account for the unequal distribution.

3 Study Figure 8.21. Then use an atlas to:
   a help describe the pattern of HIV prevalence in Russia
   b try to explain this pattern.

4 Study the information about Russia on pages 287–288. Then copy and complete the following table to show the impacts of HIV/AIDS there.

	Short-term	Medium-term	Long-term
Social			
Economic			

Internet research

Conduct some further research into the social, economic (and possibly environmental) impacts of HIV/AIDS in India. Use the latest country report from the UNAIDS website – http://www.unaids.org/en/default.asp – to research:

- the number of cases and their distribution
- HIV prevalence rates
- whether the epidemic is urban or rural
- which sex and age groups are most affected, and why
- how the disease is being spread
- attempts to manage the epidemic through treatment and prevention.

In this section you will learn about:

- the causes of cancer
- cancer's global distribution and increasing incidence
- the distribution of cancer in the USA, and its impacts

What is a non-communicable disease?

A **non-communicable disease** is one that, unlike HIV/AIDS or the common cold, cannot be passed from one person to another. Examples of non-communicable diseases include: cardiovascular disease, cancer, chronic respiratory disease, and diabetes.

What is cancer?

Cancer is the name given to a range of diseases that result from cells in the body growing abnormally and multiplying out of control (Figure 8.25). There are more than 200 different types of cancer, and each one has different causes and symptoms.

What causes it?

Cancer is caused by damaged or faulty genes, which tell the body's cells what to do. Genes are encoded with deoxyribonucleic acid (DNA), so anything that damages DNA can increase the risk of cancer. However, a number of genes within the same cell need to be damaged before it becomes cancerous.

Most forms of cancer are caused by DNA damage that occurs over a person's lifetime, e.g. too much long-term exposure to the sun can increase the risk of skin cancer. Inherited cancers – caused by specific faults in a gene inherited from a parent – are rare, but everybody has subtle variations in their genes that may increase or

decrease their risk of cancer by a small amount. Cancer, therefore, is not 'all in the genes', and it's not 'all down to lifestyle' either – it's a combination of both.

The global distribution and increasing incidence of cancer

On average, about 60% of all deaths worldwide are caused by non-communicable diseases. That figure rises to about 86% for deaths in HICs and MICs, but drops to about 50% for LICs (where diseases like cholera and HIV/AIDS still cause many deaths, especially in Africa).

One in eight deaths worldwide are currently caused by cancer. This proportion is higher than deaths from AIDS, tuberculosis (TB), and malaria *combined*! Every year, cancer claims about 7 million lives worldwide (Figure 8.24) – with about 11 million new cases being diagnosed.

Figure 8.24 *A Worldmapper map showing the distribution of deaths from all cancers in 2002. Each country has been sized in proportion to the number of deaths there caused by cancer.* ▼

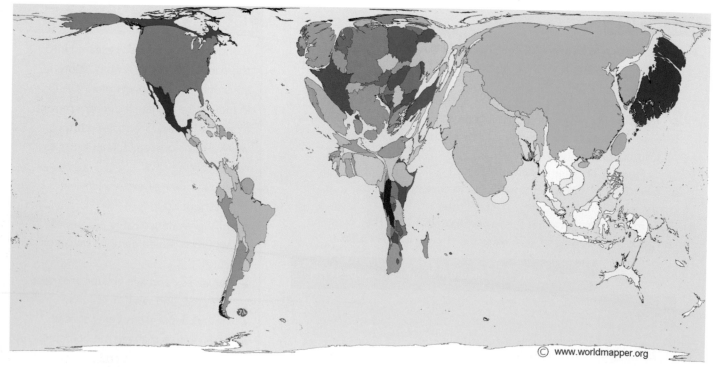

© www.worldmapper.org

However, cancer is now increasing in incidence. By 2020, it is expected that there will be around 12 million cancer deaths a year, with 15 million new cases being diagnosed annually. At least 70% of those deaths will be in LICs. This is because of lifestyle changes, such as changing diets and an increase in the number of smokers in those countries. Cancer survival rates in LICs are also just 20-30%, which is much lower than in HICs – where survival rates can be as high as 60%.

Throughout the world, about 25 million people are currently living with some form of cancer. A lot of these people suffer a reduction in their quality of life due to physical pain, mental anguish and economic hardship.

The economic impacts of cancer

Macroeconomic impacts

According to the Lance Armstrong Foundation (Figure 8.26), the global economic cost of new cancer cases in 2009 was $305 billion. This consisted of $217 billion in medical and non-medical costs, $19 billion for research, and $69 billion in lost productivity due to ill health. The medical costs included the cost of diagnoses, in-patient treatment and care, outpatient treatment and care, and drugs. The non-medical costs included the costs of transportation to and from medical providers, the costs of alternative and homeopathic treatments and care, and the value of time associated with informal care-giving.

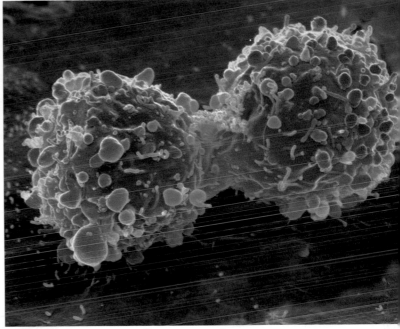

Figure 8.25 *A lung cancer cell dividing into two* ▲

Microeconomic impacts

The World Health Organisation says that a diagnosis of cancer in an adult family member doesn't just lead to the loss of their direct income. It can also cause extra problems for the family if they spend much of their remaining income and resources seeking treatment for the patient. Many families can spend large sums of money on treatments that often fail to prolong the life of a family member with advanced cancer. In addition to this, they can often feel abandoned by the formal State health care system – and may spend any remaining resources they have in seeking help from 'unscrupulous individuals who falsely promise to help'.

Pain is temporary. It may last a minute, or an hour, or a day, or a year, but eventually it will subside and something else will take its place. If I quit, however, it lasts forever.

Figure 8.26 *Lance Armstrong, seven-times consecutive winner of the gruelling Tour De France cycle race – after surviving testicular cancer* ▶

Cancer in the USA

Every year, about 560 000 Americans die of cancer – that's more than 1500 people a day. Cancer is the second most common cause of death in the USA (after heart disease). It accounts for nearly a quarter of US deaths.

In 2006, Yongping Hao of the American Cancer Society conducted some research into the geographic patterns of cancer death rates in the USA (Figure 8.27). The two maps show that the highest death rates for both men and women occurred in the eastern USA, with the lowest in the Midwest.

- In men, the death rate ranged from 186.3 per 100 000 in Utah, to 343.7 per 100 000 in the District of Columbia (where Washington DC is).
- In women, the death rate ranged from 123.4 per 100 000 in Utah, to 217.4 per 100 000 in Pennsylvania.
- The patterns for all cancers combined (as shown in Figure 8.27) were very similar to those for lung cancer:
  - Lung cancer death rates for men ranged from 35.7 per 100 000 in Utah, to 130.3 per 100 000 in Kentucky.
  - For women, they ranged from 14.8 per 100 000 in Utah, to 57.9 per 100 000 in Kentucky.
- The burden of death from all cancers, and particularly from lung cancer, highlighted the importance of tobacco control measures in the USA – as well as the need to provide localised cancer treatment and associated services.

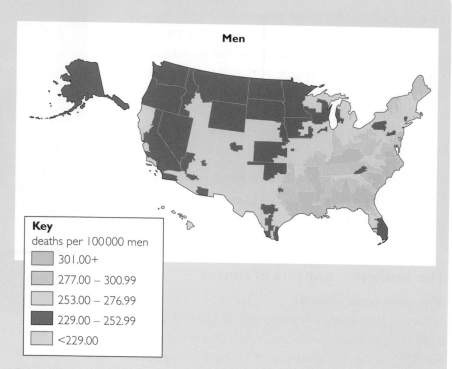

Men

Key
deaths per 100 000 men

- 301.00+
- 277.00 – 300.99
- 253.00 – 276.99
- 229.00 – 252.99
- <229.00

Figure 8.27 *The distribution, by gender, of cancer death rates per 100 000 people in the USA (from all cancers combined), 1990-2001* ▲ ▼

Women

Key
deaths per 100 000 women

- 187.00+
- 177.00 – 186.99
- 167.00 – 176.99
- 157.00 – 166.99
- <157.00

The economic impacts of cancer in the USA

The USA has by far the highest economic cost of cancer in the world:

- In 2009, the US National Institutes of Health estimated that the overall cost of cancer to the USA in 2008 was $228.1 billion (nearly 69% of the estimated total global cost).
- The $228.1 billion broke down into: $93.2 billion for direct medical costs, $18.8 billion for lost productivity due to illness, and $116.1 billion for lost productivity due to premature death.

The economic costs of cancer are high for both the person with cancer and for society as a whole. It is estimated that the loss of productivity due to cancer deaths is equivalent to approximately 1% of American GDP every year.

One of the major costs of cancer is cancer treatment. However a lack of health insurance and other barriers to health care (see pages 278-279) prevents many Americans from receiving even basic health care – let alone expensive cancer care. In 2008, the US National Health Interview Survey estimated that:

- about 24% of Americans aged 18 to 64 had no health insurance for at least part of the past year
- about 13% of children in the USA had no health insurance for at least part of the past year.

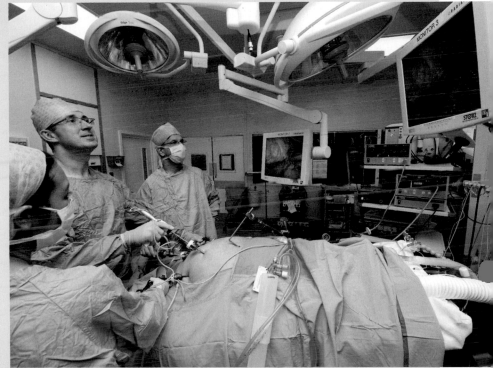

Figure 8.28 *Cancer treatment in the USA is too expensive for many Americans without health insurance* ▲

Evidence from the USA suggests that individuals with no health insurance – and those with Medicaid insurance – are more likely to be diagnosed with advanced forms of cancer (because they don't have easy access to a doctor for early diagnosis). This then leads to higher medical costs later on and poorer survival rates/higher death rates.

ACTIVITIES

1 Prepare a factfile for cancer: What is it? How do you get it? What causes it? What are its symptoms? How does it affect people? How is it treated? How can it be reduced?

2 Study Figure 8.24.
  a Describe the global distribution of all cancer deaths.
  b Suggest some of the factors that might account for the unequal distribution.

3 Compare the impacts of cancer with those of HIV/AIDS (Section 8.3). Which disease has the greatest impacts and why?

Internet research

Research some of the socio-economic issues involving cancer in the USA. Use reports from the US National Cancer Institute website – http://www.cancer.gov/ – to research:

- the Top 10 types of cancer
- trends in the numbers of cases
- which sex and age groups are most affected and why
- how people are treated (**a**) in the private health sector, and (**b**) in the public health sector – and the issues arising from this.

In this section you will learn about:
- the global spread of malnutrition
- the impacts of malnutrition on health
- the impacts that malnutrition are having on people in Ethiopia and Mexico

Stuffed and starved

Malnutrition means 'bad nourishment'. It includes having too much food (over-nutrition, leading to obesity), as well as not having enough (under-nutrition, leading to starvation). In 2007, Raj Patel (of the Center for African Studies at the University of California) wrote a book called *Stuffed and starved,* in which he considered the problems of the current food system. A system that fails to feed one billion people adequately each year, but also manages to overfeed about 800 million people worldwide. Patel called this 'Our Big Fat Contradiction'.

Stuffed – the global over-nutrition problem

Affluent consumers in HICs and MICs are enjoying a much richer and more diverse diet than previous generations could ever have dreamt of. They are also paying a comparatively low price for it – not just for food staples (like rice, maize and potatoes), but also for foods that were once considered 'luxury items' (like oysters, avocados and shrimps).

One result of this trend has been the emergence of a global obesity pandemic (Figure 8.29). The combination of a big increase in the global consumption of energy-dense meals (high in fat and protein, and low in carbohydrates) – coupled with a more sedentary lifestyle (with motorised transport, labour-saving devices at home and work, and leisure time dominated by physically undemanding pastimes like watching TV and playing computer games) – has led to a massive growth in obesity, and an increase in 'lifestyle' diseases like diabetes and heart disease.

Figure 8.29 *The geography of obesity in 2005* ▼

Key
obese adults in population %

- 30 – 40%
- 20 – 30%
- 10 – 20%
- 5 – 10%
- 0 – 5%
- No data

CASE STUDY

Mexico – fighting the flab

In 1989, fewer than 10% of Mexican adults were considered to be overweight. But, by 2008, that figure had risen to 71% of women and 66% of men. Mexican children are also following the trend, with more than 25% now thought to be overweight – an increase of 40% since 2000.

- Many Mexicans are now living more sedentary lives – just like the populations of many other HICs and MICs, particularly in Europe and the USA.
- Studies also show that many Mexicans are eating more fatty and processed foods, and fewer whole grains and vegetables (Figure 8.30).
- Foods that were once unavailable in Mexico, can now be purchased easily at modern supermarkets. In some parts of the country, it is now easier to get a soft drink than a class of clean water.

Figure 8.30 Overweight Mexicans in a Mexico City park full of junk food stands ▲

The health consequences of obesity in Mexico have led to an increased rate of diabetes, high blood pressure and heart disease. The Mexican Diabetes Federation estimates that between 6.5 and 10 million Mexicans are currently suffering from diabetes (out of a population of about 111 million). More than 70 000 Mexicans die each year from diabetes-related conditions. The rapid growth of diabetes is starting to overburden Mexico's already strained health services. If this trend continues, there are fears that the country's health care system could become bankrupt within a decade.

Starved – the global under-nutrition problem

In 2009, the UN's Food and Agriculture Organization (FAO) identified that about 1.02 billion people were suffering from under-nutrition worldwide (about a sixth of the people on the planet). The FAO went on to say that the numbers of undernourished people rose dramatically between 2006 and 2009 – and that 2009 was a 'devastating year for the world's hungry, marking a significant worsening of an already disappointing trend in global food security since 1996'. The FAO noted that there were increases in hunger in all of the world's major regions.

The geography of global under-nutrition is a complex issue, but what is clear is that patterns of hunger correlate with patterns of power – those without power tend to suffer first, while those who enjoy access to power seldom experience hunger.

According to Figure 8.31, the highest proportion of undernourished people – as a percentage of the population – live in sub-Saharan Africa. However, in actual number terms, most hungry people live in Asia – where China, India, Pakistan and Bangladesh account for most of the world total. This is because the populations of these Asian countries are so much larger than those in Africa.

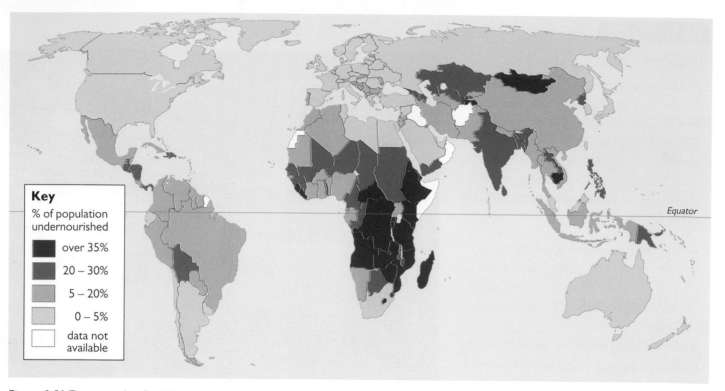

Figure 8.31 *The geography of world hunger – undernourishment by country* ▲

The health impacts of under-nutrition

Going without proper food can lead to serious health implications. The following conditions are just some of the possible health issues that can develop when the body is starved of sustenance:

- **Kwashiorkor** occurs in children, due to a lack of protein. It causes swelling of the feet and abdomen, and also thinning hair, loss of teeth, skin depigmentation, and dermatitis.

- **Anaemia** is the result of a lack of iron in the diet. It leads to a decrease in the number of red blood cells in the body, and also causes brittle or rigid fingernails, cold intolerance, and behavioral disturbances in children.

- **Beri-beri** is caused by a lack of vitamin B1. Its symptoms include severe lethargy and fatigue, together with complications affecting the cardiovascular, nervous, and muscular systems.

- **Scurvy** is caused by a lack of vitamin C. It leads to the formation of spots on the skin, spongy gums, and bleeding from the mucous membranes. Advanced scurvy causes the teeth to fall out. This disease often affected sailors in the past.

- **Rickets** is caused by a lack of vitamin D. It leads to a softening of children's bones – potentially leading to fractures and deformity. Children with rickets often have bow legs as an obvious sign.

CASE STUDY

Ethiopia in 2009 – food crisis or famine?

Former US President George W. Bush declared a policy of 'No famine on my watch'. The main official difference between a 'food crisis' and a 'famine' is whether there is enough food aid to keep starving people alive. Therefore, the effectiveness of this distinction depends largely on the generosity of the rich world. However, for many Ethiopians in 2009, playing with words like this was not very important – they were still hungry!

In 2009, millions of impoverished Ethiopians faced the threat of undernourishment – and possibly starvation – as the country faced its worst food crisis for decades (Figure 8.32):

- The number of Ethiopians who needed emergency food aid rose steadily throughout the year – from 4.9 million in January, to 5.3 million in May, and 6.2 million by June.
- Another 7.5 million received aid in return for work on community projects – as part of the National Productive Safety Net Program, for people whose food supplies are chronically insecure. This brought the total being fed to 13.7 million.

Figure 8.32 *Famine in Ethiopia in 2009* ▲

ACTIVITIES

1 Explain what Raj Patel on page 294 meant by the phrase 'Our Big Fat Contradiction'.

2 Study Figure 8.29.
   a Describe the geographical distribution of obesity.
   b Suggest some of the factors that might account for the unequal distribution.
   c The map shows that India and China have low rates of obesity. Explain why this might not be an accurate interpretation of the map.

3 In pairs, discuss the possible reasons why over-nutrition is becoming an increasing problem. Consider different income groups and different countries.

4 Study Figure 8.31.
   a Describe the geographical distribution of world hunger.
   b Suggest some of the factors that might account for the unequal distribution.
   c In pairs, consider reasons why hunger is still found in countries like the USA and the UK.

5 a Describe some of the health impacts of under-nutrition.
   b In pairs, discuss the differences between a food crisis and a famine? Draw a spider diagram to identify the differences.

In this section you will learn about:
- what constitutes a health care system
- the ideologies that define health care systems
- how health care systems are funded
- the health care systems of China and France

24/7 health care in the UK

Access to health care is important. In the UK, we expect NHS health care to be available 24 hours a day, 7 days a week. So what happens if people become ill when their GP's surgery is closed – at night or at weekends?

A report in July 2010 suggested that more and more people faced with that situation are turning to hospital accident and emergency (A&E) departments as an alternative (Figure 8.33). This is because many people do not understand how to access their local out-of-hours doctor service, so they either go straight to A&E or dial 999. The number of A&E admissions has increased by 12% since many GPs stopped offering out-of-hours services at evenings and weekends.

Over the past 5 years, the number of people attending A&E departments has risen by 1.35 million – costing the NHS an extra £330 million a year.

Even though people in the UK expect to have access to medical treatment at all times, this is a service that is not available to all people in all parts of the world. Even in some of the world's most affluent countries, like the USA, access to health care is not guaranteed. Despite this, people are now beginning to demand better access to health care facilities in some of the world's least affluent countries.

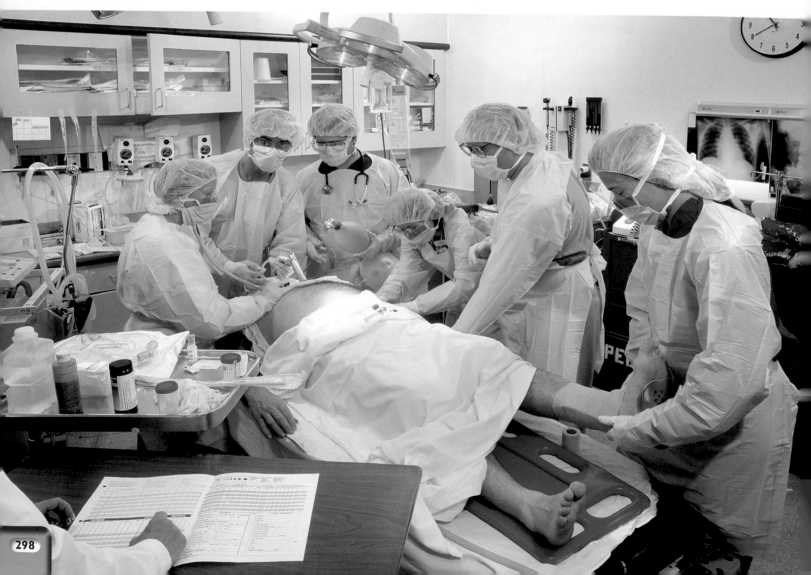

Figure 8.33 *A busy night at an A&E department, but many people now arrive with only minor ailments* ▼

Health care systems

A health care system includes all of the activities that promote, restore or maintain health. This includes all formal health delivery services, as well as the services of traditional 'healers', and the home care of sick people.

Health care systems are organised in different ways in different countries. The organisation of a health care system includes: social and political factors, the physical infrastructure of health facilities, and the needs of individuals. However, it is usually political factors that dominate the Headlines and influence the way in which health care is provided by different countries. Health care systems often reflect the political ideologies of the countries where they operate. In many countries, the government has a critical role in health care provision – to ensure that there is some form of fair distribution of health care to meet the needs of the whole population, as the panel on the right explains.

Funding health care systems

In 2006, global spending on health care was $4.7 trillion – nearly 9% of global income.

- The highest **public** health care spending per person is in Western Europe, North America (through Medicaid and Medicare), and Japan (Figure 8.34). The highest-spending countries per person are Luxembourg, Norway and Iceland (which all have relatively small populations).

- **Private** health care spending per person is highest in North America and lowest in Central Africa (Figure 8.35). The level of spending on private health care is, of course, related to the ability to pay – not the need for health care.

The key components of a health care system

An effective health care system should ensure that there is some form of equality for the people using it. This can be achieved through five key aspects:

- Equality of public expenditure. Every person in a country should have a 'fair' share of the available money. However, different individuals have different needs – with some people being healthier than others – so this ends up causing inequalities, with healthy people not receiving the same level of health spending as unhealthy people.
- Equality of 'final real income'. Final real income is when health care benefits are considered alongside a person's salary. Therefore, to ensure 'equality', those people who earn lower salaries should receive a higher proportion of health care provision.
- Equality of access. The aim of health care provision should be to ensure that all people have equal access to health care services, according to their needs.
- Equality of cost. People should make some contribution to top-up any State spending on their health care – either through insurance or direct payments.
- Equality of outcome. Resources should be provided to ensure equality of health, not just the treatment of ill health.

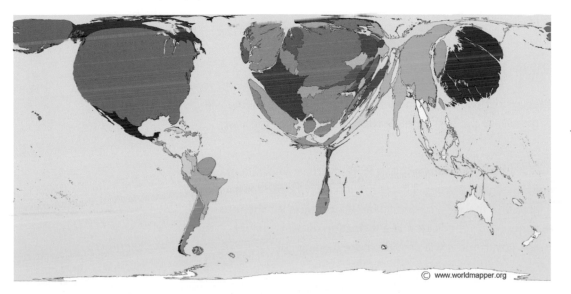

◀ **Figure 8.34**
A Worldmapper map showing global public spending on health. Public health spending includes government spending on health care (e.g. through the NHS), plus money from grants, social insurance, and non-governmental organisations

© www.worldmapper.org

The five main ways of funding health care systems are through:

- taxation
- social health insurance
- voluntary, or private, health insurance
- out-of-pocket payments (where payment is made at the point of accessing health services)
- donor contributions.

Most HICs fund their health care systems through general taxation – or compulsory social health insurance contributions. By contrast, LICs often depend more on out-of-pocket payments.

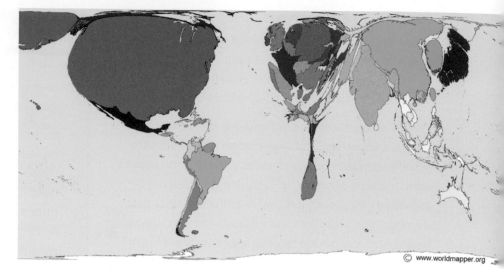

© www.worldmapper.org

Figure 8.35 *A Worldmapper map showing global private spending on health. Private health spending is when individuals or companies pay for health care directly (e.g. through private health care providers like Bupa).* ▲

CASE STUDY

Health care in China

China is developing an ambitious new health care programme, which aims to introduce health care insurance for the entire population of nearly 1.5 billion Chinese by 2020. The 'Healthy China 2020' plan intends to create a universal health care system.

A key aspect of China's new health care programme is the New Rural Co-operative Medical Care System (NRCMCS), which provides annual health insurance – at a cost of 50 yuan (US$7) per person – for those Chinese living and working in rural areas (the majority of the population). Of the 50 yuan insurance premium, 20 yuan is paid by central government, 20 yuan by the provincial government, and the final contribution of 10 yuan is made by the patient. Nearly 80% of the rural population was signed up by September 2007 (about 685 million people).

The NRCMCS provides tiered funding for patients, depending on where they receive their medical care:

- If they go to a small hospital or clinic in their local town, the scheme will cover 70-80% of their bill.
- If they go to a county hospital, the percentage of the cost being covered falls to 60%.
- However, if they need specialist help in a large modern city hospital, the scheme will only cover about 30% of the bill, so patients would have to bear most of the cost themselves.

The Chinese government also wants to prevent hospitals from overcharging patients, and doctors from prescribing unnecessary drugs. Therefore, it has proposed that hospitals will be uncoupled from pharmacies – to end financial incentives for them to over prescribe or prescribe unnecessary medicines. In anticipation of the resultant decline in hospitals' income, the government will provide more funding.

CASE STUDY

Health care in France

France operates a health care system that is mainly funded by the government, but administered through a number of social insurance schemes. These provide cover for almost the entire population. In 2004, over 80% of French people were covered by the main State-regulated insurer. The system works as follows:

- Individuals must pay a compulsory health insurance of 0.75% of their earnings, which is deducted from their salary.

- Their employer then makes a contribution of 12.8%.

- About 85% of the population also pay a voluntary top-up premium of 2.5% of their income – to ensure that their health costs are fully reimbursed.

- Recent health reforms have introduced a system of universal health coverage (*couverture maladie universelle*, CMU). Those people earning less than €6600 do not have to make any health insurance payments and are covered by the State.

Medical services are provided by general physicians, and there are no restrictions on where doctors can set up their practices. Individuals have the choice of using more than one general physician. Access to hospitals and specialist services does not depend on a referral by a general physician (which is what happens in the UK). Some specialist services (such as gynaecology) have community based specialist units.

Figure 8.36 *The French medical system was deemed the best in the world by the World Health Organisation (WHO) in 2000* ▲

ACTIVITIES

1 Define what is meant by a health care system.
2 Study Figures 8.34 and 8.35. In pairs, discuss and draw up a spider diagram to show why some countries prefer a private health care system, while others prefer a system of national care.
3 a Compare the health care systems in China and France under the headings:
- Equality of public expenditure
- Equality of 'final real income'
- Equality of access
- Equality of cost
- Equality of outcome

   b Which country's system seems the fairest? Explain your answer.
4 In pairs, discuss whether you think that 24/7 medical care in the UK is 'a right' or 'a privilege'? Feed back your ideas.

Internet research

Carry out further research into the UK's NHS, using the same headings as Activity 3. How does the NHS compare with China and France in terms of:
- advantages
- disadvantages?

In this section you will learn about:

- how pharmaceutical companies are trying to tackle the HIV virus
- the range of strategies that have taken place to tackle HIV

Living with HIV

Silvia Petretti (Figure 8.37) was 30 years old when she was diagnosed with HIV. She thought that her life had come to an end. However, she is now 44 and responsible for an Internet blog called 'Speaking Up!'. The *Independent* newspaper featured her in an article in August 2010, where she discussed her life living with HIV.

Silvia recalled that she was 'devastated, paralysed and terrified' when she found out that she was HIV positive. She thought that she was 'going to die a horrible death'. However, there was one small silver lining – Silvia was diagnosed with HIV just as the first antiretroviral drug treatments (ART) were becoming available.

Silvia began her treatment by taking 18 pills a day, which caused dry skin, nausea, tingling all over her body, and diarrhoea. However, dramatic advances in medical science in the 14 years since she was diagnosed, mean that today Silvia only has to take four pills a day, and suffers no side-effects.

Combating the threat of HIV/AIDS

The global HIV/AIDS pandemic has seen a surge in pharmaceutical developments to combat the disease. Although there is currently no preventative vaccine – or cure – antiretroviral drug developments continue to be the leading way of countering the effects of HIV, by slowing down the progress of the infection. The next page provides some examples of developments by pharmaceutical companies to tackle HIV/AIDS.

The HIV drugs market was worth an estimated $9.3 billion in 2007, and is expected to grow to $15.1 billion by 2017 – largely due to the increasing prevalence of HIV worldwide and the longer life expectancy of patients receiving treatment. The Californian biotech company Gilead is the current market leader. Its ART drug 'Truvada' earned the company $1.5 billion in 2007. GlaxoSmithKline has been the market leader for much of the history of ART – with a range of eight products on the market. However, increasing competition and the aging of its brands have meant that GSK has lost the top spot.

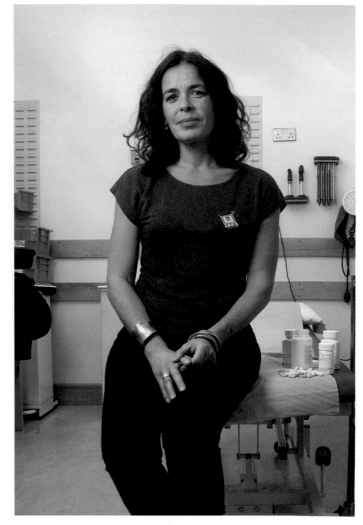

Figure 8.37 *Silvia Petretti, author of the Internet blog 'Speaking Up!', which can be found at: http://hivpolicyspeakup.wordpress.com/* ▲

Four HIV drugs into one pill

In 2009, Gilead started trials of a new four-drugs-in-one pill, which combined all of its anti-HIV drugs together. The 'Quad' pill combines tenofovir and emtricitabine (already marketed as a two-drug pill called Truvada) with a new antiretroviral drug now in Phase 3 studies (called elvitegravir). This drug belongs to a new class of antiretroviral drugs, called integrase inhibitors, and will be combined with another new drug (called GS9350), which boosts levels of elvitegravir and suppresses the HIV virus.

New HIV drug trials in the UK

A new drug called raltegravir (produced by Merck Sharp & Dohme), marketed under the name Isentress, is being made available to an estimated 73 000 HIV sufferers in the UK. The drug works by blocking an enzyme that is essential for HIV to be able to replicate itself. It means that doctors will now have a further treatment option for sufferers who have built up resistance to existing drugs. A clinical trial of raltegravir found it to be effective in patients who had been taking regular antiretroviral HIV drugs for about 10 years. Drug treatment in the UK has been helping people to live longer, but resistance to commonly used drugs has been a growing problem that raltegravir looks set to tackle.

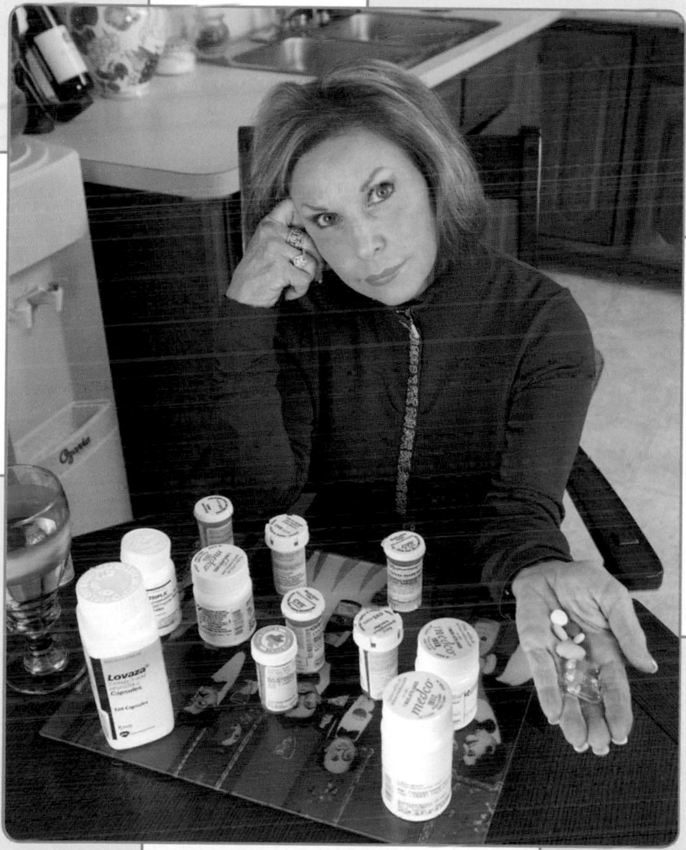

Nanoparticle technology

Johnson and Johnson are working with Tibotec to develop their antiretroviral drug Rilpivrine – currently only available as a daily oral tablet – into a slow-release drug that could be injected once a month, or even less. This would give HIV sufferers freedom from the daily chore of taking a cocktail of tablets. The drug is being developed using nano-technology (a nanometre is one millionth of a centimetre), which allows the drug to be injected into muscle tissue and to be released slowly over a period of up to six months.

Figure 8.38 *Treatment of HIV can result in sufferers taking a cocktail of drugs in order to maintain a 'normal' life* ▲

Oral vaccine technology to tackle HIV

The UK Department of Trade and Industry awarded a grant of £1.1 million to Cobra Biomanufacturing PLC to work with researchers from Cambridge and Royal Holloway Universities to develop an oral HIV vaccine. The technology stems from research that was used to produce an oral vaccine for both anthrax and plague. If it proves successful, it could pave the way for trials of the first HIV vaccine.

Action is also taking place to improve the quality of life for people living with HIV:

Eliminating HIV from semen

With HIV infection no longer a death sentence, due to antiretroviral drugs, many people who are HIV positive are able to live relatively normal lives – and make plans for the future. This has led to an increase in the number of couples made up of HIV positive men and HIV negative women who want medical help in order to have a healthy uninfected child.

Scientists in Japan have now developed a process that washes semen and eliminates the HIV virus from it. They have been successful in extracting HIV-free sperm from the washed semen and using it to artificially inseminate the female partner. The process has yet to go beyond the testing phase, but offers the possibility of starting a family for many couples where the man is HIV positive.

Internet research

In groups of 2-3, research how the world's poorest countries are now increasingly able to obtain HIV treatments. Use the websites for UNAIDS (http://www.unaids.org/en/default.asp) and AVERT (http://www.avert.org/), and a search engine, to research the following:

- 2000 – the Millennium Development Goals
- 2004 – the US Government PEPFAR
- Generic drugs
- The work done by NGOs such as the Elton John AIDS Foundation

Produce a report of 500-700 words to present to the class entitled: 'The HIV/AIDS pandemic is within sight of being brought under control.'

Transnational corporations and the treatment of HIV

Antiretroviral drugs (ARVs) are expensive. Researching a new drug costs millions of pounds, but the financial rewards for a successful treatment are huge. In the UK, a year's treatment for HIV in the 1990s could cost up to £10000 per patient. For the world's richer countries, this amount is affordable if you have government health care or private insurance. But what about the world's poorer developing nations?

A number of transnational corporations are now providing ARVs and medical services for their employees (and their families) in some of the countries most affected by HIV/AIDS. The diamond mining company De Beers, and its subsidiary Debswana, was the first company in the world to provide a programme for HIV patients in South Africa, Botswana and Namibia. These three countries have amongst the world's highest HIV prevalence rates. De Beers provides its employees, their partners and their children with a health programme, including health advice, doctors, counselling, HIV testing, nutrition supplements, and ARVs. The company argues that this benefit keeps its employees healthy, reduces the amount of working time lost to sickness, and enables employees who are HIV-positive to work for longer and continue to support their families.

ACTIVITIES

1 In pairs, discuss and draw up a spider diagram to show what issues patients with HIV face when obtaining treatment in different parts of the world.

2 a Use the information in this section to outline the technological developments being made in the treatment of HIV.

b To what extent are the world's poorest countries likely to be able to afford these latest developments?

c How well do you see the HIV pandemic being brought under control in your lifetime? Explain your answer.

In this section you will learn about:

- the global distribution of smokers
- how the tobacco industry is still big business
- the health risks of smoking
- how different countries are tackling the smoking issue

Up in smoke

Every day, about 15 billion cigarettes are sold worldwide – which works out at about 10 million every minute! In total, there are nearly 1.3 billion smokers around the world.

- Almost 1 billion of the 1.3 billion smokers are men (43% of all men aged over 15). However, they are not distributed evenly – 35% of men in HICs smoke, compared to nearly 50% in MICs and LICs (Figure 8.39).

- Women smokers account for a much smaller share of the global total – nearly 300 million (10% of all women aged over 15). Again, the distribution varies – with smokers amounting to 22% of women in HICs, and 9% in MICs and LICs (Figure 8.40). However, many women in south-east Asia chew tobacco, rather than smoke it, so they are not counted in the figures.

- Globally, cigarette smoking for men has peaked and is in decline. The same is true for women smokers in many HICs – such as the UK, Australia, Canada, and the USA. However, smoking is still on the increase amongst women in parts of eastern Europe and south-east Asia.

The tobacco industry

In 2009, the global tobacco industry produced more than 5400 billion cigarettes. The four largest international tobacco companies account for around 46% of the total global market (Figure 8.41).

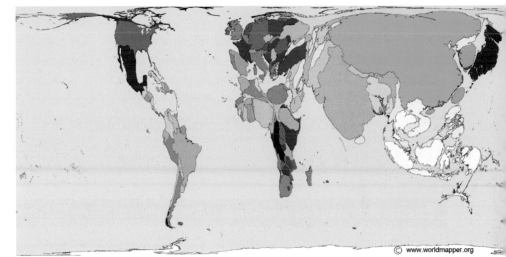

Figure 8.39 *A Worldmapper map showing male smokers. The size of the country indicates the proportion of men who smoke who live there.* ▲

Figure 8.40 *A Worldmapper map showing female smokers. The size of the country indicates the proportion of women who smoke who live there.* ▼

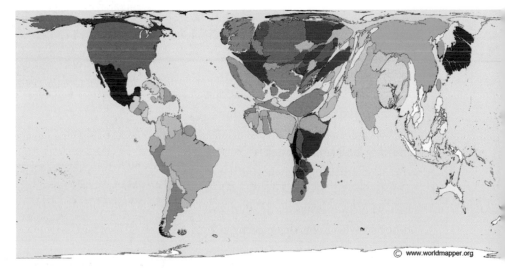

Figure 8.41 *The market share of the four main tobacco companies in 2009* ▶

Phillip Morris International	16%
British American Tobacco	13%
Japan Tobacco	11%
Imperial Tobacco	6%

> **Did you know?**
>
> Nearly a third of all cigarettes are smoked in China.

British American Tobacco

British American Tobacco (BAT) is the second largest tobacco company in the world – employing over 60 000 people. In 2009, it had a turnover of more than £40 billion (generating profits of around £4 billion).

BAT sells tobacco brands in more than 180 countries. It is the market leader in 50 of them, and has over 250 different brands in its portfolio. Its leading global brands include: Dunhill, Kent, Lucky Strike and Pall Mall. BAT claims not to encourage new smokers, but instead to meet 'the preferences of adults who have chosen to consume tobacco', and to use marketing to differentiate its brands from those of its competitors.

BAT accepts that its products 'pose health risks', but also points out that – in 2009 – its tobacco sales contributed more than £26 billion in taxes to governments worldwide (almost nine times more than its profits!). The company expects that in future a smaller proportion of people will smoke, and that individual smokers will consume fewer cigarettes. However, BAT also expects its global sales to remain unchanged in volume, because the number of adults aged over 20 is expected to grow by 11% globally over the next ten years!

The health risks of smoking

Smoking causes serious risks to health, including:

- **heart attacks and strokes.** Smokers are five times more likely to have a heart attack than non-smokers. Smoking also increases the risk of having a stroke.
- **lung problems.** Men who smoke increase their chances of dying from lung cancer by more than 22 times. For women, the figure is 12 times.
- **cancer.** Smoking increases the risk of oral, uterine, liver, kidney, bladder, stomach, and cervical cancers, and leukaemia.
- **smoking and young people.** People who start smoking in their youth – aged 11 to 15 – are three times more likely to die a premature death than someone who takes up smoking at the age of 20.

Figure 8.42 *A normal lung on the left and the diseased lung of a smoker on the right* ▲

Figure 8.43 *Smoking by parents of babies and very young children has been linked to sudden infant death syndrome (or 'cot death'), and also to higher rates of infant respiratory illness, e.g. bronchitis, colds, and pneumonia.* ▶

The global clamp down on smoking

USA

Smoking bans in the USA are being adopted by individual city and state authorities:

- California introduced a ban on smoking in public buildings in 1993. Smoking is also banned in bars, restaurants, beaches, and enclosed workplaces.
- New York banned smoking in bars, clubs and restaurants in 2003.

Figure 8.44 *In 2006, Chicago banned smoking in most public spaces (this sign is at the airport)* ▶

India

A ban on smoking in all public places was adopted in October 2008 – with a penalty fine of 200 rupees (about £3) for anyone caught. Cigarette sales to children were also banned. The Indian government is desperate to reduce India's 900 000 smoking-related deaths every year.

Figure 8.45 *Anti-smoking laws in India are often ignored* ▶

China

In May 2008, smoking was banned in most public buildings in Beijing, because of the Olympic Games. It was part of a move to encourage a healthier lifestyle in the run up to the Games, and to discourage some of China's 350 million smokers (nearly a quarter of the population).

Australia

Smoking is now banned in most public places, including airports, health clinics and workplaces. Most restaurants and shopping centres are also smoke-free areas. In Sydney, smoking has been banned at Manly and Bondi beaches. The cities of Freemantle and Perth have banned smoking in all outside dining areas, and it is expected that this will become commonplace across the whole of Australia.

ACTIVITIES

1 Use a blank outline map of the world to annotate different patterns of smoking in different countries. Use a highlighter to identify those areas of the world where smoking is (**a**) increasing and (**b**) decreasing.
2 In pairs, discuss possible reasons why:
   a rising affluence in developing countries is leading to greater smoking
   b the most affluent areas of the world have falling trends in smoking.
3 **a** What big advantage does British American Tobacco claim to bring to the countries where it operates?
   b Is this a fair justification for the actions of tobacco companies? Explain your views.
4 Using examples from this section, complete a table to show:
   a the arguments in favour of greater legislation to control and reduce smoking
   b the arguments against such legislation.

Internet research

Tobacco companies are targeting women in south-east Asia as a growing market. Research possible reasons why this particular market is such an opportunity for tobacco companies. Feed back your findings.

In this section you will learn about:
- inequalities in health and health care provision in the UK
- measuring inequalities in health
- the impacts that health inequalities can have on people living in different places

The spirit level

In March 2009, epidemiologists Richard Wilkinson and Kate Pickett published a book called *The Spirit Level: Why More Equal Societies Almost Always Do Better*. In it, they argue that inequalities within a country cause 'shorter, unhealthier and unhappier lives'. This is something that geographers have long been interested in – inequalities over space (within and between countries), and how this affects people.

Health inequalities in the UK have been linked up with geography ever since Dr John Snow first plotted the distribution of cholera deaths in Soho, London, on a street map in 1854. Snow was then able to use his mapped data to relate the spread of cholera to a contaminated water pump in Broad Street (Figure 8.46).

Identifying health inequalities in the UK

There are two key ways in which health inequalities in the UK can be assessed:

- Epidemiologists have long understood that diseases are often tied to locational factors, or the influence of environmental conditions – such as John Snow's study of cholera in London. Therefore, geographers can look at inequalities in the spread of a disease across the UK (e.g. deaths from heart disease) to consider why some places are more prone to ill health than others.
- A study of different aspects of health care provision across the country can also highlight inequalities in the availability and effectiveness of health care between regions. This could involve inequality in the distribution of health care services (e.g. the number of doctors available, or the location of hospitals), and/or inequalities of access to those facilities (e.g. travel times, transport availability, and cost).

Some health statistics commonly used to highlight geographical inequalities in the UK are life expectancy at birth (Figures 8.47 and 8.50), causes of death (e.g. the geographical distribution of 'lifestyle' diseases, such as heart disease and lung cancer), and death rates. For instance, in 2003, there was a higher death rate in Scotland than anywhere else in the UK (Figure 8.48).

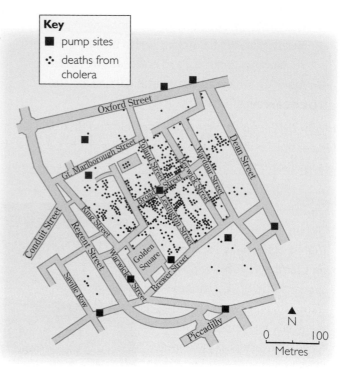

Figure 8.46 *John Snow's completed map of cholera cases in Soho, London, in 1854* ▲

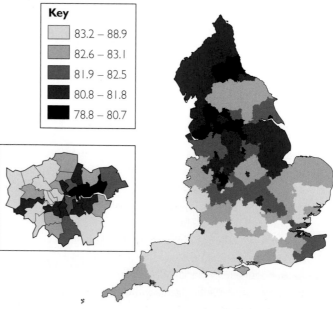

Figure 8.47 *Geographical inequalities in female life expectancy at birth for England, 2006-2008* ▲

	Death rate per 1000 population	Infant mortality rate per 1000 population	NHS spending per person
England	10.1	5.3	£1085
Wales	11.5	4.1	£1186
Scotland	11.6	5.1	£1262
Northern Ireland	8.5	5.3	£1214
United Kingdom	10.4	4.9	£1187

◀ **Figure 8.48** Key health statistics for the UK in 2003

Economic inequalities and UK life expectancy

In July 2010, researchers from Sheffield and Bristol Universities published the results of their studies into health inequalities in the UK. They found that the poorest people in Britain are twice as likely to die before the age of 65 than the richest people. The researchers' studies went right back to the 1920s, and they concluded that today's economic life expectancy gap in the UK is the highest since the Great Depression of the 1930s.

The researchers' key findings were that:

- between 1921 and 1930, for every 100 deaths before the age of 65 in the richest tenth of areas, there were 191 deaths in the poorest tenth
- between 1931 and 1939, for every 100 deaths before the age of 65 in the richest tenth of areas, there were 185 deaths in the poorest tenth
- between 1999 and 2007, for every 100 deaths before the age of 65 in the richest tenth of areas, there were 212 in the poorest tenth
- inequalities in premature death between different areas of Britain continued to rise steadily during the first decade of the twenty-first century
- life expectancy is predicted to fall for the first time in some of the poorest areas of the UK.

Incomes, occupations and health inequalities

Occupation and income are powerful influences on how people live. They count highly among a range of factors that influence UK health inequalities. These factors are often referred to as **determinates**, because they strongly determine – or influence – health. A number of these factors/determinates have been included in Figure 8.49 on page 310.

Determinates consist of three broad types:
- wider determinates, e.g. occupation, education, income, housing, etc.
- lifestyle factors, e.g. smoking, diet, alcohol, drug misuse, etc.
- preventative health care, e.g. immunisation

None of these determinates operates in isolation. They may influence each other:
- *within* categories – for instance, low educational achievement influences employment and income prospects, or
- *between* categories – e.g. high-income earners tend to enjoy much healthier diets.

Therefore, it is important to look at determinates in combination, rather than in isolation. Generally, there are many linkages between the different indicators and how they impact on health. Changing trends in one determinate might be caused by, or have implications for, other indicators.

It is possible for external actions to reduce health inequalities. For example:
- *The Clean Air Act* (of 1956) has had a major effect on the quality of urban air in the UK. The Act forbids the burning of fossil fuels in urban areas, and restricts industrial use to cleaner fuels (e.g. coke or gas, instead of coal). Before the Act, cases of bronchitis and other major lung infections were far higher in urban areas than in rural ones.
- The health of people in inner London and areas to the east was much improved after the construction of the north London sewer outfall in the late nineteenth century, which piped all raw sewage away from the city. John Snow's map (Figure 8.46) proved that contaminated water could badly affect health. Piped sewage prevented contamination of the water supply and improved health in inner London as a result.

The Health Profile of England

Each year, the Department for Health publishes a Health Profile of England. The 2009 report highlighted some improvements in health. There was:

- a declining death rate from cancers
- declining infant mortality
- increasing life expectancy – now at its highest ever level.

But, for geographers, the 2009 report highlighted continuing **geographical inequalities** in England (Figure 8.49). This conclusion is part of a regional pattern of inequalities which has persisted since the nineteenth century.

- London and south-east England has always been the wealthiest part of the UK. London's role as the capital city attracts many professional and high-salary occupations in government, companies (in their headquarters), and industry. From the nineteenth until the late twentieth century, London was the world's largest port – and east London the UK's most productive industrial area. Even though those traditional employment sectors have now been massively reduced in London, the new 'knowledge economy' has now replaced the old industries with jobs that generate even higher incomes, such as banking and finance and the creative industries (e.g. advertising and the media).

- North-east England suffered serious industrial decline throughout the twentieth century. Although its traditional industries of coal mining, steel manufacture, engineering and shipbuilding boomed at times, a sustained decline since the 1980s has led to high unemployment in the region. In spite of some new investment (e.g. car production for Nissan, and a growth of call centres in the region), employment alternatives have been mostly lower paid and in the public sector, where incomes are lower.

INDICATOR	Measure	England as a whole	North-east England	South-east England
Deprivation	%	**19.9**	33.7	5.9
Children in poverty	%	**22.4**	26.0	15.4
Statutory homelessness	Per 1000	**2.8**	3.2	1.5
Children's and young people's health				
Smoking in pregnancy	%	**14.7**	22.2	13.7
Breastfeeding initiation	%	**71.0**	52.5	75.7
Physically active children	%	**90.0**	90.5	89.5
Obese children	%	**9.6**	10.7	8.3
Teenage pregnancy (under 18)	Per 1000	**41.2**	50.6	33.3
Adults' health and lifestyle				
Adults who smoke	%	**24.1**	29.1	21.8
Binge drinking adults	%	**18.0**	26.5	16.2
Healthy eating adults	%	**26.3**	18.5	30.4
Physically active adults	%	**10.8**	10.6	11.4
Obese adults	%	**23.6**	25.2	22.0
Disease and poor health				
Over 65s 'not in good health'	%	**21.5**	27.5	16.9
Incapacity benefits for mental illness	Per 1000	**27.7**	40.2	19.8
Hospital stays for alcohol related harm	Per 100000	**1472.5**	2045.6	1161.4
Drug misuse	Per 1000	**9.8**	9.4	5.6
People diagnosed with diabetes	%	**4.1**	4.1	3.7
New cases of tuberculosis	Per 100000	**15.0**	5.6	8.0
Hip fracture in over 65s	Per 100000	**479.8**	552.3	467.5
Life expectancy and causes of death				
Male life expectancy	Years	**77.7**	76.3	78.9
Female life expectancy	Years	**81.8**	80.4	82.7
Infant deaths	Per 1000	**4.9**	4.9	4.0
Deaths from smoking	Per 100000	**210.2**	268.8	183.0
Early deaths – heart disease and strokes	Per 100000	**79.1**	92.7	66.3
Early deaths – cancer	Per 100000	**115.5**	134.3	108.0
Road injuries and deaths	Per 100000	**54.3**	42.8	54.5

Figure 8.49 *Extracts from the Health Profile of England Report for 2009, produced by the Department of Health* ▲

CASE STUDY

London – life expectancy and health issues

London – as a whole – has a very similar average life expectancy to England. However, there are wide inequalities within London (Figure 8.50).

- Residents of the Royal Borough of Kensington and Chelsea can expect to live longer than the national average life expectancy. According to the Focus on London report (2007) male residents there live to an average age of 83.7, and women to 87 (compared to the England averages of 77.7 and 81.8 years, respectively).
- However, some London boroughs have the shortest average life expectancy in England, e.g. Islington and Newham, with an average life expectancy of 76.6, and Tower Hamlets with just 76.55.

The health of Londoners also shows a mixed picture. London has the highest incidence of tuberculosis (TB) in England, and a much higher proportion of children living in poverty than elsewhere. However, it also has the lowest rate of smoking in pregnancy, and Londoners are more likely to adopt the approved option of breastfeeding.

Figure 8.50 *Male life expectancy at birth, by London borough, 2005-2007* ▶

London faces a number of particular health challenges:

- There is a wide variation in teenage pregnancy rates across the London boroughs. For example, the rates in Lambeth and Southwark are over three times higher than the rate in Richmond upon Thames.
- London has the highest rates of new diagnoses for chlamydia, gonorrhoea and syphilis in the country.
- It also accounts for approximately 40% of TB cases in the UK, plus 52% of the UK's known cases of HIV.

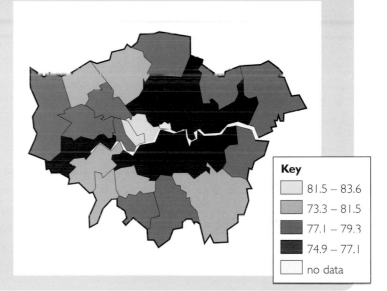

Key

	81.5 – 83.6
	73.3 – 81.5
	77.1 – 79.3
	74.9 – 77.1
	no data

ACTIVITIES

1 Explain the two ways in which health inequalities can be assessed.

2 Study Figure 8.48.
   a Suggest some possible reasons why Scotland had a higher death rate than other parts of the UK in 2003.
   b According to Figure 8.48, Scotland had the highest NHS expenditure per person in the UK, as well as the highest death rate. Explain how these two factors might be connected.

3 Explain how determinates can impact on people's health outcomes.

4 a Draw up a table to classify the indicators in Figure 8.49 into (i) wider determinates and (ii) lifestyle factors.
   b Select five wider determinates and draw a mind map to show how they are linked together.
   c Add to your mind map to show how these determinates can have profound influences on people's health and welfare.

5 Should governments try to influence (a) wider determinates, (b) lifestyle choices? Discuss the issue in pairs and draw up a table to show the main arguments in favour and against.

Internet research

In pairs, carry out research to show links between environmental quality and health. Topics could include a study of the effects of the following on health:

- an industrial incident, e.g. Bhopal (India) chemical explosion in 1984
- air pollution in UK cities
- water pollution, e.g. river pollution and drinking water in China
- government environmental legislation, e.g. the UK's *Clean Air Act* 1956.

In this section you will learn about:
- geographical variations in access to health care in the UK
- the factors that contribute to good health and long life expectancy

The gap gets ever wider

The UK's National Health Service is unique. No other country has a health care system which is free at the point of delivery to all of its citizens – regardless of wealth. The NHS provides a service that, despite its imperfections, is much admired. By contrast, before the creation of the NHS in 1948, everybody had to pay for doctors' visits, medicines or hospital stays – so the poor often went without proper health care.

But, despite its broad availability to everyone, the NHS still contains health inequalities. The House of Commons reported in 2009:

Health inequalities can be found in many aspects of health. For example, poor people not only live less long than rich, but also have more years of poor health. Access to health is also uneven. The old and disabled receive worse treatment than the young and able-bodied.

In an attempt to reduce these health inequalities, government funding for Primary Care Trusts (PCTs) – who provide hospitals and GPs surgeries – varies between trusts. Budgets for health care are allocated to PCTs according to need – with large differences in funding per person between them. For example, in 2009-10, Mid-Essex PCT (in prosperous south-east England) received £1269 per person, whereas PCTs in deprived areas received much more funding per person, e.g. City and Hackney Teaching PCT (£2136) and Liverpool PCT (£2031).

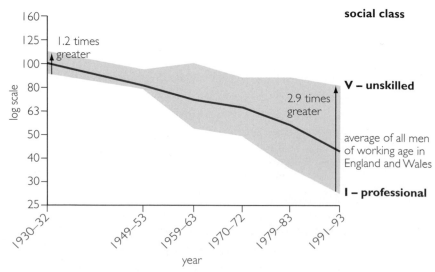

Figure 8.51 *St Bartholomews Hospital (Barts) provides services for the City and Hackney PCT in east London. Barts is in one of the UK's most deprived areas (see Section 5.7), so it receives more funding than other, wealthier, areas.* ▲

However, despite actions like differences in funding, health inequalities have been increasing for a long time – so it's a hard trend to reverse. Figure 8.52 shows the gap in mortality rates for men of different employment classes between 1930 and 1993. 100 is the average mortality rate for working men in 1930-32. The 'average of all working age men' line shows that the mortality rate fell overall up to 1993. However, on either side of that line, the mortality rate for unskilled workers (V) fell less fast than for professionals (I). In 60 years, the gap between the two groups widened considerably. This is not because the poor are getting less healthy – in fact, in 2009, the life expectancy of the poorest 20% was as high as that of the richest 20% in 1979. It is simply that richer people are getting healthier more quickly than poorer people.

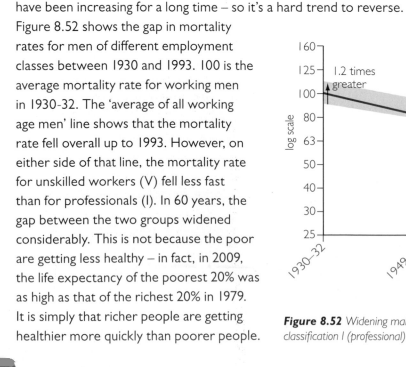

Figure 8.52 *Widening male health inequalities in the UK (1930-1993), according to job classification I (professional) to V (unskilled)* ▲

Income and inequalities I – diet

Health inequalities are due mainly to differences in income. One health indicator is diet. Healthy diets are influenced by income, because healthy foods tend to be more expensive than 'value' items in supermarkets (Figure 8.53). Many 'value' items are higher in carbohydrates, fat and salt – and lower in fibre, vitamins and protein. This affects the diet of low-income families (and many pensioners on fixed incomes), who are more likely to eat unhealthily as a result (Figure 8.54) – creating greater health problems.

Poor nutrition in deprived areas is nothing new. But a more-recent problem is that of too many calories from an imbalanced diet. As a result, obesity is rising more rapidly in low-income groups.

Item	Healthy alternative products and price	Value foodstuff and price
Bread	Large wholemeal loaf £1.32	Large white sliced 'value' 47p
Meat and fish	Free range chicken breasts £13.19 per kg Salmon fillets £16.67 per kg Lean mince £4.79 per kg	Frozen pack 'value' burgers £2.64 per kg Sausages – cheapest £2.09 per kg Fish fingers £1.88 per kg pack
Butter / spread	Flora Pro-Active £3.50 per 500g Butter £5.00 per 500g	'Value' spread £1.75 per 500g
Cereal	Organic muesli £4.50 per kg	Value cornflakes £1.88 per kg
Milk	Organic low fat 76p per litre	Standard low fat 55p per litre
Vegetables	Fresh broccoli £1.59 per kg	Value baked beans 69p per kg
Potatoes	Baking potatoes £1.47 per kg	Value frozen chips 48p per kg
Cooking oil	Extra virgin olive oil £6.50 per litre	Value vegetable oil £1.20 per litre

Figure 8.53 Comparing healthy foodstuffs with 'value' products. Prices are from a major supermarket chain in October 2010. ▲

Figure 8.54 The proportion of people eating five portions of fruit and vegetables a day, by employment group (according to the British Heart Foundation) ▶

Fruit and vegetable consumption	Job type				
	Managerial and Professional jobs	Intermediate jobs (e.g. supervisors)	Small employers and self-employed workers	Lower supervisory and technical jobs	Semi-routine and routine jobs
% of men who eat no fruit and vegetables daily	5	8	8	9	12
% of men who eat five portions of fruit and vegetables daily	28	24	25	22	18
% of women who eat no fruit and vegetables daily	–	4	6	6	8
% of women who eat five portions of fruit and vegetables daily	35	25	27	26	21

Income and inequalities 2 – exercise

Health inequalities in the UK are also reflected in the provision of leisure facilities, which are of two types – public and private:

Figure 8.55 Truro Leisure Centre in rural Cornwall. This is a public access leisure complex, which is run by Carrick Leisure Limited (a not-for-profit company) for Cornwall County Council. ▼

Public sector facilities are owned and run by local councils, e.g. sports centres and swimming pools. These are run as a service for local people, and are often located in local schools or close to local housing. Their admissions prices are lower than those of private facilities. Visitors are generally charged per visit, with entry to a swimming pool costing about £3–5. Those on low incomes or benefits receive low-cost or free entry. However, public facilities like the leisure centre in Figure 8.55 are increasingly being transferring to private companies, who run them on a not-for-profit basis for local councils.

Private sector leisure centres are owned and run by companies such as David Lloyd or Virgin Active. These are often larger than public facilities, offer a wider range of activities, and are run for profit. They are located in large urban centres (e.g. London, Manchester, Leeds), or in major concentrations of population. There are very few in rural areas, because the demand and profit are not high enough. Membership is normally by monthly payment. It can range from £30 for off-peak membership with limited hours, to over £150 per month for full membership in London's prestige health clubs.

CASE STUDY

Healthy lifestyles in west Cornwall?

Rural areas are often healthy places, with fresh food and air. But access to health facilities can be poor. West Cornwall is one of the UK's most remote areas. It has an elderly population, and one of the UK's lowest average incomes. In 2007, West Cornwall was identified as one of the UK's most deprived areas for health, as measured by:

- a shorter life expectancy
- a greater likelihood of serious illness and disability
- the proportion of adults under 60 who were suffering anxiety disorders, periods in hospital, and/or claiming health benefits.

The Royal Cornwall Hospital in Truro provides a wide range of health treatments. But Cornwall is a long county and – for those living in Penzance – a 26-mile journey to Truro could mean life or death in the case of a severe heart attack or stroke. In the tourist season, journeys take even longer and traffic congestion can be a problem.

Figure 8.56 *A new doctors' surgery in Cornwall – but one of the problems is that it's only accessible by car for most patients* ▲

- Only 38% of villages in west Cornwall have a doctor's surgery – and, even then, the surgery hours vary. St Just has a surgery every weekday, but the villages of Sennen, St Buryan and Polgigga have surgeries just one morning a week.
- Access to transport is also difficult. Buses to Penzance and St Ives operate in 70% of villages, but there may be only three or four a day. People increasingly rely on cars or taxis.
- Penzance and St Ives each have a Leisure Centre with a wide range of facilities, but access by public transport is difficult and costly. An average five-mile bus trip to Penzance Leisure Centre is £3.50 return, which can double the cost of a visit.

ACTIVITIES

1 In pairs or small groups, discuss whether PCTs in deprived areas should be funded differently from those in wealthier areas. Explain your thinking.

2 a Explain how the foods listed in Figure 8.53 could affect health, both positively or negatively.

  b Explain the significance of the data in Figure 8.54.

3 Draw up a table to compare advantages and disadvantages in location, facilities and cost of public and private health and leisure facilities, such as those in Figure 8.55.

4 In pairs, draw up a table showing the health issues facing isolated rural communities. Also provide suggestions about how these issues could be addressed. Feed back your ideas in class.

AIDS Acquired Immune Deficiency Syndrome. Better referred to as the conditions brought about by the deterioration in the body's immune system caused by HIV – known as AIDS-Related Conditions (ARC)

behavioural disease A disease created or spread by human behaviour, e.g. conditions caused by smoking

cancer The name given to a range of illnesses that result from one of the cells in the body growing out of control

case mortality rate The number dying from a disease divided by the number of those diagnosed with the disease

chronic disease A disease that is long-lasting or recurrent and normally incurable. Examples include: cardiovascular disease, cancer, chronic respiratory disease, and diabetes

crude rates Measure the basic statistics of any population, such as birth or death rates per 1000

development of disease The way in which a disease actually develops, e.g. the development of the malarial parasite in a mosquito or human host

epidemic A widespread growth in cases of a disease

epidemiology The study of the spread of disease in medicine

health care system A system created by governments to provide health care across the population

health insurance An amount paid to pay for health treatment. This is usually private, unlike public contributions which are funded through taxation

HIV The Human Immuno-Deficiency Virus, the virus which causes AIDS

incidence rate How many people develop a condition within a period of time, e.g. one year

infant mortality rate The number of deaths of children before their first birthday expressed as a rate per 1000 population

malnutrition Hunger, or insufficient calories

morbidity Illness, or any diseased state, disability, or condition of poor health from any cause. It can be used to refer to the existence of disease, or the extent that a health condition affects a patient

mortality Deaths. It is the condition of being mortal, or susceptible to death

mortality gap The gap in death rates between wealthier and poorer income groups or social classes

mortality rate The number of deaths (in general, or from a specific cause) in a population

nano-technology Micro-technology, used for example in drug treatments where tiny amounts can be administered

notifiable disease Any disease that is required by law to be reported to the government

obesity A condition caused by over-consumption of calories

pandemic A global infection, found in (almost) every country in the world

parasitic infection An infection acquired through a parasite, e.g. amoeba which can cause dysentery

pharmaceutical Linked to drugs and their development

prevalence rate An indication of the number of people in a population suffering from a condition at any one time

primary care trusts (PCTs) UK administrative areas which provide hospitals and doctors' surgeries. Likely to be abolished after 2010

private sector leisure centres Owned and run by companies for profit

public sector Owned and run by local councils as a service

spread of disease The way in which a disease moves from once person or place to another

under-nourishment A shortage of protein or calories

under-nutrition Lacking adequate minerals and vitamins

vaccine An injection designed to prevent an infection

1 (a) Study Figure 8.14 (page 283). Explain one additional data source that could help you to explain differences in numbers of cases between the countries shown. *(4 marks)*

> **(a)** Think of one other source, e.g. health care spending.

(b) Outline one reason why infectious diseases such as malaria spread easily in some countries. *(4 marks)*

> **(b)** Focus on factors that encourage the spread of disease.

(c) Explain two actions that can be taken to reduce or prevent a 'disease of affluence' that you have studied. *(7 marks)*

> **(c)** Think of what, for instance, governments can do.

> **(d)** The key word here is 'effectiveness'. Think of a project you have studied which either is or is not effective, and why.

(d) Assess the effectiveness of one attempt to control the spread of one disease that you have studied. *(15 marks)*

2 (a) Describe the differences between 'mortality rate' and 'morbidity'. *(4 marks)*

> **(a)** Two separate definitions are less important than picking out the differences between them.

(b) Study Figure 8.18 (page 286). Describe the variations in the global distribution of HIV-AIDS. *(4 marks)*

> **(b)** Naming areas of the world might help you to describe where cases are high or low.

> **(c)** Think of one factor, e.g. health care funding or international aid.

(c) Explain why some countries are better placed to treat diseases (e.g. HIV-AIDS) than others. *(7 marks)*

> **(d)** Use different impacts, e.g. social, economic, or environmental. Make use of case studies.

(d) Assess the impacts of one disease you have studied upon a country or countries. *(15 marks)*

3 (a) Study Figure 8.48 (page 309). Describe the variations in public health spending shown. *(4 marks)*

(b) Explain one factor that may make it difficult for some countries to fund public health care. *(4 marks)*

(c) Briefly outline the arguments in favour of funding health care publicly. *(7 marks)*

(d) Assess the effectiveness of the health care system in one country you have studied. *(15 marks)*

The aim of this chapter is to give you a summary of the Geographical Skills requirement in the AQA AS specification. Remember that each skill is incorporated into the main body of this book where appropriate.

The AS specification requires you to develop:

- a variety of basic, investigative, cartographic (maps), graphical, applied ICT, and statistical skills;
- a critical awareness of the appropriateness and limitations of different skills and resources.

In your exams, you should take with you a basic mathematical set (ruler, protractor, and so on) and a calculator.

The Skills boxes

The Skills boxes in chapters 1 and 5, the two core topics, explain and develop the skills required for Unit 2.

Skills boxes can be found on the following pages: 11, 16, 24, 29, 40, 44, 50, 60, 160, 169, 170, 179, 188, 191.

There are also opportunities for skills practice in the activities throughout the book.

Basic Skills

In addition to the basic skills of literacy (spelling, punctuation, grammar, and the correct use of sentences and paragraphs), you will need to be able to annotate illustrative material such as maps, diagrams, graphs, and photographs.

An **annotation** is a label that might offer a detailed description or some form of explanation. It might draw out links or interconnections between features and processes. Figure 9.1 is an example of an annotated photograph of a meandering river.

You may be asked to use an overlay in an exam. This simply adds another layer of information. Just be sure to follow the instructions to ensure that you position the overlay correctly on the base map, photo, or diagram.

Woodford Stream – a meandering river on Exmoor

◄ *Figure 9.1 An example of an annotated photo*

Highest velocity causes undercutting on the outside bend of the meander, forming a river cliff

River bank damaged by cattle drinking from this place

Slip-off slope on the inside bend of the meander where the velocity is low and deposition takes place

Riffle – shallow water in a wide channel

Investigative Skills

These are the skills that are employed when you carry out a geographical investigation. They form the sequence of an investigation, and reflect its different stages:

1 Identification of aims, geographical question, or hypothesis, and justification of the investigation (e.g. the geographical context).

2 Conducting a risk assessment.

3 Identification, selection, and collection of quantitative and qualitative evidence to investigate the chosen question or hypothesis. This includes the selection of appropriate sampling strategies. Data will be collected from primary and secondary sources.

4 Processing, presentation, analysis, and interpretation of evidence.

5 Drawing conclusions, and showing an awareness of the validity of the conclusions.

6 Evaluation (reliability, accuracy, and so on) and identification of opportunities for further research.

Cartographic Skills

These are skills involving the use of maps. In your exams you may be asked to use and interpret a variety of maps, including:

- atlas maps
- base maps
- OS maps (at different scales, but most commonly 1:25 000 or 1:50 000)
- sketch maps
- maps with proportional symbols (e.g. squares, circles)
- maps showing movement (flow lines, desire lines, and trip lines)
- choropleth, isoline, and dot maps (Figure 9.3)

There are a few things to remember to help you interpret maps correctly and achieve high marks:

- Take time to read the key carefully. Make sure you are clear what the map is showing and what the colours or symbols mean.
- When **describing**, put into words what you can see. Look for patterns (e.g. linear or radial), concentrations, and trends (e.g. an increase from east to west). Use facts and figures from the map to support your statements. Look for any anomalies (exceptions) and identify them clearly. Be as detailed as you can.
- When **explaining** or giving reasons, try to suggest why certain patterns or trends exist. If possible refer to other information available to you, such as maps or text (e.g. steep slopes – shown by closely spaced contours – might help to explain why a flood has occurred). Don't be afraid to have a go at a possible explanation – as long as your argument makes sense, you will be credited.

Look at Figure 9.3, which shows an isoline map with accompanying description and attempted explanation.

◀ *Figure 9.2 A glossary of map types*

Map type	Description
Sketch map	Sketch maps show selected information chosen by the drawer. They may be based on a map such as an OS map or drawn to show spatial details at a field locality. They should include an orientation and approximate scale.
Map with proportional symbols	Proportional symbols (such as squares or circles) positioned in appropriate places on a base map, e.g. tourist numbers visiting countries in the Caribbean shown by proportional symbols such as circles or suitcases, positioned on the countries.
Flow lines	Proportional arrows (width drawn in proportion to the value being shown) used to show movement from one place to another, e.g. oil, tourists. The arrow needs to be positioned with its base at the source and its arrow head at the destination. Traffic movement at a local scale can also be shown using flow lines that map actual routes.
Desire lines and trip lines	Desire lines are straight lines on a map drawn from the point of origin to the destination, e.g. origin of shoppers using a supermarket. Regular travel lines, e.g. tourist flows, are called trip lines.
Choropleth map	A map using a gradation of shading densities or colours to show grouped values for areas, such as countries or counties, e.g. population densities, GNP. Ideally, there should be between four and eight equally-sized categories. When interpreting a choropleth map, remember that the often sharp differences that may exist at the boundaries of two adjacent areas on the map can be misleading and probably don't occur on the ground.
Isoline map	Isolines are lines that join points of equal value, e.g. contour lines join points of equal height. They are usually drawn on to a base map to show relative location, e.g. pedestrians in a town centre.
Dot map	Dots are used to represent a particular value or number, e.g. a population of 1 million, and are located accurately on a map. It is the density (spread) of the dots that creates the visual impact of the map. Dot maps give powerful impressions, but they are not easy to extract accurate information from.

▼ *Figure 9.3 An isoline map showing the distribution of exotic plant species in Britain*

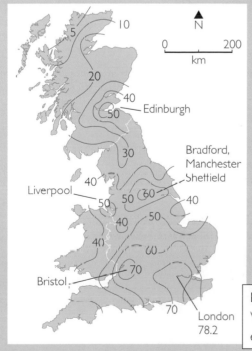

Description and explanation

The map shows a great variation in the percentages of exotic plant species in Britain. The highest values are in southern England (60-70%), centred on London (78.2%) and Bristol. High values elsewhere coincide with the location of urban areas (e.g. Liverpool 50%, Bradford, Manchester, and Sheffield 60%, and Edinburgh 50%). The lowest values (10-20%) are in northern Scotland.

Exotic plant species were introduced through major ports, which explains the high values for London, Bristol, Liverpool, and Edinburgh. Once introduced, these species then spread out from the ports and urban areas. Southern England has the highest values due to its proximity to continental Europe, the souce of many of the exotic species. Remote regions far away from ports and urban areas, such as Wales and northern Scotland, have lower percentages of exotic species.

Key

Values (e.g. 60, 70) are percentages of the total British exotic plant species that can be found in a 50x50km quadrat

Graphical Skills

There are a variety of graphs and diagrams that you need to be able to draw and interpret, including:

- line graphs – simple, comparative, compound, and divergent
- bar graphs – simple, comparative, compound, and divergent
- scattergraphs (with use of best-fit line)
- pie charts and proportional divided circles
- triangular graphs
- radial diagrams
- logarithmic scales
- dispersion diagrams (see below)

When constructing a graph or diagram, remember to write detailed labels on the axes and give it a full title. Add labels and annotations to describe and explain what the graph or diagram shows.

When interpreting graphs and diagrams, you should consider the same points made earlier about description and explanation. Remember to refer to facts and figures to support the points you make, and make sure you use the correct units.

◀ *Figure 9.4 A glossary of graphs and diagrams*

Graph / diagram	Description
Line graph	Used to show continuous data, most commonly changes taking place over time. It is possible to sub-divide the area below a line graph to show different proportions of the total. This creates a compound line graph.
Bar graph	Bars are commonly used to show data that is unconnected, e.g. quantities of different items bought from a shopping centre. Such graphs are conventionally drawn with a small gap between each bar. A histogram is a type of bar chart that shows data that is derived from a single population, e.g. separate categories of pebble sizes from a single beach sample. Histograms are drawn with the bars touching. A bar graph can be divided up to form a compound bar graph.
Scattergraph	A scattergraph is used to consider a possible relationship between two data sets. It involves the plotting of individual points and the drawing of a best-fit line to show the main trend of the points.
Pie charts and proportional divided circles	A pie chart is a circle divided into segments to show the proportions of a total population, e.g. percentages of different vegetation types in a quadrat. They can be drawn proportionally whereby the size of the circle is determined by a single factor, e.g. population size.
Triangular graphs	These graphs have three axes. They are well suited to certain investigations involving three variables, e.g. soil texture that comprises percentages of clay, sand, and silt.
Radial diagrams	Radial diagrams are graphs with axes originating from a central point. They are well suited to show orientation, e.g. the orientation of corries. When complete, the points on each of the axes can be joined together to form a polygonal shape.
Logarithmic scales	Logarithmic scales are useful when trying to represent a large range of data on a single graph, e.g. GNP of countries.

ICT Skills

You should be familiar with the application of various ICT skills, including:

- use of remotely-sensed data such as photographs and digital images such as those captured by satellites. Remote sensing involves measurement or recording at a distance – by, for example, satellites, aircraft, or spacecraft. A good example is the use of satellites and radar to observe the weather. Information collected digitally can be processed to focus on certain aspects, for example vegetation or infrared heat.
- use of databases (e.g. census data). Databases most commonly involve the use of computer programs such as Excel. Here information can be tabulated and manipulated by the user to produce graphs and diagrams.
- use of GIS (Geographical Information Systems). GIS essentially involves the use of layered data on a spatial base map. An ordinary OS map is a good example, as is the Environment Agency's 'floodline' maps that use shades of blue to represent flood risks on an OS base map.
- presentation using ICT (text, diagrams, and so on). ICT can be used widely in the presentation of information, both in fieldwork investigations and in general geographical research. This can include the use of tables, annotated maps and photos, graphs, and diagrams.

Statistical Skills

At AS, you should be familiar with the application and interpretation, including the application of significance level in inferential statistical results, of the following statistical methods:

- measures of central tendency (mean, mode, and median)
- measures of dispersion, interquartile range, and standard deviation
- Spearman's rank correlation test

Measures of central tendency

Mean – this is often referred to as the 'average'. It involves adding up the individual values for a set of data and dividing it by the number of values. So, if 10 pebbles have a combined total size of 35 cm, the average pebble size is 35/10 = 3.5 cm.

Mode – this is the number that appears most frequently in a data set. This measure of central tendency is only really valid when there is a substantial data set – otherwise it can be misleading.

Median – this is the central or midway value in a data set. There should be an equal number of values above and below the median. If there is an odd number of values (e.g. 23), the median value is the middle value when all the values have been ranked (i.e. the twelfth). If there is an even number of values, the median is the average of the two middle values.

Measures of dispersion

Dispersion diagram – this describes the dispersion of a data set, or its spread. It is most commonly displayed using a dispersion diagram (Figure 9.5).

Interquartile range – to calculate the interquartile range you need to divide the ranked data set into four equal quarters (Figure 9.5). Notice that the top of the interquartile range is marked by the upper quartile line and the bottom is marked by the lower quartile line. The interquartile range can be used to compare dispersions between two or more sets of data (Figure 9.5).

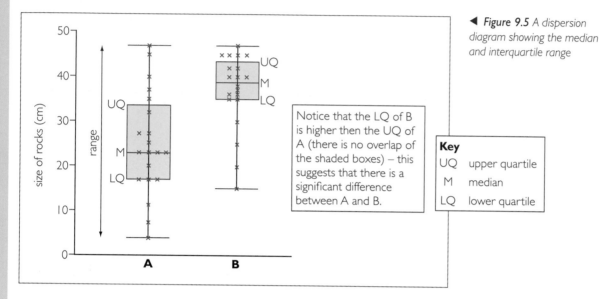

◀ **Figure 9.5** *A dispersion diagram showing the median and interquartile range*

Notice that the LQ of B is higher then the UQ of A (there is no overlap of the shaded boxes) – this suggests that there is a significant difference between A and B.

Key

UQ   upper quartile
M   median
LQ   lower quartile

Standard deviation – this is a statistical measure of the spread of values around the mean. It is calculated using the equation shown in the box on the right.

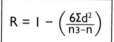

$$R = 1 - \left(\frac{6\Sigma d^2}{n3-n} \right)$$

A high standard deviation indicates that the data is widely spread around the mean value. A low standard deviation indicates that the data is clustered closely around the mean.

Spearman's rank correlation test

The Spearman's rank correlation test is a statistical method that can be used to assess the strength of an apparent relationship between two values. It is usually employed if a scattergraph suggests a causal relationship between two variables, where one factor seems to affect another. Turn back to pages 31-32 to see how the test is applied to a set of data.

In common with all statistical tests, it is the interpretation of the result that is critical. By using a table of significance it is possible to judge the strength of an apparent relationship. This is given as a percentage likelihood that the result did not occur by chance. In geographical investigations, a 95% threshold value is usually acceptable.

It is important to remember that:
- a conclusion is likely to be less secure if only a few values are used to calculate Spearman's rank. Ideally there should be over 10 sets of values;
- it is up to you as a geographer to make a causal link between two sets of data. A causal relationship is not an automatic conclusion of the test.

If you want to be successful then you need to know and understand the geography that you have been studying, but you also need to know how you will be examined, what kinds of questions you will come up against in the exam, how to use what you know, and what you will get marks for. That is where this chapter can help.

What is the AQA AS Geography specification about?

AQA Geography is a new specification written for the twenty-first century. It includes some new content, e.g. 'Health issues', and offers a new twist to old geographical favourites, e.g. 'Rivers, floods and management' and 'Population change'. Along with A2, it provides a balance between your own physical, human, and/or environmental interests and key geographical topics that provide you with the knowledge, understanding, and skills so that you are ready for further study in higher education or for employment.

What does the AS specification include?

The AS specification consists of two Units, which are broken down into topics.

- **Unit 1: Physical and Human Geography** consists of two compulsory topics looking at some of today's big issues – one physical, one human.

In terms of **Physical Geography**, *Rivers, floods and management* is the compulsory topic. It looks at a range of natural processes which occur along rivers, and which have led to some characteristic landforms. It also focuses upon flooding as a threat to people and places, and how flood threats might be managed.

In terms of **Human Geography**, *Population change* is the **compulsory topic**. It looks at today's rapid growth in global population and the challenges that population issues present for different countries, and for local areas. These challenges include an examination of the links between numbers of people and economic growth, and the resources available, and what happens when populations rise rapidly, or slowly.

In addition to these compulsory topics, you must study at least **one** of three Physical options and at least **one** of three Human options. These are:

Physical topics

- *Cold environments* explores the landform processes that occur in glacial regions and their margins (known as periglacial areas). It shows how these processes have created unique landforms. It investigates how traditional lifestyles in these regions are changing rapidly and whether they face a sustainable future or not.
- *Coastal environments* looks at a range of processes which occur along coasts, and which have led to unique landforms. It also focuses on coastal flooding and erosion as threats to people and places, and how these might be managed.
- *Hot desert environments and their margins* explores the landform processes that occur in the world's arid regions, and how these processes have created unique landforms. It investigates how population pressures in these regions are occurring as a result of desertification, and how these might be managed in future.

Human topics

- *Food supply issues* investigates the causes and consequences of food surpluses and shortages in different parts of the world, and how far food problems can partly be solved by technology, e.g. by genetic modification.
- *Energy issues* focuses on the world's dependency on oil and other fossil fuels and the issues that arise from energy surpluses in some parts of the world and shortages in others. It explores how far the world can continue using up finite energy reserves, or whether sustainable supplies might replace them.
- *Health issues* focuses on global health issues, and ways in which disease and infection vary between different countries in the world. It looks at different health challenges, such as HIV and cancer. It explores how inequalities of wealth – globally or locally – can affect human health significantly.

- **Unit 2 Geographical Skills** will allow you to develop a range of geographical skills in terms of geographical research, data handling, and analysis, including your ability to apply your knowledge and skills to unseen information and resources and fieldwork. These skills are to be taught as an integral part of Unit 1 Physical and Human Geography (both compulsory and option topics) and not as a separate unit. These skills have therefore been integrated into every chapter in this book.

How will you be assessed?

If you understand how you will be assessed, and how you can pick up marks, you should find it easier to do well in your exams.

There are two exams for AS Geography (and no coursework). The table shows what the exams are like:

Unit	Question paper format	Marks and method of marking
Unit 1 Physical and Human Geography	This exam is 2 hours long and consists of two sections, A and B. Each section has four questions. The questions match the topics in Unit 1. You have to answer **four** questions in total: • **Question 1** (*Rivers, floods and management*) which is compulsory. • **One** other question from Section A – either Question 2 (*Cold environments*), Question 3 (*Coastal environments*), or Question 4 (*Hot desert environments and their margins*). • **Question 5** (*Population change*) – which is compulsory. • **One** other question from Section B – either Question 6 (*Food supply issues*), Question 7 (*Energy issues*) or Question 8 (*Health issues*). All questions use resource materials, e.g. maps, data tables, photographs, and are of two types: • those requiring **short answers**. These usually carry between 3 and 7 marks. • those requiring **longer essay answers**. These generally require some kind of argument to be developed, e.g. 'Using examples, discuss how successful different flood management strategies have proved.' These usually carry 15 marks.	Each question carries 30 marks. This means there is a total of 120 marks for this exam – i.e. one mark a minute! You need to know your stuff and be able to write quickly! **Short answers** (up to about 4 marks) are 'point marked'. You need to state a point, or give an example, and so on, to get a mark. So 3 marks will require 3 points. **Longer answers** carrying 5 marks or more are 'level marked'. Level 1 is the lowest. The more detail and explanation you can give, the higher level you will achieve. • Questions carrying up to 8 marks are marked in **two** levels. They generally require explanations with examples. • Questions carrying 15 marks are marked in **three** levels. The highest level – Level 3 – needs named examples, and an answer to the main question.
Unit 2 Geographical Skills	This exam is 1 hour long and consists of two sections. There may be a separate item, such as an OS map, but this varies. There are two questions which carry a variety of marks. They are based around data (e.g. tables of statistics, graphs, OS maps, photographs) and test the skills that you will have been taught in Unit 1. You will be expected to use the resource booklet and your own ideas from fieldwork and research that you have done. You cannot take any materials into the exam.	Each question carries 25 marks, making a total of 50 marks for this exam. Timing is a little more relaxed over 60 minutes, but this is to allow you time to perform calculations and plot data. You will have no time to spare! Like the exam for Unit 1, **short answers** (up to about 4 marks) are point marked. **Longer answers** carrying 5 marks or more are level marked.

Know your exam papers

You will do better in exams if you know what the questions will look like, how to interpret them, and how to respond to the question.

What will the questions be like?

The table above explains what type of questions you will get in the exams. Here are some examples.

Sample question for Unit 1 Question 1 (Rivers, floods and management)

1 Study Figure 3.

(a) Explain two possible causes of the flooding shown in Figure 3. *(4 marks)*

(b) Suggest the likely impacts of this flood on Cockermouth. *(5 marks)*

(c) Suggest ways in which changing land use can contribute to the likelihood of floods such as these. *(6 marks)*

(d) Referring to examples, discuss how effective hard engineering can be in controlling floods. *(15 marks)*

(Total 30 marks)

▶ *Figure 3 Flooding in Cockermouth, Cumbria, November 2009*

Sample question for Unit 2 Question

7 Study Figure 5.

(a) Name the two counties with the highest dependency ratio. *(2 marks)*

(b) Referring to both Figure 4 and Figure 5, suggest possible reasons why the percentage employed in tertiary employment varies. *(4 marks)*

(c) Referring to the data, describe how you would try to identify the least deprived county in Figure 5. *(5 marks)*

(d) Referring to the data, justify what you believe to be the **two** most significant factors in determining deprivation in East Anglia. *(6 marks)*

(e) Suggest how a programme of fieldwork could help to investigate further the population and employment patterns in **one** of these counties. *(8 marks)*

(Total 25 marks)

▲ *Figure 4 Map showing the counties of East Anglia*

	Essex	Herts	Beds	Cambs	Suffolk	Norfolk	UK
Annual population growth %	0.5	0.5	0.6	1.1	0.4	0.5	0.3
% aged 0–15	20	20	22	21	20	19	20
% aged 65 or more	16	16	13	14	18	19	16
% in Primary employment	5	5	5	7	8	8	6
% in Secondary employment	24	24	31	25	26	26	26
% in Tertiary employment	70	71	64	60	66	66	69
% unemployed, male/female	7/3	6/3	7/3	6/2	5/2	6/3	8/3
Average weekly incomes	£539.90	n/a	£548.00	£560.50	£455.20	£446.20	£522.60
% earning less than £250 per week (male)	8.2	n/a	8.1	6.9	12.4	10.4	11.3
% earning less than £250 per week (female)	24.1	21.3	23.8	17.4	27.2	32.3	24.4

▲ *Figure 5 Population and employment data from counties in East Anglia, UK*

Command words

One of the first things to do when you look at an exam question is check out the command word. This tells you what the examiner wants you to do. This table gives you some of the most commonly used command words.

Command word	What it means	Example question
Account for	Explain the reasons for – you get marks for explanation rather than description.	Account for the changes in Cornwall's employment structure.
Analyse	Identify the main characteristics and rate the factors with respect to importance.	Analyse the economic effects of Objective One funding in Cornwall.
Assess	Examine closely and 'weigh up' a particular situation, e.g. strengths and weaknesses, for and against.	Assess the success of coastal management schemes using named examples.
Comment on	This is asking you to assess a statement. You need to put both sides of the argument.	Comment on the view that desertification is caused by human rather than physical factors.
Compare	Identify similarities and differences between two or more things.	Compare the causes and impacts of hazards in California and the Philippines.
Contrast	Identify the differences between two or more things.	Contrast the impacts of hazards in California and the Philippines.
Define	Give a clear meaning.	Define river discharge.
Describe	Say what something is like; identify trends.	Describe the trends shown in the graph.
Discuss	Similar to assess.	'Global warming is a myth.' Discuss.
Evaluate	The same as assess.	Evaluate the success of flood management schemes.
Examine	You need to describe and explain.	Examine the attempts by governments to control population growth.
Explain	Give reasons why something happens.	Explain, using examples, whether you think hazards are really 'natural'.
How far?	You need to put both sides of an argument.	How far is Kiribati a nation facing multiple hazards?
Illustrate	Use specific examples to support a statement.	Illustrate the ways that debt makes development difficult in some developing countries.
Justify	Give evidence to support your statements.	Suggest three ways in which Botswana could attempt to reduce poverty. Justify your suggestions.
List	Just state the factors, nothing else is needed.	List the different methods of coastal protection.
Outline	You need to describe and explain, but more description than explanation.	Use the graphs to outline the trends in global disasters.
To what extent?	The same as 'How far?'	'The Sydney Olympics – economic and environmental benefits, but at a social cost.' To what extent is this true?

Interpreting the questions

You have probably already been told by your teacher to read the exam questions carefully, and answer the question set, not the one you think it might be. That means you need to interpret it to work out exactly what it is asking. So, look at the key words. These include:

Command words – these have distinct meanings which are listed opposite

Theme or topic – this is what the question is about. The examiner who wrote the question will have tried to narrow the theme down, and you need to spot how so that you do not write everything you know about the theme.

Focus – this shows how the theme has been narrowed down.

Case studies – look to see if you are asked for specific examples.

Here is an example of a question that has been interpreted using the key words above – these questions are part of a 30-mark question:

• Command words

In (c), 'Suggest ways in which…' means you must give more than one way to show the effects of changing land use.

In (d), 'Referring to examples…' means you are being asked for either a case study or named examples of hard engineering.

In (d), 'discuss how effective…' means that you need to weigh up the good and bad points about hard engineering.

• Theme or topic

This question is for Rivers, floods and management, looking specifically at the causes of flooding and flood management.

Study Figure 3.

(c) Suggest ways in which changing land use can contribute to the likelihood of floods such as these. *(6 marks)*

(d) Referring to examples, discuss how effective hard engineering can be in controlling floods. *(15 marks)*

• Case studies

Choose one named area only.

• Focus

Part (c) asks you to talk about 'ways'. You need to be specific in suggesting how (for example) urbanisation or deforestation can contribute to the likelihood of flooding, not everything you know about floods.

Part (d) asks you to discuss the effectiveness of hard engineering. You must name particular methods of hard engineering.

How to use case studies in the exam

Case studies are in-depth examples. There are lots of case studies in this book; some are one or two pages long, others are six or eight pages. You will need to use these to answer some of the exam questions. The following questions are examples in which you would be expected to use case studies to answer a question on *Population change*.

> 1 Referring to named examples, explain why fertility rates are high in low income countries. *(15 marks)*
>
> 2 Using examples, explain how population change can bring both benefits and challenges. *(15 marks)*

Notice that these questions do not ask you to write down everything you know about a case study. They ask you to focus on two things:

- Each question asks for **examples**. Make sure you refer to more than one case study – you could limit your marks if you restrict yourself to just one named example.
- Each question asks you to **explain**. Make sure your examples prove a point. In question 1, you need to explain the reasons why women have more children in LICs, so don't just describe in general terms. Focus on examples – e.g. '*In Uganda, fertility rates are high because…*'

There are things you can do to help yourself learn and use case studies – these are outlined below.

Have a portfolio of case studies

Build up a 'portfolio' of case studies, so that you know which part of the specification and content the case studies fit. A grid like this might help:

Topic name:		
Subject content	Relevant theory	Case studies

The grid works like an index – or pigeon hole. If you fill it in, you will know exactly what case study and theory is appropriate for each part of a topic. It also helps to make links between topics.

How to remember your case study

Look again at questions 1 and 2 above about *Population change*. There are several case studies in this book that you could use to answer these questions.

Question 1 is a prime question for using the case study of Uganda (section 5.2). Here is one way of remembering your case study – draw a spider diagram with all the key information on it. This diagram is only a start, and you would need to fill in more details – for example, about lack of education amongst girls, the effects of poverty on fertility rates, and so on.

Poverty
- Schooling isn't free – families have to pay for secondary education.
- Families can't afford to send children to school.
- 24% of Ugandan families are undernourished – so infant mortality is high.

Poor health care in rural areas
- Skilled health workers attend only 20% of births.
- Most births have no health workers in attendance.

Why fertility rates are high in Uganda

Rural isolation
- Difficult for some children to get to school – it might be 5-10 km to the nearest village.

Lack of education among girls
- Girls often forced to stay at home to help with domestic work.
- Uneducated girls marry early (about 13-15 years old).
- Fertility is high as they have a long period of producing children.
- Educated women have careers and marry later, and so have reduced fertility.

Using your case study

Once you have revised and learned your case study you can use it to answer exam questions. Look again at the question on fertility rates. It asks you for named examples – you can use Uganda. It asks you to explain (give reasons) why fertility rates are high. The reasons are linked to:

- lack of education among girls
- poverty
- rural isolation
- poor health care in rural areas.

How to use fieldwork in the Unit 2 exam

Fieldwork and research skills are a key feature of Unit 2 Geographical Skills. Whatever fieldwork you have carried out, you will be expected to use it in the exam. You will have to answer a question something like this:

> (c) Explain how **analysis of fieldwork data** could be used to help you understand the theory or concept on which your fieldwork was based. *(8 marks)*

Step 1   Understand the question

This is a complex question and it would be easy to write a lot, but gain few marks. To avoid this, two skills are essential:

- Read the question. It is not 'tell the examiner all you know about fieldwork' – it's about how you would **analyse data** about what you have investigated.
- Underline the command words so that you do what the question asks. So, explain, and remember to focus on the theory or concept on which your fieldwork was based.

Use your pen or pencil to highlight the key words.

Step 2   Focus the question

For the sake of this example, the question will be focused on rivers fieldwork, which set out to investigate whether hard management strategies had proved effective along a stretch of river.

- You **must** mention a named area – examiners only award a maximum of about two-thirds of marks for generalised answers that do not refer to places.
- Remember, it must focus on hard management – 'soft' methods will not do.

Step 3   Answer the question

To answer the question, you need to refer to your fieldwork, and particularly to the data. You would refer to patterns that you found in velocity along the river.

- If velocity increased, you could show why this occurred. It could be because channel shape became more efficient, or because gradient increased. You could therefore refer to theories about how and why river channels influence velocity.
- If velocity decreased, it could be because gradient decreased, or because the river channel became broader and shallower – therefore probing theories about hydraulic radius.
- If velocity varied along the channel, a combination of factors may be having an effect, like those outlined above. You could show how this disproved general theories about what happens to velocity, but also how individual places helped to illustrate specific ways in which river channels behave.

You will get credit for:

- Naming a stretch of river that you have studied.
- Describing your fieldwork data, and what general patterns emerged.
- Explaining how far the patterns of velocity varied and **why**. For example: *'I found that velocity varied between the five locations where I collected data. For the first three places, an increase in gradient from two to five degrees led to an increase in velocity, just as my textbook said it would. But then velocity slowed down a lot at the next two places, even though gradient stayed the same – in spite of textbooks telling me that velocity ought to increase downstream. The reason why this happened was because the river channel was choked with sediment as there had been a recent flood, so the river spread out and lost energy. Although this didn't fit what I'd learnt about velocity, it fitted perfectly what I'd found about hydraulic radius – the broader and shallower the channel, the slower velocity is.'*

Following these three steps can help you write about fieldwork on other topics. Here are examples of Unit 2 questions for particular topics:

> **Coastal environments:** Explain how a programme of fieldwork has helped you to understand the effects of human interference in natural landscapes.
>
> **Population change:** Explain how the conclusions you drew from your fieldwork have helped you to understand how changes can occur in populations.

How are exam papers marked?

An examiner will mark your exam papers and will have clear guidance about how to do it. They must mark fairly, so that they mark the first candidate's exam paper in exactly the same way as the last one. You will be rewarded for what you know and can do, rather than being penalised for what you have left out. If your answer matches the best qualities in the examiner's mark scheme then you will get full marks.

If you look at the table at 'How will you be assessed?' on page 324, it tells you which parts of the exam are point marked and which parts are level marked. Are you clear about what that means?

Point marking

Look at this question:

> (b) Outline the characteristics of one policy of migration control. *(4 marks)*

There are four marks for this question and you have to outline one policy of migration control. So, you get one mark (or point) for either outlining four characteristics (or features) OR for developing an explanation which expands a point you are making. Follow these guidelines:

- Remember that 'outline' means 'sketch out'. Detail is not expected.
- A range of examples could be correct for this answer. So you could write about Australian skills-based migration or about open-door policies in the EU.
- As well as giving four marks for four characteristics, examiners will award marks if you expand on a statement, perhaps with examples or reasons. For example, you could say: 'Australia has a skills-based migration system [1 mark], *which awards points depending on education and qualifications* [1 mark] *or competence in spoken English, and helps to maintain a skilled population* [1 mark].'

Level marking

Look at this question:

> Study Figure 7.
> (a) Assess the economic and political consequences of international migration. *(15 marks)*

There are 15 marks for this question and the examiner will award these using the following mark scheme:

Level	Mark	Descriptor
Level 1	1-6	Little structure in answer. Points made are simple and random. Describes trends with little explanation. May refer to reasons for migration. Geographical terminology is rarely used. Frequent written language errors.
Level 2	7-12	Some structure. Begins to focus content with some purpose – seeks to explain reasons for migration. Shows understanding of a few consequences of migration. Geographical terminology is used with some accuracy. Lower end descriptive. There are some written language errors.
Level 3	13-15	Clear explanation of reasons for migration. Structured detailed account of several consequences of migration, and of its consequences politically, using appropriate geographical terms and exemplification to show understanding. Written language errors are rare.

So, to gain the most marks, your answer needs to meet the criteria for Level 3.

How to gain marks, and not lose them

Gaining marks instead of losing them is just a matter of technique. You might get a lower mark than you really ought to, not because you do not understand the material, but because you do not know how to use it. Here are some helpful hints based on choosing the right question, case studies, the type of question you are answering, and communication.

Do not be seduced by an attractive cartoon or photo – how difficult is the whole question?

If you choose the wrong question, have the guts to change in the first 5–6 minutes. If you find yourself running out of time, bullet points are better than no points.

Choose the right question

Read all the questions carefully and then choose. At this stage, you might want to sketch out a quick few points in pencil to make sure you include examples you think you might forget.

Always choose questions from the part of the specification you have studied – the other questions only look easy because you haven't studied them.

Choose the right case study – don't put down something just because you have revised it.

Get the length right – look at the mark allocation.

Case studies

Get a balance between breadth and depth. Does the question ask for one or more case studies or examples?

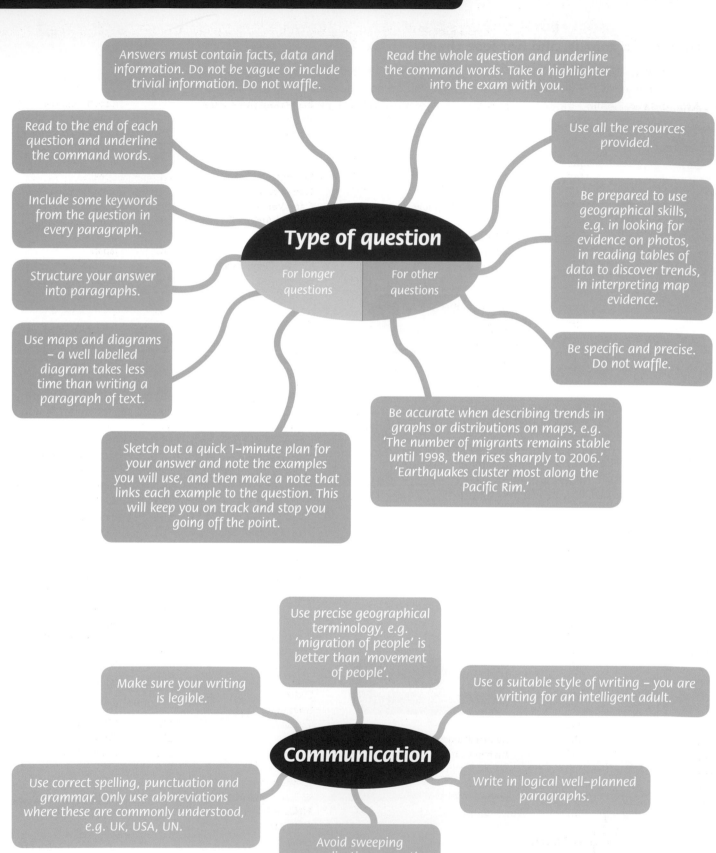

Answers must contain facts, data and information. Do not be vague or include trivial information. Do not waffle.

Read the whole question and underline the command words. Take a highlighter into the exam with you.

Read to the end of each question and underline the command words.

Use all the resources provided.

Include some keywords from the question in every paragraph.

Be prepared to use geographical skills, e.g. in looking for evidence on photos, in reading tables of data to discover trends, in interpreting map evidence.

Structure your answer into paragraphs.

Type of question

For longer questions

For other questions

Use maps and diagrams – a well labelled diagram takes less time than writing a paragraph of text.

Be specific and precise. Do not waffle.

Sketch out a quick 1-minute plan for your answer and note the examples you will use, and then make a note that links each example to the question. This will keep you on track and stop you going off the point.

Be accurate when describing trends in graphs or distributions on maps, e.g. 'The number of migrants remains stable until 1998, then rises sharply to 2006.' 'Earthquakes cluster most along the Pacific Rim.'

Use precise geographical terminology, e.g. 'migration of people' is better than 'movement of people'.

Make sure your writing is legible.

Use a suitable style of writing – you are writing for an intelligent adult.

Communication

Use correct spelling, punctuation and grammar. Only use abbreviations where these are commonly understood, e.g. UK, USA, UN.

Write in logical well-planned paragraphs.

Avoid sweeping generalisations, e.g. the South of England is rich and the North is poor.

List of LICs, MICs, and HICs

Based on the World Bank classification of economies by income, 2010

LIC – low income country   MIC – middle income country   HIC – high income country

East Asia and the Pacific

Cambodia	LIC
China	MIC
Fiji	MIC
Indonesia	MIC
Kiribati	MIC
North Korea	LIC
Laos	LIC
Malaysia	MIC
Marshall Islands	MIC
Micronesia	MIC
Mongolia	MIC
Burma (Myanmar)	LIC
Palau	MIC
Papua New Guinea	MIC
Philippines	MIC
Samoa	MIC
Solomon Islands	MIC
Thailand	MIC
Timor-Leste	MIC
Tonga	MIC
Vanuatu	MIC
Vietnam	LIC

Europe and Central Asia

Albania	MIC
Armenia	MIC
Azerbaijan	MIC
Belarus	MIC
Bosnia & Herzegovina	MIC
Bulgaria	MIC
Georgia	MIC
Kazakhstan	MIC
Kosovo	MIC
Kyrgyzstan	LIC
Latvia	MIC
Lithuania	MIC
Macedonia	MIC
Moldova	MIC
Montenegro	MIC
Poland	MIC
Romania	MIC
Russia	MIC
Serbia	MIC
Tajikistan	LIC
Turkey	MIC
Turkmenistan	MIC
Ukraine	MIC
Uzbekistan	LIC

Latin America and the Caribbean

Argentina	MIC
Belize	MIC
Bolivia	MIC
Brazil	MIC
Chile	MIC
Colombia	MIC
Costa Rica	MIC
Cuba	MIC
Dominica	MIC
Dominican Republic	MIC
Ecuador	MIC
El Salvador	MIC
Grenada	MIC
Guatemala	MIC
Guyana	MIC
Haiti	LIC
Honduras	MIC
Jamaica	MIC
Mexico	MIC
Nicaragua	MIC
Panama	MIC
Paraguay	MIC
Peru	MIC
St Kitts and Nevis	MIC
St Lucia	MIC
St Vincent and the Grenadines	MIC
Surinam	MIC
Uruguay	MIC
Venezuela	MIC

Middle East and North Africa

Algeria	MIC
Djibouti	MIC
Egypt	MIC
Iran	MIC
Iraq	MIC
Jordan	MIC
Lebanon	MIC
Libya	MIC
Morocco	MIC
Syria	MIC
Tunisia	MIC
Yemen	LIC

South Asia

Afghanistan	LIC
Bangladesh	LIC
Bhutan	MIC
India	MIC
Maldives	MIC
Nepal	LIC
Pakistan	MIC
Sri Lanka	MIC

Sub-Saharan Africa

Angola	MIC
Benin	LIC
Botswana	MIC
Burkina Faso	LIC
Burundi	LIC
Cameroon	MIC
Cape Verde	MIC
Central African Republic	LIC
Chad	LIC
Comoros	LIC
Congo, Dem. Rep. of	LIC
Congo, Rep. of	MIC
Cote d'Ivoire	MIC
Eritrea	LIC
Ethiopia	LIC
Gabon	MIC
Gambia, The	LIC
Ghana	LIC
Guinea	LIC
Guinea-Bissau	LIC
Kenya	LIC
Lesotho	MIC
Liberia	LIC
Madagascar	LIC
Malawi	LIC
Mali	LIC
Mauritania	LIC
Mauritius	MIC
Mozambique	LIC
Namibia	MIC
Niger	LIC
Nigeria	MIC
Rwanda	LIC
Sao Tome and Principe	MIC
Senegal	LIC
Seychelles	MIC
Sierra Leone	LIC
Somalia	LIC
South Africa	MIC
Sudan	MIC
Swaziland	MIC
Tanzania	LIC
Togo	LIC
Uganda	LIC
Zambia	LIC
Zimbabwe	LIC

HICs – OECD* countries

Australia
Austria
Belgium
Canada
Czech Republic
Denmark
Finland
France
Germany
Greece
Hungary
Iceland
Irish Republic
Italy
Japan
South Korea
Luxembourg
Netherlands
New Zealand
Norway
Portugal
Slovakia
Spain
Sweden
Switzerland
UK
USA

HICs – others

Andorra
Antigua and Barbuda
Aruba
Bahamas, The
Bahrain
Barbados
Bermuda
Brunei
Cayman Islands
Croatia
Cyprus
Equatorial Guinea
Estonia
Israel
Kuwait
Liechtenstein
Malta
Monaco
Oman
Puerto Rico
Qatar
San Marino
Saudi Arabia
Singapore
Slovenia
Taiwan
Trinidad and Tobago
United Arab Emirates

This 2010 listing is according to 2008 GNI per capita:
LIC – US$975 or less
MIC – US$976 - US$11 905
HIC – US$11 906 and above

The World Bank actually uses four groups: 'low income', 'lower middle income', 'upper middle income', and 'high income' – this book combines the two middle income groups into a single 'MIC' group to allow a more convenient three-tier division of world economies.

* OECD – Organisation for Economic Co-operation and Development: an international organisation of 34 countries with the aim of stimulating economic progress and world trade; most members are HICs.